Advances in Functional Polymers and Composites

Advances in Functional Polymers and Composites

Guest Editors

Bing Wang
Chenglong Guan
Lihua Zhan

MDPI

Basel • Beijing • Wuhan • Barcelona • Belgrade • Novi Sad • Cluj • Manchester

Guest Editors

Bing Wang
School of Mechanical
Engineering and Automation
Fuzhou University
Fuzhou
China

Chenglong Guan
School of Mechanical
Engineering and Automation
Fuzhou University
Fuzhou
China

Lihua Zhan
Light Alloys Research
Institute
Central South University
Changsha
China

Editorial Office

MDPI AG
Grosspeteranlage 5
4052 Basel, Switzerland

This is a reprint of the Special Issue, published open access by the journal *Polymers* (ISSN 2073-4360), freely accessible at: https://www.mdpi.com/journal/polymers/special_issues/83B891UHR8.

For citation purposes, cite each article independently as indicated on the article page online and as indicated below:

Lastname, A.A.; Lastname, B.B. Article Title. *Journal Name* **Year**, *Volume Number*, Page Range.

ISBN 978-3-7258-5309-0 (Hbk)
ISBN 978-3-7258-5310-6 (PDF)
https://doi.org/10.3390/books978-3-7258-5310-6

Contents

About the Editors

Bing Wang

Bing Wang is a Postdoctoral Fellow at the University of Cambridge, having recieved his Ph.D. and postdoctoral training at the University of Hull. He is active both nationally and overseas, currently serving at the School of Mechanical Engineering and Automation, Fuzhou University, as a Minjiang Distinguished Professor and Ph.D. supervisor. Prof. Wang is a Young Committee Member of the Materials Branch of the Chinese Mechanical Engineering Society, an Executive Council Member of the Fujian Mechanical Engineering Society, and a Life Member of Clare Hall, University of Cambridge. He serves as Guest Editor for four SCI-indexed journals, is a Youth Editorial Board Member and Guest Editor of *Applied Mechanics*, and has served as the Conference Chair/Co-chair of several well-known international conferences. He has acted as a reviewer or outstanding reviewer for more than fifty SCI journals, including *Nature*. His long-term research focuses on the multi-scale adaptability, functional integration, and actuation reliability of bistable deployable structures. According to Clarivate's Web of Science database, he ranks first worldwide in the field of "bistable deployable structures". In the past five years, he has presided over five major projects, including three funded at the national level, such as grants from the National Natural Science Foundation of China and provincial key projects, and has played a leading role in seven other projects, including those supported by the UK Aerospace Technology Institute and Fujian Provincial major science and technology programs. He has published over 60 SCI papers as the primary or corresponding author, including more than 30 in Q1 journals, with three ESI highly cited papers and a single publication cited over 550 times.

Chenglong Guan

Chenglong Guan obtained his Ph.D. from Central South University and is a C-Class High-Level Talent in Fujian Province. He is currently a Qishan Scholar at Fuzhou University, working as an Associate Researcher and Master's Supervisor at the School of Mechanical Engineering and Automation. Dr. Guan serves as Guest Editor for the SCI journals *Polymers* and *Computers, Materials & Continua* and has been appointed as the Chair of the CoMEA2025 sub-session and Vice Chair of the Organizing Committee of the 18th Pan-Pearl River Delta Plastic Engineering Academic Conference. He is also a member of the Organizing Committee for the EI-indexed conference ICMTAE2025. Dr. Guan's research focuses on the collaborative manufacturing of lightweight, high-strength aerospace and space structures, specifically in the areas of advanced composite component molding processes, thermal/mechanical field control, structural–functional integration design, and multi-physics field coupling simulation. As the principal investigator, he has led several projects, including one funded by the Fujian Provincial Natural Science Foundation, an open project from a national key laboratory, and the Qishan Scholar Talent Program at Fuzhou University. He has also contributed significantly to projects within the National Key R&D Program and National Natural Science Foundation-funded initiatives. Dr. Guan has authored over thirty academic papers, including twenty-one SCI-indexed papers as the first or corresponding author, with fourteen published in Q1 journals and one ESI highly cited paper. He has contributed to a group standard approved by the China Ordnance Society (ranked 2/5). His research achievements have been recognized with his receipt of the Excellent Doctoral Dissertation Award from the Nonferrous Metals Society of China and the Gold Medal in the 3rd National Postdoctoral Innovation and Entrepreneurship Competition (Fujian Province Selection, ranked 2/4).

Lihua Zhan

Lihua Zhan is a nationally recognized high-level talent and a beneficiary of the State Council Special Allowance. She currently serves as Dean of the School of International Education at Central South University and holds several academic and professional appointments, including Vice Chairman of the Plastic Engineering Branch of the Chinese Mechanical Engineering Society, Vice Chairman of the Changsha Association for Science and Technology, and Deputy Director of the Precision Forming Engineering Committee of the China Ordnance Society. Prof. Zhan has long been engaged in the study of integrated forming and performance manufacturing theories and technologies for lightweight, high-strength, large-scale components. She has led more than thirty national-level projects, including the National 973 Program, the National Key R&D Program, the National Natural Science Foundation of China (Key and General Programs), and major projects supported by the State Administration of Science, Technology and Industry for National Defense. Her breakthroughs in integrated manufacturing technologies for large and complex thin-walled components enabled the high-quality fabrication of the world's largest 10-meter-class cryogenic propellant tank domes for heavy-lift rockets. This achievement was reported by the China National Space Administration and Nature and was showcased at the 4th Military–Civil Integration High-Tech Equipment Exhibition. As first or corresponding author, Prof. Zhan has published over 250 papers in leading journals, including the *International Journal of Plasticity*. She holds 110 granted invention patents, one of which contributed to an industrial technology transfer valued at over RMB 200 million as part of six core patents.

Preface

The rapid development of polymers and polymer-based composites has established these materials as one of the most significant and dynamic areas in advanced materials science and engineering. By combining polymeric matrices with functional fillers or reinforcing phases, they provide unique advantages, including low density, high specific strength, design flexibility, and adaptability in manufacturing. Beyond their intrinsic mechanical attributes, continuous innovations in material design and processing have enabled polymers and composites to achieve multifunctionality, integrating properties such as electromagnetic shielding, thermal insulation, corrosion resistance, and bio-compatibility. These advances have expanded their applications across aerospace, transportation, sustainable energy, construction, healthcare, and many other strategic sectors.

This Reprint is dedicated to a comprehensive exploration of advanced functional polymers and composites, with particular emphasis on the interplay between material systems, structural design, processing technologies, and performance evaluation. The chapters encompass a wide spectrum of topics, including novel polymer and nanocomposite development, innovative molding and curing processes, machining and assembly techniques, defect detection methodologies, and multi-scale characterization. In addition, attention is devoted to emerging directions such as multifunctional integration, environmental sustainability, and the translation of advanced composites into practical engineering applications.

The motivation for compiling this Reprint stems from the dual imperatives of scientific advancement and industrial demand. On the one hand, the pursuit of new knowledge in material chemistry, mechanics, and processing continues to stimulate fundamental discoveries. On the other hand, the increasing reliance on high-performance, lightweight, and sustainable materials in critical industries imposes stringent requirements for larger-scale, integrated, and multifunctional composite structures. These converging drivers not only present formidable challenges in design and manufacturing but also provide unprecedented opportunities for interdisciplinary innovation and technological breakthroughs.

This Reprint seeks to provide an authoritative reference for researchers, engineers, and graduate students engaged in this field. By integrating theoretical foundations with recent advances and application-oriented perspectives, we aim to foster cross-disciplinary dialog, stimulate novel research directions, and contribute to the broader dissemination of knowledge in polymers and composites. Ultimately, it is our hope that this work will not only advance the scientific understanding of functional polymers and composites but also promote their widespread adoption in high-performance equipment and emerging technologies, thereby supporting progress across multiple disciplines, including materials science, mechanical engineering, aerospace, chemistry, and beyond.

Bing Wang, Chenglong Guan, and Lihua Zhan

Guest Editors

polymers

MDPI

Editorial

Advanced Polymer Composites and Applications

Bing Wang [1], Lihua Zhan [2,3] and Chenglong Guan [1,*]

1 Fujian Provincial Key Laboratory of Terahertz Functional Devices and Intelligent Sensing, School of Mechanical Engineering and Automation, Fuzhou University, Fuzhou 350108, China; b.wang@fzu.edu.cn
2 State Key Laboratory of Precision Manufacturing for Extreme Service Performance, Central South University, Changsha 410083, China; yjs-cast@csu.edu.cn
3 Light Alloys Research Institute, Central South University, Changsha 410083, China
* Correspondence: guancl@fzu.edu.cn

Polymer composite materials, engineered by combining polymeric matrices with functional fillers or reinforcing phases, represent a frontier in advanced materials science driven by the dual imperatives of performance enhancement and sustainable development. The production of polymer matrix composite components encompasses a multifaceted process involving structural design, preform shaping, curing and forming, machining and assembly, and inspection and analysis. It is imperative to recognize that each of these steps has a direct and substantial impact on the overall performance and functionality of the resultant structure. In light of the escalating adoption of polymer and composite materials in high-end applications across various industries, composite components are evolving towards larger scale, integral structures, increased complexity, and multi-functions. These developments inevitably give rise to heightened challenges in the design and manufacturing of advanced functional polymers and composites.

In this Special Issue, we have collected the most recent advances in functional polymers and composites, including 10 original research papers and two review articles, co-authored by 71 scientists and engineers from 35 institutions and 10 countries. The research topics mainly covered the development of functional polymers, structural design, molding processes, and engineering applications.

With the increasingly widespread application of advanced polymer composites in aerospace, satellite communications, automobiles, and other fields, higher demands are now placed on their functionalities—such as electromagnetic shielding performance and thermal insulation performance—while ensuring that the materials possess excellent fundamental mechanical properties. The development of nanocomposites based on filler modification technology provides a solution for achieving functional integration in these materials. Khalil et al. [1] synthesized Zn-Al ferrite/polypyrrole nanocomposites and conducted a comprehensive characterization. X-ray diffraction analysis confirmed the presence of ZnO, $AlFeO_3$, and Fe_2O_3 phases, and high-resolution transmission electron microscopy (HR-TEM) revealed a distinctive core–shell morphology. Variations in aluminum content not only affect magnetization, but also alter the dielectric permittivity and relaxation time, thus demonstrating that the Zn-Al ferrite/polypyrrole nanocomposite is a promising candidate material for advanced microwave absorption applications. Li et al. [2] proposed an effective and universal filler modification and nanocomposite preparation method. Specifically, the montmorillonite (MMT) surface was coated with polydopamine (PDA) to obtain DMMT, and then DMMT gel was compounded with solid ethylene propylene diene monomer rubber (EPDM) via the gel compounding method. Compared with the unmodified MMT filler EPDM, the EPDM/DMMT nanocomposite showed much fewer filler aggregates in the matrix, and the highest tensile strength of the composites reached 6.5 MPa with 10 phr DMMT, almost 200% higher than that of pure EPDM.

Moreover, attributes such as low cost, recyclability, and environmental friendliness also constitute crucial factors that need to be considered during the R&D process of novel polymers and composites. Jasinski et al. [3] explored the possibility of the utilization of coniferous bark as a filler in wood–polymer composites (WPCs), and discovered that bark-filled composites exhibit lower thickness swelling and water absorption, as well as lower water contact angles and surface free energy, compared to analogous composites filled with coniferous sawdust. Ail et al. [4] aimed to utilize waste black tea bags (BTBs) and date palm surface fibers (DPSFs) efficiently by developing new composite materials for thermal insulation and sound absorption. Hybrid boards prepared by bonding BTBs and DPSFs with polyvinyl acetate resin were tested and found to possess favorable thermal conductivity, noise reduction performance, and thermal stability, along with high mechanical properties. This material not only shows potential for competing with synthetic counterparts derived from petrochemicals in building insulation, but also helps reduce the number of landfills and the level of environmental pollution.

The development of novel material systems forms the foundation for achieving multifunctional integration, while geometric structure design and advanced manufacturing techniques are core to ensuring both dynamic and static mechanical properties, along with the forming quality, of polymer matrix composite components; they are also pivotal for determining whether these components can achieve practical engineering application. Huang et al. [5] investigated the shrinkage behavior of core height in honeycomb sandwich structures during the secondary bonding process and found that the out-of-plane compression deformation of aramid honeycomb cores was primarily caused by dehumidification, pressurization, and creep. By employing the viscoelastic Burgers mechanical model and applying the nonlinear surface fitting method, the total height shrinkage deformation behavior of the aramid honeycomb core during the curing process can be accurately predicted. Addressing the critical role of bone scaffolds in tissue engineering, Bakhtiari et al. [6] fabricated four types of polylactic acid (PLA) scaffolds with distinct structures via 3D printing. They investigated the influence of pore architecture on the scaffolds' mechanical properties under quasi-static and cyclic compression, revealing that increased strut thickness correlated with higher compressive strength, while enhanced fatigue performance across different topologies was associated with the minimum cross-sectional area of the scaffolds. However, fatigue damage was observed in all structures under higher strains. Zhang et al. [7] investigated the influence of vibration parameters on the porosity, fiber weight fraction, and mechanical properties of polymer composite components fabricated via the vibration pretreatment–microwave curing process. Utilizing an orthogonal experimental design to optimize the process parameters, they found that vibration acceleration alters the escape pathways of pores within the components, which is the dominant factor affecting their mechanical properties. Furthermore, a positive correlation was demonstrated between the interlaminar shear strength and impact strength of the components.

It is precisely the development of these novel material systems, coupled with the introduction of new structures and processes, that have enabled polymer matrix composites to not only maintain their intrinsic attributes—such as high mechanical strength, low density, and processability—but also integrate functional properties including corrosion resistance, friction resistance, electromagnetic shielding, thermal insulation, and freeze resistance. This has significantly expanded their applications in traditional fields like aerospace, marine, and automotive engineering, while demonstrating promising potential in emerging domains such as healthcare, food processing, and construction. Barros et al. [8] produced recycled substrates via hot compression molding with different proportions of pure CuO and sludge incorporated onto surfaces, enhancing their antibacterial properties through surface functionalization. This approach not only enables the recycling of abandoned fishing nets into novel substrates for algae cultivation, but also offers a viable pathway for reusing sludges typically disposed of in landfills.

In the context of the biomedical applications of polymer matrix composites, Wilk-Kozubek et al. [9] delve into the intriguing phenomenon of low-temperature thermochromism, with a particular focus on its applications in temperature-sensitive fields like medical storage. This study highlights recent advances in polydiacetylenes (PDAs)—a class of conjugated polymers engineered through structural alterations of monomers to achieve irreversible color transitions at specific low temperatures—and identifies them as ideal materials for constructing reliable temperature indicators that ensure the integrity of thermally sensitive products. Giuliani et al. [10] developed electrospun composites based on polycaprolactone (polymer matrix) and tungsten powder for radiation shielding applications in high-dose environments such as aerospace and healthcare. They demonstrated that the radiation shielding effectiveness of these composites increased with thickness and/or the number of stacked layers, effectively reducing radiation-induced apoptosis.

In traditional industrial sectors such as construction and machinery, polymer coatings emerge as a more efficient and less wasteful alternative when considering component durability, manufacturing costs, and environmental sustainability—particularly through the use of adhesive polymers to ensure coating durability and functional effectiveness. Torskaya et al. [11] applied thin (several hundred microns thick) UHMWPE coatings to formed rubber rings and conducted friction tests on both the coated samples and pure UHMWPE under varying loads and velocities. Their results demonstrated that this polymer combination imparts antifrictional properties and wear resistance to the surface layer while preserving the damping performance of the rubber substrate. Kolya et al. [12] provided a comprehensive examination of eco-friendly polymer nanocomposite coatings, including their synthesis, characterization, and performance. By integrating nanoparticles—such as nano-clays, graphene oxide, and metal oxides—into biopolymer matrices, these nanocomposites demonstrate improved thermal stability and char formation properties. This significantly mitigates the flammability of wood substrates, supporting the expanded utilization of wood in sustainable construction practices and aligning with global initiatives toward achieving carbon neutrality.

We hope that this Special Issue will contribute to disseminating the latest progress in functional polymers and composites, as well as stimulating the interest of its audiences to work in this important and vibrant area to promote and benefit the multidisciplinary scientific communities. Owing to the word limit on this Editorial, audiences are recommended to refer to the original papers for further information on their specific interests.

Funding: This work was funded by the National Natural Science Foundation of China, grant numbers 52175373 and 52475152; the Natural Science Foundation of Fujian Province, grant number 2024J01240; and the Open Research Fund of the State Key Laboratory of Precision Manufacturing for Extreme Service Performance, grant number Kfkt2024-06.

Conflicts of Interest: The authors declare no conflicts of interest.

References

1. Khalil, H.F.; Elsharkawy, S.G.; AL-Harby, N.F.; El-Batouti, M. Zn-Al Ferrite/Polypyrrole Nanocomposites: Structure and Dielectric and Magnetic Properties for Microwave Applications. *Polymers* **2024**, *16*, 2432. [CrossRef] [PubMed]

2. Li, Z.; Chen, Z.; Sun, W.; Liu, Y.; Wang, X.; Lin, J.; Wang, J.; He, S. Properties of EPDM Nanocomposites Reinforced with Modified Montmorillonite. *Polymers* **2024**, *16*, 2381. [CrossRef] [PubMed]
3. Jasinski, W.; Auriga, R.; Lee, S.H.; Adamik, L.; Borysiuk, P. Coniferous Bark as Filler for Polylactic Acid-Based Biocomposites. *Polymers* **2024**, *16*, 2669. [CrossRef] [PubMed]
4. Ali, M.; Almuzaiqer, R.; Al-Salem, K.; Alshehri, H.; Nuhait, A.; Alabdullatif, A.; Almubayrik, A. New Eco-Friendly Thermal Insulation and Sound Absorption Composite Materials Derived from Waste Black Tea Bags and Date Palm Tree Surface Fibers. *Polymers* **2024**, *16*, 2989. [CrossRef] [PubMed]
5. Huang, X.; Hu, X.; Guo, J.; Zhang, D.; Yao, S.; Zhan, L.; Ma, B.; Huang, M.; Zhang, L. Aramid Honeycomb Cores under Constant Pressure: Unveiling the Out-of-Plane Compression Deformation. *Polymers* **2024**, *16*, 1974. [CrossRef] [PubMed]
6. Bakhtiari, H.; Nouri, A.; Tolouei-Rad, M. Fatigue Performance of 3D-Printed Poly-Lactic-Acid Bone Scaffolds with Triply Periodic Minimal Surface and Voronoi Pore Structures. *Polymers* **2024**, *16*, 2145. [CrossRef] [PubMed]
7. Zhang, D.; Zhan, L.; Ma, B.; Guo, J.; Jin, W.; Hu, X.; Yao, S.; Dai, G. Effect of Vibration Pretreatment–Microwave Curing Process Parameters on the Mechanical Performance of Resin-Based Composites. *Polymers* **2024**, *16*, 2518. [CrossRef] [PubMed]
8. Barros, D.; Nobre, L.; Antunes, J.; Bessa, J.; Cunha, F.; Mota, C.; Gomes, F.; Henriques, M.; Fangueiro, R. Upcycling Fishing Net Waste and Metal Oxide from Electroplating Waste into Alga Cultivation Structures with Antibacterial Properties. *Polymers* **2024**, *16*, 3415. [CrossRef] [PubMed]
9. Wilk-Kozubek, M.; Potaniec, B.; Gazinska, P.; Cybinska, J. Exploring the Origins of Low-Temperature Thermochromism in Polydiacetylenes. *Polymers* **2024**, *16*, 2856. [CrossRef] [PubMed]
10. Giuliani, C.; De Stefano, I.; Mancuso, M.; Fiaschini, N.; Hein, L.A.; Gattia, D.M.; Scatena, E.; Zenobi, E.; Del Gaudio, C.; Galante, F.; et al. Advanced Electrospun Composites Based on Polycaprolactone Fibers Loaded with Micronized Tungsten Powders for Radiation Shielding. *Polymers* **2024**, *16*, 2590. [CrossRef] [PubMed]
11. Torskaya, E.; Shkalei, I.; Stepanov, F.; Makhovskaya, Y.; Dyakonov, A.; Petrova, N. Using Thin Ultra-High-Molecular-Weight Polyethylene Coatings to Reduce Friction in Frost-Resistant Rubbers. *Polymers* **2024**, *16*, 2870. [CrossRef] [PubMed]
12. Kolya, H.; Kang, C.W. Eco-Friendly Polymer Nanocomposite Coatings for Next-Generation Fire Retardants for Building Materials. *Polymers* **2024**, *16*, 2045. [CrossRef] [PubMed]

polymers

MDPI

Article

Upcycling Fishing Net Waste and Metal Oxide from Electroplating Waste into Alga Cultivation Structures with Antibacterial Properties

Daniel Barros [1,*], Luís Nobre [1], Joana Antunes [1,*], João Bessa [1], Fernando Cunha [1], Carlos Mota [2], Fernanda Gomes [3,4], Mariana Henriques [3,4] and Raul Fangueiro [1,5]

1 Fibrenamics Association, University of Minho, 4800-058 Guimarães, Portugal; luisnobre@fibrenamics.com (L.N.); joaobessa@fibrenamics.com (J.B.); fernandocunha@det.uminho.pt (F.C.); rfangueiro@dem.uminho.pt (R.F.)
2 Beyond Composite, 4410-309 Canelas, Portugal; carlosmota@beyondcomposite.pt
3 Centre of Biological Engineering, Laboratório de Investigação em Biofilmes Rosário Oliveira, University of Minho, 4710-057 Braga, Portugal; fernandaisabel@ceb.uminho.pt (F.G.); mcrh@deb.uminho.pt (M.H.)
4 LABBELS—Associate Laboratory, 4800-122 Braga, Portugal
5 Department of Textile Engineering, University of Minho, 4800-058 Guimarães, Portugal
* Correspondence: danielbarros@fibrenamics.com (D.B.); joanaantunes@fibrenamics.com (J.A.)

Abstract: Plastic waste, especially discarded fishing nets, and electroplating sludges pose significant environmental challenges, impacting marine ecosystems and contributing to pollution. In alga cultivation, invasive microorganisms often hinder growth, necessitating strategies to combat these issues. This study aimed to develop recycled substrates for alga cultivation by repurposing fishing nets and enhancing their surfaces with antibacterial properties using copper oxide (CuO). Additionally, it explores the reuse of CuO from electroplating sludge, providing a sustainable solution that addresses both marine and industrial waste while supporting healthy alga development. Recycled substrates were produced, with different proportions of pure CuO and sludge (1 and 2 wt%) incorporated on the surface. These compositions were processed by hot compression molding and then the antibacterial activity was characterized using a qualitative and quantitative method. The results indicate the possibility of recycling fishing net into new substrates to alga cultivation and the functionalization of their surface using CuO as an antibacterial agent. The antibacterial tests showed a better activity for pure CuO compared to the residual sludge, and better for the higher surface concentration of 2 wt%. Despite the limited bacterial inhibition observed, there is an opportunity for reusing these sludges, typically disposed of in landfills, to obtain specific antibacterial agents that can be applied to the surface of substrates for algal growth.

Keywords: recycling; antibacterial activity; electroplating waste; sustainability; algae; eco-composite

1. Introduction

Plastic waste generated by modern mass consumption practices is a significant contributor to marine pollution, affecting ocean ecosystems on a global scale. This waste comes from various sources, including household trash, fishing activities, discarded ropes, and other plastic materials. Once in the ocean, plastics pose a severe threat to marine life and ecosystems, disrupting their balance and causing widespread damage [1–3].

Among the various sources of marine plastic pollution, discarded fishing nets, often referred to as "ghost nets", are particularly problematic. These nets, abandoned at the end of their useful life, not only pollute the marine environment but also entrap and harm marine species, leading to biodiversity loss [4,5]. The recycling and proper waste treatment of these nets represent a viable approach to reducing their environmental impact and preventing further damage to marine ecosystems [6–8]. An innovative strategy involves repurposing these waste materials to create recycled substrates for algae cultivation, providing a dual benefit of waste reduction and resource reuse.

Algae are highly diverse aquatic organisms that inhabit a variety of environments, including oceans, rivers, and lakes, around the world [9]. In natural settings, algae attach to substrates such as rocks, corals, and even artificial surfaces like docks or ropes. This adaptability makes alga cultivation feasible in coastal and offshore areas, although these environments often present significant challenges. Successful alga cultivation depends on factors such as adequate water quality, sufficient light, and nutrient availability, as well as the use of suitable substrates to support alga growth [10,11]. While ropes are traditionally employed as substrates for alga production, there is potential to develop alternative materials that improve seeding, cultivation, and harvesting efficiency.

The cultivation of alga offers numerous environmental, economic, and social benefits. Algae play a vital role in maintaining aquatic ecosystems by contributing to the biological balance and producing oxygen through photosynthesis. Additionally, algae are rich in proteins, minerals, and vitamins, making them valuable for nutritional, medicinal, and cosmetic applications. They can also be used to generate organic matter and fertilizers, contributing to sustainable agricultural practices [9–11].

Despite these advantages, alga cultivation faces challenges, particularly due to external factors that can hinder growth and reduce production. Microorganisms such as parasites, bacteria, fungi, and viruses can compete with algae for resources, inhibit their growth, and even cause diseases that lead to cell damage or death [12,13]. For example, *Alphaproteobacteria*, *Gammaproteobacteria*, *Amoebophrya*, and *Labyrinthula* are known to cause cellular damage to algae, while diatoms primarily compete for space and nutrients, further exacerbating cultivation difficulties [14–18]. Addressing these challenges requires exploring innovative methods to mitigate the impact of harmful organisms and enhance alga productivity.

One promising solution is the application of antibacterial finishes to substrate surfaces. These finishes can prevent bacteria and other harmful microorganisms from adhering to the substrates, thereby creating a more favorable environment for alga growth. Such approaches are already widely used in hospital settings, where metallic coatings are employed to reduce the spread of infections by eliminating harmful bacteria and viruses upon contact.

Metals like copper, silver, and zinc, along with their oxides, have demonstrated effective antimicrobial properties and are commonly used in hygienic applications [19]. For instance, copper oxide (CuO) is particularly effective in maintaining its antimicrobial properties even when oxidized. In this project, copper oxide was chosen as the antimicrobial agent for functionalizing recycled substrates used in alga cultivation [20,21]. This CuO is sourced from the reuse of electroplating sludge, an innovative approach that not only addresses alga cultivation challenges, but also contributes to waste management and sustainability.

Electroplating is a vital industrial process used to coat metal parts, but it generates significant waste, particularly in the form of sludge containing heavy metals like copper (Cu), nickel (Ni), zinc (Zn), and chromium (Cr) [22,23]. The improper disposal of this waste can lead to severe environmental contamination, affecting water and soil quality [24–26]. Therefore, developing effective methods for treating and reusing this sludge is crucial for mitigating its environmental impact.

This study aimed to develop innovative substrates for alga cultivation using recycled fishing net waste as the base material. To enhance these substrates, their surfaces were modified with antibacterial agents, such as metal oxides, to inhibit the growth of harmful microorganisms. Additionally, the research explores the potential for reusing metal oxides, specifically those extracted from electroplating sludge, as a sustainable source of antibacterial agents. By combining the recycling of ocean plastics with the valorization of industrial waste, this approach not only mitigates environmental pollution, but also contributes to the advancement of sustainable materials for alga cultivation.

2. Experimental Procedures

2.1. Materials

To produce the substrates for alga cultivation, waste fishing nets were acquired from the entity Peniche Ocean Watch, Portugal. To produce the substrates with antimicrobial properties, pure CuO microparticles were used and acquired from Sigma-Aldrich, Lyon, France, along with residual sludge containing about 10 wt% of copper oxide, originating from the electroplating process, supplied by an entity located in the Braga district. These materials allowed for producing antibacterial substrates to develop alga cultivation substrates.

2.2. Sample Development

Different combinations of recycled fishing nets functionalized with CuO were produced via the compression molding process, using fishing net waste, pure CuO and residual sludges from the electroplating processes. The samples were developed by adding 1 and 2 wt% of CuO/sludge, that was added on the recycled substrates previously, produced by hot-pressing. The samples were produced at 220 to 230 °C (melting temperatures of polyamide, composition material of the fishing nets), and the functionalized coatings were added applying it into the mold surface before the hot-pressing process.

In Figure 1, a schematic representation of the production cycle for different sample combinations is presented.

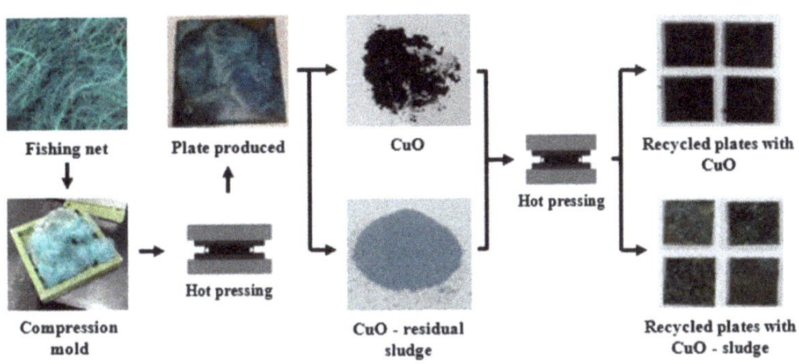

Figure 1. Schematic representation of the sample production.

The addition of antibacterial elements was performed on each plate surface in 1:99 and 2:98 weight percentage (1% and 2%, respectively), as listed in Table 1. Then, each combination of plates was cut in 2 × 2 cm specimens to perform antibacterial analysis.

Table 1. Produced samples.

Sample	Antibacterial Agent	Sample Reference
Recycled fishing net (only)	-	Sub_ref
Antibacterial powder	CuO	CuO_ref
	CuO sludge	CuO_slu_ref
Recycled fishing net	1% CuO	Sub_1%CuO
	2% CuO	Sub_2%CuO
	1% CuO sludge	Sub_1%CuO_slu
	2% CuO sludge	Sub_2%CuO_slu

2.3. Antibacterial Activity

The antibacterial activity test was carried out using two methods, a qualitative test and a quantitative test based on international standards: modified JIS L 1902:2008 standard (Testing for antibacterial activity and efficacy on textile products) [27] and ISO 22196 standard (Plastics—Measurement of antibacterial activity on plastic surfaces) [28], respectively.

Samples measuring 2 × 2 cm of various compositions were considered for testing, with reference samples comprising substrate without antimicrobials and the different antibacterial agents used (powder). All samples were sterilized by ultraviolet radiation for 30 min on each side.

Both of the testing methods used a Gram-positive and a Gram-negative bacterial strain: *Staphylococcus aureus* ATCC 6538 and *Escherichia coli* ATCC 25922, respectively. This analytical method was selected to evaluate the antibacterial activity of the additive agents applied to the surface during the substrate production process. According to the literature, these two bacteria are commonly used to assess antibacterial capacity.

The halo test, also known as the disk diffusion test, was performed as follows. An inoculum was incubated in a tryptic soy broth (TSB) overnight at 37 °C and 120 rpm. Then, 1 mL from the inoculum with 1×10^6 cells/mL was added to 15 mL of tryptic soy agar (TSA) warmed at 45 °C. This solution was disposed in a sterilized Petri dish. After medium solidification, the samples were placed over the agar and incubated for 24 h at 37 °C. For powdered antimicrobials, a 10 mg sample was used. The evaluation of the antibacterial activity was made based on the observation of the zone of inhibition, around and under the samples.

The quantitative test was performed as follows. With 1×10^6 cells/mL, 200 μL of inoculum was placed onto the surfaces of the coated reference samples. The bacterial inoculum was covered with a plastic cover film (1 × 1 cm). The samples were incubated for 24 h at 37 °C. After this period of incubation, the cells were collected by adding 10 mL of a washing solution (TSB + 7.0 g/L of Tween 80). Serial dilutions were made and then plated on TSA for 24 h at 37 °C. The number of colony-forming units (CFUs) was determined.

Antibacterial activity (R) was determined by calculating according to Equation (1):

$$R = Ut - At \quad (1)$$

where "Ut" is the average of the logarithm of the number of viable bacteria (N), in cells/cm^2, taken from the control samples after 24 h and "At" is the average of the logarithm of the number of viable bacteria (N) in cells/cm^2, taken from the functionalized samples after 24 h. Qualitative and quantitative analysis comprised three independent assays performed in duplicate.

2.4. Mechanical Characterization

The mechanical characterizations carried out were tensile and flexural tests, performed in an universal test machine (Hounsfield H100KS). For these tests, a 2.5 kN load cell and a crosshead speed of 2 mm/min were used in accordance with the ISO 527-4 (Plastics—Determination of tensile properties) [29] standard for tensile tests and the ISO 178 (Plastics—Determination of flexural properties) standard for flexure test [30].

2.5. Fourier Transform Infrared Spectroscopy (FTIR) Characterization

The chemical composition of the sample production was analyzed using Fourier transform infrared spectroscopy (FTIR) coupled with the attenuated reflection (ATR) technique using SHIMADZU—IRAffinity-1S, Kyoto, Japan. All the spectra were obtained in the transmittance mode with 45 scans over a wavenumber range of 4000–400 cm^{-1}. This technique will allow for comparing the chemical compositions of the produced fishing net recycled substrates with and without antibacterial agent.

2.6. Scanning Electron Microscopy (SEM) and Elemental Analyses

The specimen's surface morphology evaluation was made using the Scanning Electron Microscopy (SEM); each is a technique widely used in scientific research to visualize the morphology of materials at the micro- and nanoscales. It employs an electron beam to generate high-resolution images of the sample surface. In this study, SEM analysis was conducted into specific area of different specimens using an SEM instrument, FEI Nova, 200, Hillsboro, OR, USA, to evaluate the surface differences of the fishing net recycled surface with and without antibacterial agent. Additionally, by performing an elemental analysis of the developed samples and the antibacterial agents used, it is possible to evaluate their composition and determine the elements they are made of. These elemental analyses were made using an EDS instrument—Pegasus X4M (EDS/EBSD).

2.7. Fourier Transform Infrared Spectroscopy (XRD) Characterization

X-ray diffraction (XRD) is a powerful analytical technique used to investigate the crystalline structure of materials by examining the diffraction pattern produced when X-rays interact with the sample's atomic planes. This technique was conducted using an XRD instrument—Bruker D8 Discover. By analyzing the diffraction patterns, it was possible to assess the structural integrity and determine the crystalline phases present in each sample, providing insights into the effectiveness and purity of CuO obtained from different sources.

2.8. Surface Roughness Characterizaiton

Surface roughness testing was performed on the recycled fishing net substrates, in accordance with ISO 21920-2 standards [31]. This testing utilized a Mitutoyo SJ-210, Kanagawa, Japan, which allowed for the measurement of the arithmetic average roughness (Ra), a parameter that represents the average surface profile deviations from the mean line. This evaluation enabled a comparison of the effects of different antibacterial agents on surface roughness.

2.9. Water Absorption Characterizaiton

Water absorption tests were conducted on samples of recycled fishing nets with and without antibacterial agents, aiming to evaluate the impact of these agents on the water absorption properties of the recycled substrates. The test was conducted following the ASTM D570 standard [32], where the samples were fully immersed in distilled water for a 24 h, and the percentage of water absorbed was calculated according to Equation (2):

$$Water\ absorption\ (\%) = \frac{W_d - W_w}{W_w} \times 100 \quad (2)$$

where

W_d = dry mass of the sample;
W_w = wet mass of the sample.

3. Results and Discussion

3.1. Measurement of Antibacterial Activity on Recycled Fishing Net Substrate Surfaces

The antibacterial activity of recycled fishing nets functionalized with pure CuO and sludge was evaluated against standard strains, including the Gram-positive bacteria *Staphylococcus aureus* and the Gram-negative bacteria *Escherichia coli*.

In Figure 2, the results of the antibacterial activity are presented according to the qualitative method, where visually and through the halo test, it was possible to assess the substrate's ability to inhibit bacterial growth, providing a comprehensive understanding of the antibacterial effectiveness of each sample.

When considering specifically the recycled network substrate (Sub_ref), a significant absence of antibacterial activity against the tested bacterial strains was observed. The lack of antibacterial efficacy may be attributed to the nature of the recycled material, which showed to not possess intrinsic properties capable of inhibiting bacterial growth. Also, for the CuO_slu_ref sample, no halo formation was visible.

In contrast, the pure antibacterial agent CuO (CuO_ref) demonstrated robust antibacterial activity, evidenced by the formation of clear halos around the areas of contact with the bacteria (*S. aureus*: 16 mm and *E. coli:* 11 mm), indicating an effective inhibition of bacterial growth.

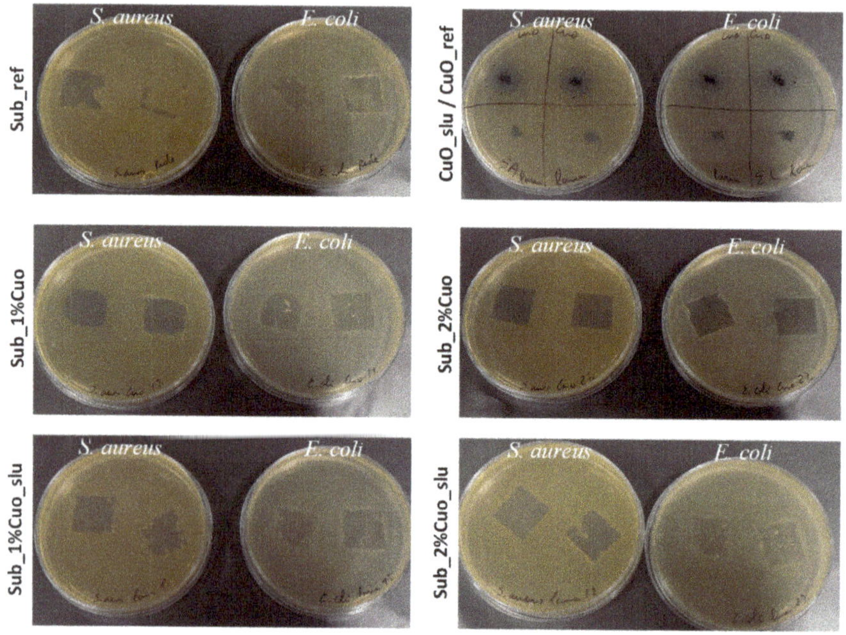

Figure 2. Qualitative halo test results for reference and coating samples, and antibacterial powders.

When these antibacterial agents were added to the surface of the recycled network substrate at 1 and 2 wt% concentrations, it was found that substrates previously without any or low antibacterial activity exhibited some activity. Comparing the two concentrations considered, a higher activity was observed for the 2 wt% concentration, as it contains a higher concentration of the antibacterial agent, leading to better efficacy in eliminating bacteria. From the results of the qualitative assays, a greater antibacterial efficacy was observed for the CuO_ref and sub_2% CuO samples.

The results obtained using the quantitative method, which determined the antibacterial activity (R), are presented in Table 2.

Table 2. Antibacterial activity (R) of functionalized substrates with 1 and 2 wt% of pure CuO and sludge against *Escherichia coli* and *Staphylococcus aureus*.

Functionalization	*E. coli*	*S. aureus*
Sub_ref	0	0
Sub_1%CuO	0	0.7 ± 0.1
Sub_2%CuO	0.7 ± 0.64	1.3 ± 0.55
Sub_1%CuO_slu	0.3 ± 0.02	0.2 ± 0.05
Sub_2%CuO_slu	0.4 ± 0.09	0.3 ± 0.11

Based on the results from Table 1, and although $R < 2$ (the minimum threshold to consider antibacterial activity occurs), it was observed that substrates functionalized with 2 wt% pure CuO exhibited indeciduous of some bacterial inhibition, whereas non-functionalized and CuO_Slu substrates showed no activity ($R = 0$). With the increase in antibacterial agent concentration on the surface from 1 to 2 wt%, an increase in the R was observed, especially for the samples functionalized with pure CuO. This is attributed to a greater exposure of these agents to bacteria, thereby enhancing their effectiveness.

In substrates coated with pure CuO, higher bacterial inhibition was noted against *S. aureus*. Substrates with CuO from residual sludge exhibited no bacterial inhibition against

S. aureus and *E. coli*. This must be attributed to the difference in CuO concentration in each functionalization, as pure CuO substrates have a concentration of 100%, while residual sludge substrates have approximately 10 wt% copper oxide concentration (according to the supplier). Accordingly, when comparing bacterial inhibition among the residues, better results were observed for samples with pure CuO. As for samples functionalized with residual sludge, despite exhibiting lower antibacterial activity, the results obtained may be considered as positive, giving that it is a residue from galvanoplasty processes, where copper oxide present at concentrations of approximately 10 wt%.

These results may indicate the feasibility of reusing residual sludge from galvanoplasty processes, thereby avoiding their disposal in landfills. Additionally, it is suggested that using higher concentrations of CuO may increase the R value, thereby enhancing the antibacterial capacity of the surface, or test the antibacterial activity by testing other sludges from different processes that may contain other elements capable of possessing antibacterial properties.

3.2. Mechanical Properties

For the mechanical characterization tests, tensile and flexural tests were conducted. Specimens were cut into bone and rectangular shapes according to the respective standards. The results are presented in Figure 3.

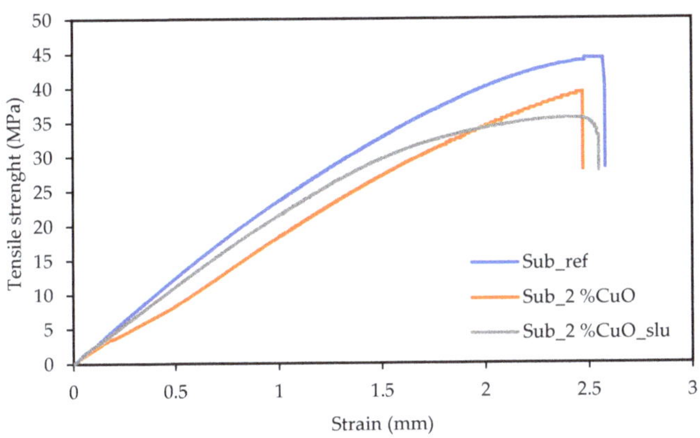

(**a**)

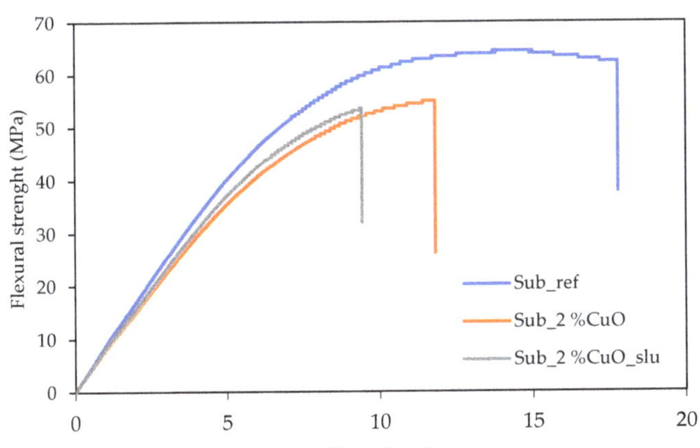

(**b**)

Figure 3. Mechanical characterization results: (**a**) tensile and (**b**) flexural.

According to the tensile results, it is possible to verify a small decrease in tensile strength when adding 2% of antibacterial agents, 11.3% to the addition of 2% CuO, and 19.6% to the 2% CuO from sludge. Like the findings of the tensile tests, there was a small decrease of 14.6% in flexural strength for the samples with an addition of 2% CuO and 16.6% to the 2% CuO from residual sludges.

For both additions of 2% antibacterial agent, a slight decrease in tensile and flexural strength was observed. This may be related to the addition of these agents to the surface of recycled fishing net substrates. During the surface functionalization process, these substrates undergo thermal reprocessing, which can lead to a loss of properties in the polymeric matrix. The samples with 2% CuO derived from electroplating sludge residues exhibited lower resistance compared to the samples with pure CuO, due to the presence of impurities.

3.3. Fourier Transform Infrared Spectroscopy (FTIR)

In this section its presented FTIR test results for the different specimens produced to evaluate the influence of adding 2 wt% of antibacterial agent in the surface. The results are presented in Figure 4.

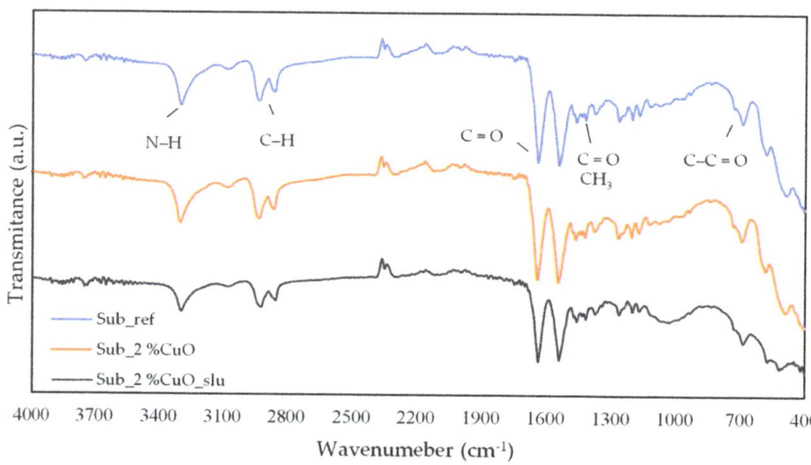

Figure 4. FTIR spectra of the fishing net substrates with and without antibacterial agent.

The FTIR spectrum of the polyamide shows characteristic bands:

- Around 3300 cm^{-1}: stretching vibration of the amine groups (at ca. 3075 cm^{-1}, the overtone of N–H bending has a signal [33];
- Around 2925 and 2854 cm^{-1} (aliphatic C–H stretching), 1636 cm^{-1} (C=O stretching), and 1475 cm^{-1} (aromatic C=C), and 1405 cm^{-1} (C–N stretching);
- Around 1634 cm^{-1} and 1535 cm^{-1} presents the (C=O) and (N–H) bands;
- peaks appearing at 1474 cm^{-1} and 1370 cm^{-1}: C-H asymmetric bending in (–CH_2) or (–CH_3–) and symmetric bending in (–CH_3–), respectively;
- At about 834 cm^{-1} and 686 cm^{-1}: C-C=O stretching vibration and N–H out-of-plane bending vibration, respectively [34];

After analyzing the FTIR results and comparing them with the spectra available in databases, it is confirmed that the obtained spectrum corresponds to that of polyamide (PA) [35].

Comparing the spectra of recycled substrates with those functionalized with antibacterial agents (both pure and residual), no significant differences were observed in the obtained spectra. The primary distinction lies in the peak intensity, which decreased when 2 wt% of the antibacterial agent was added, as its presence "hinders" signal detection, leading to reduced peak intensity.

3.4. Scanning Electron Microscopy SEM and Elemental Analyses

Using SEM/EDS analysis, it was possible to evaluate the surface of the samples with and without the addition of antioxidant agent, to assess the dispersion of these agents on the surface and evaluate the present elements in each surface composition. Firstly, is presented the results of the elemental analysis performed by EDS, exclusively for the antibacterial agents, especially to understand the composition of the electroplating residue, as presented in Figure 5, and the chemical composition in wt% of each element in Table 3.

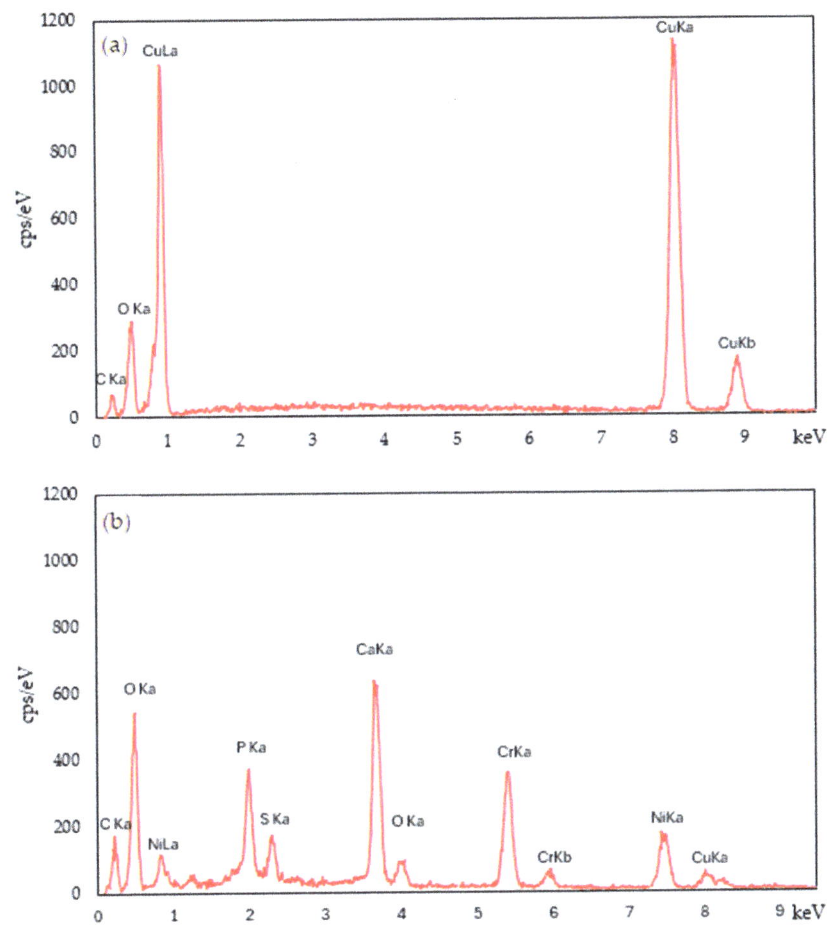

Figure 5. EDS spectra of each antibacterial agent: (**a**) pure CuO and (**b**) sludge from the electroplating process.

Table 3. Chemical composition (EDS) of pure CuO and electroplating sludge.

Sample	Element (wt%)									
	C	O	Cu	Mg	P	S	Ca	Cr	Ni	Total
CuO	8.54	13.89	77.57	-	-	-	-	-	-	100
Sludge	16.09	37.99	3.7	0.88	5.47	2.2	11.51	11.57	10.59	100

The EDS analysis confirmed that pure CuO is primarily composed of copper (Cu) at energy levels of 0.9, 8.1, and 8.9 keV, and oxygen (O) at 0.6 keV. From the obtained results, it is evident that the sludge resulting from the electroplating process contains a wide array of elements identified at various peaks: carbon (C) at 0.2 keV, oxygen (O) at 0.6 keV, nickel (Ni) at 0.9 and 7.5 keV, magnesium (Mg) at 1.3 keV, phosphorus (P) at 2 keV, sulfur (S) at 2.3 keV, calcium (Ca) at 3.7 and 4 keV, chromium (Cr) at 5.4 and 6 keV, and copper (Cu) at 8 and 9 keV.

Through the chemical analysis, the mass percentage of each element detected in both antibacterial agents was determined. In the CuO powders, a pronounced presence of Cu and O was observed, as expected due to the use of pure powder. In the electroplating sludge powder, several elements were detected, with a notable presence of Ca, Cr, and Ni, which appear in the composition as oxides, justified by the high oxygen content.

SEM results of recycled fishing net substrates, with and without 2 wt% antibacterial agents, are presented in Figure 6, along with the corresponding EDS analysis to better understand the two antibacterial agents.

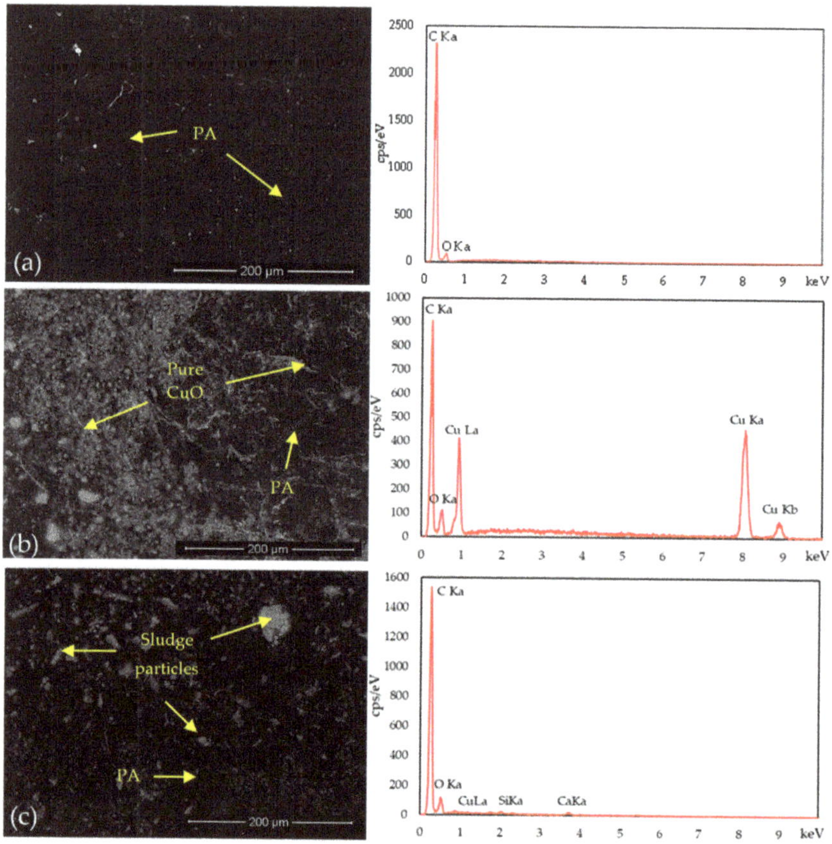

Figure 6. SEM images and EDS spectrs of the fishing net substrates: (**a**) reference, (**b**) with 2% CuO and (**c**) with 2%CuO_slu.

The morphological and elemental characterization of the reference substrate, produced exclusively from fishing nets (Figure 6a), reveals a smooth and homogeneous surface composed of polyamide (PA), the primary constituent of the nets. A few white spots are visible on the surface, likely representing residual contaminants from the fishing nets, which remained despite the washing and grinding processes. These residues persisted through compression molding and became embedded within the substrate, contributing to its final surface characteristics. Through EDS, it is possible to identify the elements carbon (C) (0.26 keV) and oxygen (O) (0.56 keV), which are characteristic of the chemical structure of polyamide.

The SEM results to the substrate surface with the addition of 2 wt% pure CuO (Figure 6b), a good distribution of particles across the surface is observed, though it is not entirely homogeneous. Certain areas exhibit a higher density of pure CuO particles, with small clusters also present. EDS analysis further confirms the presence of carbon (C) and oxygen (O) peaks, while the addition of pure CuO is evident from the appearance of copper (Cu) peaks at 0.9, 8.1, and 9 keV.

From the SEM images of the specimens with 2% CuO_slu (Figure 6c), it was also possible to observe a good distribution of particles. However, the presence of agglomerates suggests that the addition process of these agents to the surface needs optimization to achieve a more homogeneous antibacterial surface. Based on the EDS results, the presence of various elements detected in the analysis of the residual sludge alone can be confirmed. However, the identified peaks display much lower energy levels. This indicates that these elements are present on the substrate surface, but their lower concentration resulted in reduced antibacterial activity of these substrates, due to the weak presence of CuO.

SEM images further show a greater presence of CuO compared to CuO_slu on the substrate surfaces, due to the better distribution and adhesion of CuO microparticles compared to the larger CuO_slu macroparticles.

3.5. Fourier Transform Infrared Spectroscopy (XRD)

The phase composition and element determination were analyzed by the XRD pattern, shown in Figure 7.

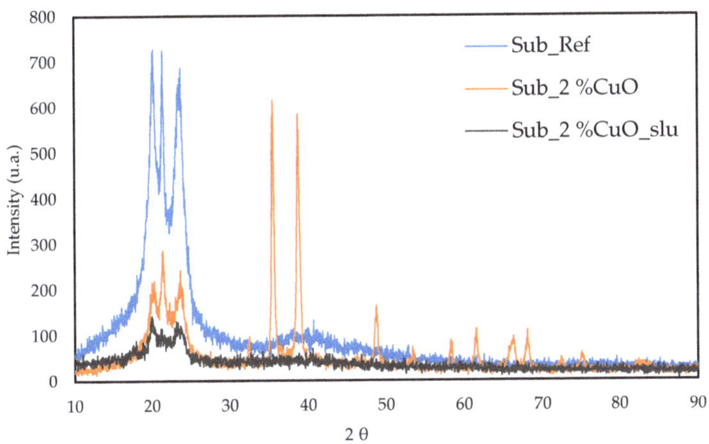

Figure 7. X-ray diffraction (XRD) of the reference fishing nets substrates, with 2% CuO and 2%CuO_slu.

The sample Sub_Ref was analyzed by X-ray diffraction (XRD) to determine its composition and crystalline structure. The results show two main peaks at $2\theta = 20.4^\circ$ and $2\theta = 23.6^\circ$, corresponding to the (200) and (002) crystallographic planes, confirming the presence of polyamide. The absence of additional peaks suggests a typical semicrystalline structure with high purity and no significant crystalline phases. These crystallographic planes indicate a characteristic polyamide organization, with well-defined crystalline regions, confirming the material's identity (X) [36,37].

The sample 2 %CuO was analyzed by X-ray diffraction (XRD) to determine its composition and crystalline structure. The results show two main peaks at $2\theta = 35.3^\circ$ and $2\theta = 38.6^\circ$, corresponding to the crystallographic planes (111) and (022), indicating the presence of CuO. Additional less intense peaks were observed at $2\theta = 48.84^\circ$, 53.5°, 58.4°,

61.6°, 66.2°, and 68°, corresponding to the crystallographic planes (202), (202), (113), (022), (220), and (222), according to the literature [38,39]. Furthermore, peaks corresponding to polyamide were also detected, resulting from readings over the substrate on which CuO was deposited during sample processing, confirming the presence of polyamide on the material's surface.

For the sample Sub_2% CuO_slu, only the characteristic peaks of polyamide were observed, while the other elements detected in the EDS analysis did not appear in the spectrum, which displayed an increase in noise. This may indicate that there is no significant crystalline material present in the analyzed plating sludge, or that if it exists, its concentration is too low, rendering the peaks undetectable. Additionally, as this is a waste material, the surface of the substrate may be contaminated, complicating peak visualization. The degradation of the material or alterations in the crystalline structure, resulting from chemical or thermal processes, may also contribute to the absence of peaks in the spectrum.

3.6. Surface Roughness

In this section, the surface roughness test results for the different specimens produced to evaluate the influence of adding 2 wt% of antibacterial agent in the surface are presented, as shown in Figure 8.

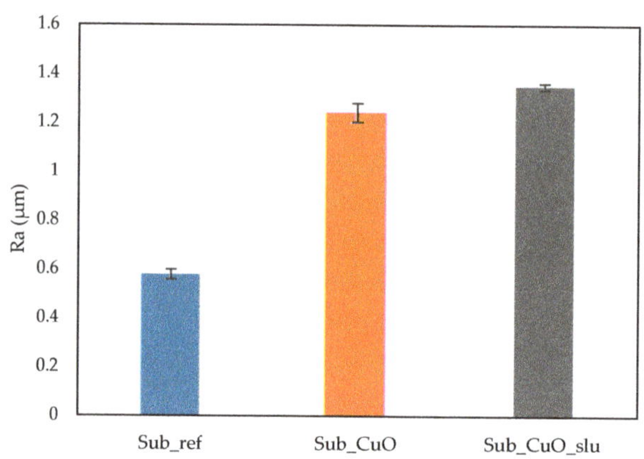

Figure 8. Surface arithmetic average roughness (Ra) of the reference fishing net substrates, with 2% CuO and 2%CuO_slu.

The results reveal a significant increase in the roughness of the additive surfaces compared to the reference surface made from fishing net waste. The surface of the Sub_ref sample exhibited an arithmetic average roughness (Ra) value of 0.578 μm, while the Sub_2%CuO sample, which includes 2% by mass of pure CuO microparticles, reached 1.242 μm. The Sub_2%CuO_slu sample, containing 2% by mass of macroparticles derived from electroplating process waste—which includes various oxides, as verified by EDS and XRD—showed an Ra of 1.349 μm.

The comparison between the additive samples indicates that the addition of macroparticles results in an even greater increase in roughness compared to the addition of CuO microparticles. This increased roughness is advantageous for algal cultivation, as rougher surfaces enhance algal adhesion, thereby promoting their growth in aquatic environments.

3.7. Water Absorption

For the water absorption tests, the samples were weighed when dry and after being submerged in water to calculate the percentage of water absorption retained. The obtained values are presented in Figure 9.

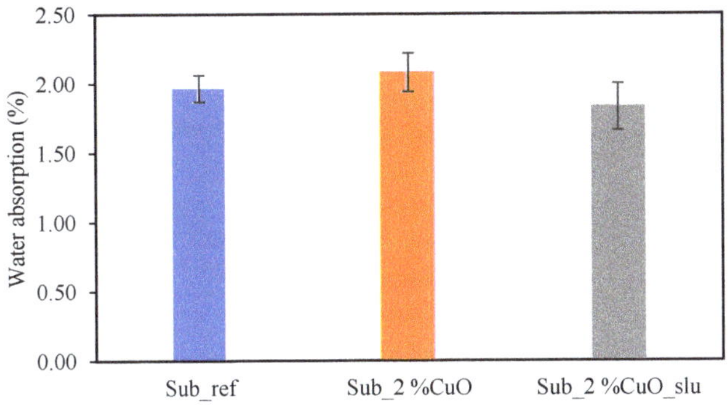

Figure 9. Water absorption results.

After the water absorption tests, no differences were observed between the substrates made with fishing nets, regardless of whether 2 wt% of the antibacterial agent was applied. The percentage of water absorption obtained was approximately 2%. The standard deviations were around 0.14, indicating high reproducibility in the tests performed, as well as in the samples produced. The addition of the antibacterial agent to the surface of the substrates made from fishing net residue did not affect water absorption in any way.

4. Conclusions

In summary, the results highlight that the surface functionalization treatment of substrates, derived from recycled fishing nets, with antibacterial agents yielded several significant findings:

- Enhanced antibacterial capacity observed in the surface-treated substrates.
- Despite the obtained values being R > 2 for the minimum threshold to be considered antibacterial activity, a greater potential of bacterial elimination efficiency exhibited by pure CuO compared to residual sludges from electroplating processes was observed.
- Surface treatment with pure CuO 2 wt% demonstrated superior capacity of possible bacterial inhibition compared to 1 wt% due to the obtained results.
- Despite the low levels of antibacterial activity, by adjusting the concentrations of the antibacterial agent, the results obtained show promising indicators for the potential reuse of residual sludges in developing solutions with antibacterial capabilities.
- A slight decrease in mechanical strength was observed in samples with 2% pure CuO and 2% CuO from electroplating sludge (CuO_slu). This reduction is due to the reprocessing of the substrates after surface modification.
- FTIR analysis confirmed that the substrates are made of polyamide (PA). The detected peaks did not change significantly, but their intensity was reduced, indicating that the substrate structure remained unchanged after the addition.
- EDS identified the elements in the antibacterial agents. SEM showed the distribution of these elements on the surface and their morphology.
- XRD confirmed the composition of the substrates. Cu and O were found in the pure CuO sample, while additional elements, including Cu and O, were identified in the sample with electroplating residues.
- Roughness tests showed a significant increase in samples with pure CuO and CuO_slu. This higher roughness benefits alga growth by providing better conditions for adhesion and development.
- Water absorption tests showed no significant differences between samples. This was expected, as the agents were added in low concentrations and only on the surface, without altering substrate porosity.

- This preliminary study allowed us to verify the promising antibacterial capacity of CuO, with the next steps being the evaluation of its antibacterial activity during alga cultivation in a controlled environment.

The results obtained for the antibacterial agents added to the surface of the polymeric substrates demonstrated their potential in eliminating bacteria that may colonize the surface. This validates the feasibility of the deposition method employed during the substrate production process at the laboratory scale. Furthermore, additional characterizations confirmed that the necessary conditions were established to develop substrates suitable for the cultivation and growth of algae. As the next steps, we are considering evaluating larger quantities of the antibacterial agent and conducting these tests using bacteria in real environmental conditions, alongside the assessment and monitoring of alga growth.

Author Contributions: Conceptualization, D.B.; methodology, L.N., C.M. and J.B.; validation, J.B., F.C. and R.F.; formal analysis, J.B., J.A., F.G. and M.H.; investigation, D.B.; resources, F.C. and R.F.; data curation, D.B.; writing—original draft preparation, D.B.; writing—review and editing, L.N. and F.G.; visualization, J.A. and F.G.; supervision, J.B. and M.H.; project administration, F.C. and R.F.; funding acquisition, F.C. and R.F. All authors have read and agreed to the published version of the manuscript.

Funding: This research was funded by the EEA Grants, "PT-INNOVATION-0109" under the program "BlueGrowt".

Institutional Review Board Statement: Not applicable.

Data Availability Statement: All data are contained within this article.

Conflicts of Interest: Carlos Mota is employed by Beyond Composite. The remaining authors declare that the research was conducted in the absence of any commercial or financial relationships that could be construed as potential conflicts of interest.

References

1. Kwon, G.; Cho, D.W.; Park, J.; Bhatnagar, A.; Song, H. A review of plastic pollution and their treatment technology: A circular economy platform by thermochemical pathway. *Chem. Eng. J.* **2023**, *464*, 142771. [CrossRef]
2. Harris, P.T.; Maes, T.; Raubenheimer, K.; Walsh, J.P. A marine plastic cloud—Global mass balance assessment of oceanic plastic pollution. *Cont. Shelf Res.* **2023**, *255*, 104947. [CrossRef]
3. Deville, A.; Vazquez-Rowe, I.; Ita-Nagy, D.; Kahhat, R. Ocean-based sources of plastic pollution: An overview of the main marine activities in the Peruvian EEZ. *Mar. Pollut. Bull.* **2023**, *189*, 114785. [CrossRef] [PubMed]
4. Karadurmuş, U.; Bilgili, L. Environmental impacts of synthetic fishing nets from manufacturing to disposal: A case study of Türkiye in life cycle perspective. *Mar. Pollut. Bull.* **2024**, *198*, 115889. [CrossRef]
5. Royer, S.J.; Corniuk, R.N.; McWhirter, A.; Lynch, H.W., IV; Pollock, K.; O'Brien, K.; Escalle, L.; Stevens, K.A.; Moreno, G.; Lynch, J.M. Large floating abandoned, lost or discarded fishing gear (ALDFG) is frequent marine pollution in the Hawaiian Islands and Palmyra Atoll. *Mar. Pollut. Bull.* **2023**, *196*, 115585. [CrossRef]
6. Mathis, J.E.; Gillet, M.C.; Disselkoen, H.; Jambeck, J.R. Reducing ocean plastic pollution: Locally led initiatives catalyzing change in South and Southeast Asia. *Mar. Policy* **2022**, *143*, 105127. [CrossRef]
7. Liotta, I.; Avolio, R.; Castaldo, R.; Gentile, G.; Ambrogi, V.; Errico, M.E.; Cocca, M. Mitigation approach of plastic and microplastic pollution through recycling of fishing nets at the end of life. *Process Saf. Environ. Prot.* **2024**, *182*, 1143–1152. [CrossRef]
8. Citterich, F.; Giudice, A.L.; Azzaro, M. A plastic world: A review of microplastic pollution in the freshwaters of the Earth's poles. *Sci. Total Environ.* **2023**, *869*, 161847. [CrossRef]
9. Goswami, R.K.; Mehariya, S.; Verma, P. Advances in microalgae-based carbon sequestration: Current status and future perspectives. *Environ. Res.* **2024**, *249*, 118397.
10. Jiang, X.; Gao, Z.; Zhang, Q.; Wang, Y.; Tian, X.; Shang, W.; Xu, F. Remote sensing methods for biomass estimation of green algae attached to nursery-nets and raft rope. *Mar. Pollut. Bull.* **2020**, *150*, 110678. [CrossRef]
11. Kim, Y.D.; Hong, J.P.; Song, H.I.; Park, M.S.; Moon, T.S.; Yoo, H.I. Studies on technology for seaweed forest construction and transplanted Ecklonia cava growth for an artificial seaweed reef. *J. Environ. Biol.* **2012**, *33*, 969.
12. Gleason, F.H.; Van Ogtrop, F.; Lilje, O.; Larkum, A.W. Ecological roles of zoosporic parasites in blue carbon ecosystems. *Fungal Ecol.* **2013**, *6*, 319–327. [CrossRef]
13. Sullivan, B.K.; Trevathan-Tackett, S.M.; Neuhauser, S.; Govers, L.L. Review: Host-pathogen dynamics of seagrass diseases under future global change. *Mar. Pollut. Bull.* **2018**, *134*, 75–88. [CrossRef] [PubMed]
14. Steinrücken, P.; Jackson, S.; Müller, O.; Puntervoll, P.; Kleinegris, D.M. A closer look into the microbiome of microalgal cultures. *Front. Microbiol.* **2023**, *14*, 1108018. [CrossRef]

15. Li, C.; Song, S.; Liu, Y.; Chen, T. Occurrence of Amoebophrya spp. infection in planktonic dinoflagellates in Changjiang (Yangtze River) Estuary, China. *Harmful Algae* **2014**, *37*, 117–124. [CrossRef]
16. Kim, S.; Park, M.G. Effect of the endoparasite Amoebophrya sp. on toxin content and composition in the paralytic shellfish poisoning dinoflagellate Alexandrium Fundyense (Dinophyceae). *Harmful Algae* **2016**, *51*, 10–15.
17. Sullivan, B.K.; Sherman, T.D.; Damare, V.S.; Lilje, O.; Gleason, F.H. Potential roles of *Labyrinthula* spp. in global seagrass population declines. *Fungal Ecol.* **2013**, *6*, 328–338. [CrossRef]
18. Żak, A.; Kosakowska, A. The influence of extracellular compounds produced by selected Baltic cyanobacteria, diatoms and dinoflagellates on growth of green algae Chlorella vulgaris. *Estuar. Coast. Shelf Sci.* **2015**, *167*, 113–118. [CrossRef]
19. Asamoah, R.B.; Yaya, A.; Mensah, B.; Nbalayim, P.; Apalangya, V.; Bensah, Y.D.; Damoah, L.N.; Agyei-Tuffour, B.; Dodoo-Arhin, D.; Annan, E. Synthesis and characterization of zinc and copper oxide nanoparticles and their antibacteria activity. *Results Mater.* **2020**, *7*, 100099. [CrossRef]
20. Bharadishettar, N.; Bhat, K.U.; Bhat Panemangalore, D. Coating Technologies for Copper Based Antimicrobial Active Surfaces: A Perspective Review. *Metals* **2021**, *11*, 711. [CrossRef]
21. Chang, T.; Babu, R.P.; Zhao, W.; Johnson, C.M.; Hedstrom, P.; Odnevall, I.; Leygraf, C. High-Resolution Microscopical Studies of Contact Killing Mechanisms on Copper-Based Surfaces. *Appl. Mater. Interfaces* **2021**, *13*, 49402–49413. [CrossRef] [PubMed]
22. Rajoria, S.; Vashishtha, M.; Sangal, V.K. Review on the treatment of electroplating industry wastewater by electrochemical methods. *Mater. Today Proc.* **2021**, *47*, 1472–1479. [CrossRef]
23. Rajoria, S.; Vashishtha, M.; Sangal, V.K. Electroplating wastewater treatment by electro-oxidation using synthesized new electrode: Experimental, optimization, kinetics, and cost analysis. *Process Saf. Environ. Prot.* **2024**, *183*, 735–756. [CrossRef]
24. Gong, J.; Tan, R.; Wang, B.; Wang, Z.; Gong, B.; Mi, X.; Deng, D.; Liu, X.; Liu, C.; Deng, C.; et al. Process and mechanism of strengthening chlorination cascade recovery of valuable metals from electroplating sludge. *J. Clean. Prod.* **2022**, *376*, 134330. [CrossRef]
25. Wang, H.; Liu, X.; Zhang, Z. Approaches for electroplating sludge treatment and disposal technology: Reduction, pretreatment and reuse. *J. Environ. Manag.* **2024**, *349*, 119535. [CrossRef]
26. Du, J.; Zeng, L.; Zhang, S.; Xiao, C.; Zhang, G.; Cao, Z.; Li, Q.; Wang, M.; Guan, W.; Wu, S. Complete recycling of valuable metals from electroplating sludge: Green and selective recovery of chromium. *Chem. Eng. J.* **2023**, *467*, 143484. [CrossRef]
27. *JIS L 1902*; Testing Antibacterial Activity and Efficacy on Textile Products. Japanese Standards Association: Tokyo, Japan, 2013.
28. *ISO 22196:2011*; Measurement of Antibacterial Activity on Plastics and Other Non-Porous Surfaces. ISO: Geneva, Switzerland, 2011.
29. *ISO 527-4:2023*; Plastics—Determination of Tensile Properties. Part 4: Test Conditions for Isotropic and Orthotropic Fibre-Reinforced Plastic Composites. ISO: Geneva, Switzerland, 2023.
30. *ISO 178:2019*; Plastics—Determination of Flexural Properties. ISO: Geneva, Switzerland, 2019.
31. *ISO 21920-2:2021*; Geometrical Product Specifications (GPS)—Surface Texture: Profile. Part 2: Terms, Definitions and Surface Texture Parameters. ISO: Geneva, Switzerland, 2021.
32. *ASTM D570-22*; Standard Test Method for Water Absorption of Plastics. ASTM: West Conshohocken, PA, USA, 2022.
33. Nandiyanto, A.B.; Ragadhita, R.; Fiandini, M. Interpretation of Fourier Transform Infrared Spectra (FTIR): A Practical Approach in the Polymer/Plastic Thermal Decomposition. *J. Sci. Technol.* **2023**, *8*, 113–126. [CrossRef]
34. Tummino, M.L.; Chrimatopoulos, C.; Bertolla, M.; Tonetti, C.; Sakkas, V. Configuration of Simple Method for Different Polyamides 6.9 Recognition by ATR-FTIR Analysis Coupled with Chenometrics. *Polymers* **2023**, *15*, 3166. [CrossRef]
35. Database of ATR-FT-IR Spectra of Various Materials—Polyamide (Nylon 6). Available online: https://spectra.chem.ut.ee/textile-fibres/polyamide/ (accessed on 15 June 2024).
36. Farias-Aguilar, J.C.; Ramírez-Moreno, M.J.; Téllez-Jurado, L.; Balmori-Ramírez, H. Low pressure and low temperature esynthesis of polyamide-6 (PA6) using Na as catalyst. *Mater. Lett.* **2014**, *136*, 388–392. [CrossRef]
37. Mahmood, N.; Islam, M.; Hameed, A.; Saeed, S. Polyamide 6/Multiwalled Carbon Nanotubes Nanocomposites with Modified Morphology and Thermal Properties. *Polymers* **2013**, *5*, 1380–1391. [CrossRef]
38. Veisi, H.; Karmakar, B.; Tamoradi, T.; Hemmati, S.; Hekmati, M.; Hamelian, M. Biosynthesis of CuO nanoparticles using aqueous extract of herbal tea (*Stachys lavandulifolia*) flowersand evaluation of its catalytic activity. *Sci. Rep.* **2021**, *11*, 1983. [CrossRef]
39. Ashok, C.H.; Rao, K.V.; Chakra, C.S. Structural analysis of CuO nanomaterials prepared by novel microwave assisted method. *J. At. Mol.* **2014**, *4*, 803–806.

polymers

MDPI

Article

New Eco-Friendly Thermal Insulation and Sound Absorption Composite Materials Derived from Waste Black Tea Bags and Date Palm Tree Surface Fibers

Mohamed Ali *, Redhwan Almuzaiqer, Khaled Al-Salem, Hassan Alshehri, Abdullah Nuhait, Abdullah Alabdullatif † and Abdulrahman Almubayrik †

Mechanical Engineering Department, College of Engineering, King Saud University, P.O. Box 800, Riyadh 11421, Saudi Arabia; ralmuzaiqer@ksu.edu.sa (R.A.); kalsalem@ksu.edu.sa (K.A.-S.); hashehri@ksu.edu.sa (H.A.)
* Correspondence: mali@ksu.edu.sa
† Undergraduate senior students.

Abstract: A tremendous amount of waste black tea bags (BTBs) and date palm surface fibers (DPSFs), at the end of their life cycle, end up in landfills, leading to increased pollution and an increase in the negative impact on the environment. Therefore, this study aims to utilize these normally wasted materials efficiently by developing new composite materials for thermal insulation and sound absorption. Five insulation composite boards were developed, two were bound (BTB or DPSF with polyvinyl Acetate resin (PVA)) and three were hybrids (BTB, DPSF, and resin). In addition, the loose raw waste materials (BTB and DPSF) were tested separately with no binder. Thermal conductivity and sound absorption coefficients were determined for all boards. Thermal stability analysis was reported for the components of the tea bag (string, label, and bag) and one of the composite hybrid boards. Mechanical properties of the boards such as flexural strain, flexural stress, and flexural elastic modulus were determined for the bound and hybrid composites. The results showed that the thermal conductivity coefficients for all the hybrid composite sample boards are less than 0.07 at the ambient temperature of 24 °C and they were enhanced as the BTB ratio was reduced in the hybrid composite boards. The noise reduction coefficient for bound and all hybrid composite samples is greater than 0.37. The composite samples are thermally stable up to 291 °C. Most composite samples have a high flexure modulus between 4.3 MPa and 10.5 MPa. The tea bag raw materials and the composite samples have a low moisture content below 2.25%. These output results seem promising and encouraging using such developed sample boards as eco-friendly thermal insulation and sound absorption and competing with the synthetic ones developed from petrochemicals in building insulation. Moreover, returning these waste materials to circulation and producing new eco-friendly composites can reduce the number of landfills, the level of environmental pollution, and the use of synthetic materials made from fossil resources.

Keywords: natural thermal insulation materials; noise reduction coefficient; agro-waste utilization; three-point bending moment; thermogravimetric analysis; moisture content

1. Introduction

Worldwide in 2020, seven million tons of tea were produced [1]. The annual growth rate of black tea production is expected to increase by 2.2% to reach 4.42 million tons in 2027 [2]. According to another source [3] the world consumes from 18 to 20 billion cups of tea daily. Negi et al. [4] have shown that 190,000 tons of tea waste were generated due to 857,000 tons of tea produced in India. Consumption of this huge amount of tea either loose or in the form of tea bags creates a tremendous amount of waste. This waste will end up in garbage bins, or landfill, or may be burnt. About 25% of those tea bags is polypropylene for bag sealing. Accordingly, this will pollute, damage, and create an environmental problem.

Another common agricultural waste is date palm surface fibers. According to the Food and Agriculture Organization of the United Nations [5], Saudi Arabia produces 1.2 million tons of dates annually. Those date palm trees develop a tremendous amount of waste during their trimming time every year. One of those wastes is date palm surface fibers, which if not utilized efficiently will have a large environmental impact, especially if burnt, as usually happens in some parts of the world [6]. Saudi Arabia generates five hundred thousand tons of date palm waste annually [7]. Among these wastes, one hundred thousand tons are from leaves and fifteen thousand tons are from the date palm surface fibers [8–10]. In addition, both black tea bags (BTBs) and date palm surface fibers (DPSFs) could have good economic benefits if utilized efficiently since they are biodegradable, eco-friendly, natural, and sustainable. Therefore, the current study is focused on using both BTBs and DPSFs as new novel raw material polymers and their composites for thermal insulation and sound absorption.

Hussien et al. [11] have studied using tea waste as an additive with different percentages in unfired clay brick mixtures to improve their physical and mechanical properties. Ibrahim et al. [12] investigated new ceramic bricks made of zeolitic tuff with a small additive percentage of tea waste to develop their physical and mechanical characteristics as building materials. A similar study has been conducted by Ozturk et al. [13]. Anjum et al. [14] have shown that incorporating tea waste with a small percentage in a clay sample reduced its thermal conductivity from 0.54 to 0.3 W/(m K). Another way of using black tea waste is to extract the microcrystalline cellulose through microwave heating instead of using conventional heating and avoiding the traditional acid hydrolysis method as explained by Debnath et al. [15]. The tea bags were not tested for acoustic control. However, three different grades of spent tea leaves were studied in terms of sound absorption coefficient (SAC) by Wong et al. [16]. They found that the SAC of all sample leaves reached a maximum above 0.7 at a frequency range of 1993–3861 Hz. In a similar study by Ersoy and Kücük [17], SAC was observed to increase at a frequency range of 500–6300 Hz to a maximum of 0.6.

Ali et al. [18] have recently developed new composites made of DPSFs and pineapple leaf fibers (PALFs) using wood adhesive as a binder. Their results showed average thermal conductivity coefficients of 0.054–0.070 W/(m K) for the composite of DPSF, PALF, and the binder. These hybrid composites also had good sound absorption coefficients greater than 0.5 for frequencies greater than 1000 HZ. On the other hand, the bound composites of DPSF and the binder had a SAC > 0.5 for almost all frequencies greater than 300 Hz. DPSF was used with polystyrene to form new thermal insulation composite materials for buildings by Raza et al. [19]. The thermal conductivity coefficient of the composite having 20% DPSF was 0.053 W/(m K). In another study, Raza et al. [20] developed four samples for thermal insulation made of DPSF with different densities using polyvinyl alcohol as a binder. The average thermal conductivity coefficients of the samples were 0.038–0.051 W/(m K). New thermal insulation boards have been developed from DPSF with different densities using cornstarch as a binder by Ali and Abdelkareem [21]. The board's average thermal conductivity coefficients were between 0.0475 and 0.0697 W/(m K). DPSF was hybridized with apple of Sodom fibers and different binders such as wood adhesive, cornstarch, and white cement to form new composites used for thermal insulation by Alabdulkarem et al. [22]. The developed boards had average thermal conductivity coefficients of 0.04234–0.05291 W/(m K). Their boards were also tested for sound absorption in the communication frequency range up to 2000 Hz. Their result showed that adding apple of Sodom fibers to the DPSF enhances the SAC over that of pure DPSF for any kind of binder. The SAC for the hybrid composite of DPSF and apple of Sodom fibers was found to be greater than 0.5 for a frequency range of 800–2000 Hz for all used samples.

On the other hand, Fouladi et al. [23] have measured the sound absorption coefficients for some natural fibers such as dry grass (SAC = 0.14 and 0.98 at 500 Hz and 2000 Hz, respectively), corn (SAC = 0.16 and 0.81 at 500 Hz and 2000 Hz, respectively), coir (SAC = 0.12 and 0.97 at 500 Hz and 2000 Hz, respectively), and sugar cane (SAC = 0.13 and 0.63 at

500 Hz and 2000 Hz, respectively) and found that these natural fiberboards could be used as alternatives for common building acoustic materials. In addition to that, Berardi and Iannace [24] have measured the noise reduction (NR) and sound absorption coefficients for some natural fibers such as sheep wool (NRC = 0.55 and 0.70 for thicknesses of 0.04 and 0.06 m, respectively), hemp (NRC = 0.4 for 0.03 m thickness), coconut (NRC = 0.5 and 0.75 for thicknesses of 0.05 and 0.1 m, respectively), kenaf (NRC = 0.55 and 0.70 for thicknesses of 0.04 and 0.06 m, respectively), cane (NRC = 0.40 and 0.50 for thicknesses of 0.04 and 0.08 m, respectively), cork (NRC = 0.30 for 0.03 m thickness), and cardboard (NRC = 0.50 for 0.10 m thickness). Their results showed that both coefficients depended on porosity, thickness, and the density of the natural fibers and were recommended for use in buildings.

As seen from the preceding literature survey, wasted tea bags have not been tested as insulation materials or for acoustic control, however, wasted DPSF was experimentally proven to be good for both thermal insulation and sound absorption as a polymer or composite with other materials. This finding motivates the current research of testing the novelty of using wasted tea bags for their thermal insulation characteristic and sound absorption as a polymer or as a hybrid with DPSF to form different composites with different densities. Therefore, laboratory experiments were designed to develop new bound and hybrid composite boards made of BTBs and DPSF with polyvinyl acetate as a resin. The thermal conductivity and sound absorption coefficients were measured for these boards with other mechanical and thermal stability analyses. The experimental results are promising for these materials to be used for thermal insulation applications in buildings since they are biodegradable and eco-friendly. Therefore, they could, in the future, be a candidate to replace fossil-fuel-based synthetic and petrochemical thermal insulation materials.

2. Materials and Methods

Used black tea bags (BTBs) were collected from the nearby cafes and the households of the authors' team. These tea bags were washed and dried using an electrical convection oven in the laboratory at 90 °C for 12 h, allowing all the water to evaporate. It should be mentioned that each BTB was kept with its label, string, and two staple pins, one to fix the string to the label and the other to the bag as shown in Figure 1a. The date palm surface fibers (DPSFs) were collected from the agricultural authority during the trimming time at a specific time of the year. During this time, huge amounts of date palm surface fibers are normally discarded. Another source of the DPSFs was the nearby farms during their trimming time. The collected DPSFs were washed and dried in the same way as the tea bags. The DPSFs were cut to approximately 10–15 cm long. Figure 1b defines the DPSF and their cut pieces.

Figure 1. Raw wasted materials; (**a**) details of a tea bag showing the string, bag, label, and two staple pins, and (**b**) date palm surface fibers and their cut pieces.

2.1. Loose Sample Polymer Preparation for Testing

The loose wasted BTBs and the DPSFs were enclosed in two wooden frame boxes with inside sizes of 28.5 × 28.5 × 4.1 cm^3 as shown in Figure 2a,b, one for the BTBs and the other for the DPSFs. This frame size was chosen to be ready for fitting inside the heat flow meter for thermal conductivity coefficient measurements as will be explained in Section 4.

Figure 2. Loose raw materials in the wooden boxes, (**a**) BTBP # 1 and (**b**) DPSFP # 3.

2.2. Composite Bound Sample Preparation for Testing

Safe and non-toxic polyvinyl acetate (PVA) resin (wood adhesive, Figure 3a) was used to bind the loose raw materials of BTBs and DPSFs shown in Figure 2a,b. The ingredients of the resin with its chemical and physical properties can be obtained from [25]. A solution of resin and water was made on a mass basis with a ratio of one part of the resin to two parts of water. The loose raw materials were completely immersed in the resin solution to be sure that each BTB or DPSF was in contact with the resin solution. Two stainless steel molds of size 30 × 30 × H cm^3 (Figure 3b,c) were used to make wetted samples, one for BTBs and the other for the DPSFs, where H is the height of each sample in the mold based on its thickness. The mold surfaces were covered with aluminum foil sheets to prevent the sample from sticking to the stainless steel mold. The mold was then moved to a hydraulic press to adjust the sample to a specified thickness from 20 to 50 mm according to each sample. After that, the stainless steel mold with the sample was moved to the convection oven for drying. The last step was to move the hot dried mold with the sample, keep it until it cooled down to the laboratory ambient temperature, remove the sample from the mold, and move it to a heat flow meter for thermal conductivity coefficient measurements. Figure 4 summarizes the sample preparation processes. Table 1 shows the specifications of the loose polymers and bound composites.

Figure 3. Resin and the stainless steel mold; (**a**) bucket of polyvinyl acetate (PVA) resin (wood adhesive), (**b**) the mold before covering with aluminum foil sheets, (**c**) after covering.

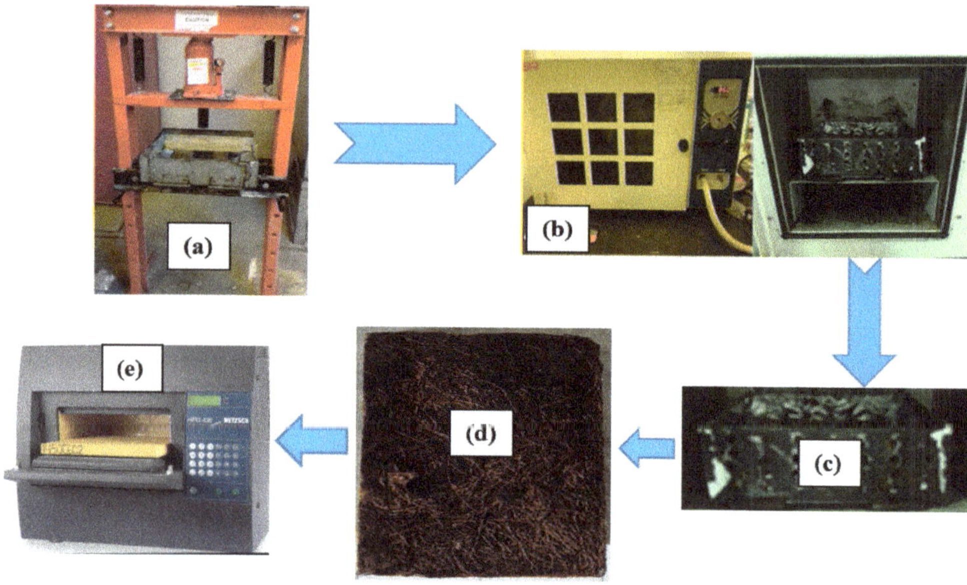

Figure 4. Sample preparation; (**a**) Pressing the sample; (**b**) convection oven; (**c**) cooling the mold with the board; (**d**) sample board; and (**e**) heat flow meter for thermal conductivity measurement.

Table 1. Complete dimensions, mass, percentage, volume, and density of the prepared samples.

Material Specifications	Sample Number								
	BTBP (# 1)	BTBC (# 2)	DPSFP (# 3)	DPSFC (# 4)	COMP (# 5)	COMP (# 6)	COMP (# 7)	COMP (# 8)	COMP (# 9)
BTB %	100	65.4	0.0	0.0	64	39.0	19.0	13.5	34.0
DPSF %	0.0	0.0	100	81.3	21	39.0	57.0	40.5	34.0
Ratio of polymerized (PVA) resin to total mass %	0.0	34.6	0.0	18.7	15.0	22.0	24.0	46.0	32.0
Thickness, δ (mm)	41.0	18.0	46.0	46.0	25.0	32.0	42.0	25.0	29.0
The sample's volume (cm^3)	3330	1620	4132	4140	2250	2880	3780	2250	2610
BTB's mass (g)	439	500	0.0	0.0	375	250	125	100	250
DPSF' mass (g)	0.0	0.0	500	500	125	250	375	300	250
Resin's mass (g)	0.0	265	0.0	115	86	148	158	341	234
Figure #	2(a)	5(a)	2(b)	5(b)	5(c)	5(d)	5(e)	5(f)	5(g)
Apparent density of dried samples (kg/m^3)	132	472	121	149	260	225	174	329	281
Total dried mass (g)	439	765	500	615	586	648	658	741	734

2.3. Composite Hybrid Sample Preparation for Testing

Hybrid composite means mixing BTBs with DPSFs with different compositions with the same binder used in Section 2.2 (PVA) using the same mold specified and shown in Figure 3. Table 1 shows the complete specifications, dimensions, percentage of each polymer in the sample, mass, volume, figure number of the sample, and the density of the samples. Figure 5 shows the real pictures of all prepared samples with the resin, either bound or hybrid composites.

Figure 5. Plan and side view of the prepared samples; (**a**) bound composite of BTBs and resin, (**b**) bound composite of DPSFs and resin, and (**c**–**g**) hybrid composite of BTBs, DPSFs, and resin, see Table 1 for more details.

3. Characterization

3.1. Mechanical Test for Bound and Hybrid Composite Samples

The bending moment test (three-point) was obtained following the standard ASTM D790-03 [26] for all composite samples # 2, 4, 5, 6, 7, 8, and 9, either bound or hybrid, as shown in Table 1. The cut specimens for the test (Figure 6) have dimensions $20 \times 5.0 \times \delta$ cm^3, where δ is the thickness as defined in Table 2. The universal testing machine (UTM, INSTRON 5984) (Instron, Norwood, MA, USA) (Figure 7) in our mechanical engineering laboratory with a crosshead speed of 2 mm/min was used to determine the deflection D, the applied force F, flexural strain ϵ_f, and flexural stress σ_f as defined by Equation (1), where E_f is the flexural elastic modulus. Table 2 specifies the other used dimensions.

$$\sigma_f = \frac{3FL}{2b\,\delta^2},\ \epsilon_f = \frac{6\,D\delta}{L^2},\ E_f = \frac{L^3 S}{4b\delta^3} \quad (1)$$

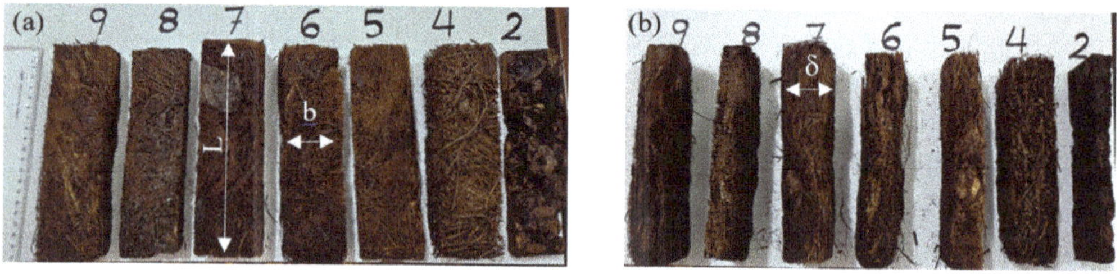

Figure 6. Cut specimens for the bending test; (**a**) Plan view and (**b**) side view.

Table 2. Bending specimens' specifications.

Specimens' Number	Thickness (δ) (mm)	Width (*b*) (mm)	Span (*L*) (mm)	Slope (*S*) (N/mm)
2	18.0	47.0	150.0	3.2
4	46.0	51.0	150.0	1.4
5	25.0	55.0	150.0	7.9
6	32.0	53.0	150.0	0.45
7	42.0	51.0	150.0	1.7
8	25.0	53.0	160.0	49.0
9	29.0	51.0	160.0	5.2

Figure 7. The universal testing machine (UTM, INSTRON 5984) used for the bending test.

3.2. *Thermal Conductivity Coefficient Measurement*

A heat flow meter (Lambda, HFM 436 (NETZSCH-Gerätebau GmbH, Wittelsbacherstraße 42, 95100 Selb, Germany, Figure 4e) was used to determine the thermal conductivity coefficient for all samples: loose polymers (Figure 2a,b) and bound or hybrid composites (Figure 5a–g). The HFM follows the ASTM-C518 [27] standard testing method. The allowable size of the sample to be used by the HFM is $30 \times 30 \times \delta$ cm^3, where δ is a variable thickness up to 10 cm. A self-automated sensor provided by the HFM measures the thickness. The thermal conductivity coefficient and the temperature were measured up to an accuracy of ±1% to 3% W/(m K) and ±0.01 °C, respectively, as provided by the manufacturer of the HFM. The thermal conductivity coefficients were measured for all samples from 20 °C to 80 °C.

3.3. *Sound Absorption Coefficient Measurement*

Two impedance tubes with inside diameters of 10 cm and 3 cm of BSWA (BSWA Technology Co. Ltd., Bejing, China) were used to measure the sound absorption coefficient (SAC). The large tube (10 cm) and the small one (3 cm) were used for a frequency range of 63–1600 Hz and 800–6300 Hz, respectively. Therefore, samples were prepared for this test with 10 cm and 3 cm diameters for bound and hybrid composites as shown in Figure 8 with specification details in Table 1. The impedance tubes' principle of working and the position of microphones at each frequency range can be obtained from our previous publication, Ali et al. [28]. The BSWA software VA-Lab IMP, Version: V1.03, conforms to ISO 10534-1 [29] and ISO 10534-2 [30] standards.

Figure 8. Samples for sound absorption coefficient measurement.

3.4. *Microstructure Analyses of the Black Tea Bags*

The black tea bags were characterized by applying scanning electron microscopy (SEM) and energy dispersive X-ray (EDX) spectroscopy analysis.

3.4.1. Scanning Electron Microscopy Analysis

Scanning electron microscopy (SEM) (JEOL; JSM7600F, Peabody, MA 01960, USA) was used to determine the surface morphology of the black tea bag sample (BTBP, # 1) at different magnifications. The tea bag sample must be oven-dried first and then coated with platinum to avoid any electrostatic charging, which may happen during the test.

3.4.2. Energy Dispersive X-Ray (EDX) Spectroscopy Analysis

The chemical composition of the black tea bag was obtained using EDX analysis at different spots. This test gives qualitative results about the composition of the tea bag. It should be mentioned that the field-emission SEM (FE-SEM) (JEOL; JSM7600F) is equipped with EDX.

3.5. *Thermal Stability of the Tea Bags*

Thermal stability analysis was performed for the tea bag components such as string, bag, and label using thermogravimetric analysis (TGA) and its differential (DTGA). The test uses the SDT Q600 V20.9 Build 20 setup (New Castle, DE, USA), a TA instrument fitted with a nitrogen purge gas. The test was also performed for the composite sample # 7. The initially used mass for the tea bag, string line, label, and composite sample # 7 are 19.45 mg, 24.96 mg, 25.42 mg, and 7.96 mg, respectively. An alumina pan was used to heat the sample mass from room temperature up to 550 °C with a heating rate of 10 °C/min. The flow rate of the nitrogen gas was 100 mL/min. It should be noted that the thermal stability of the DPSF was described in our previous paper [21].

3.6. Moisture Content Test of the Tea Bags

Figure 9 presents the samples used for moisture content determination. These samples were first oven-dried for 8 h at 100 °C and, at the end of the drying time, the masses were recorded as $m2$. Then, the samples were left in the laboratory environment of relative humidity and temperature of 51.7% and 21.6 °C, respectively. The masses were measured every 5 min as $m1$ until the readings were steady. The amount absorbed by each sample was calculated as a percentage amount of that just after drying, $m2$, following Equation (2) of the standard ASTM D2974-07A [31].

$$\% \text{ of moisture content} = \frac{m1 - m2}{m2} \times 100 \quad (2)$$

Figure 9. Samples used for moisture content determination; numbers refer to the sample number in Table 1.

4. Results and Discussion

4.1. Three-Point Bending Moment

Figure 10a,b show the force versus deflection and the flexure stress versus flexure strain, respectively, for the bound and hybrid composite samples # 2, 4, 5, 6, and 7. Table 3 shows flexure modulus, E_f, flexural stress σ_f, and flexural strain ε_f as defined by Equation (1) for the composite samples. It should be noted that σ_f presents its maximum at the end of the linear region following [32] as shown in Figure 10b. Furthermore, the flexural strain ε_f is calculated at the maximum value of σ_f. The slope S was calculated from the load–deflection curve in the linear zone (Figure 10a). By inspection of specimens # 2, 5, and 7 in Table 3, it is observed that the flexure modulus, E_f, increases as the density increases from bound composite to hybrid composite, which agrees with the results of [33,34]. The degree of coherence (percent of the polymerized (PVA) resin in the specimen) presents an important factor in the flexure modulus E_f and flexural stress σ_f. Accordingly, as this percentage increases one would expect an enhancement of both E_f and σ_f as shown in Figure 11a–d when comparing samples # 7 and # 8 as a group and # 9 and # 6 as another group, respectively. Each group almost has an equal mass of polymers but with increased resin mass. Therefore, Figure 11a–d show that the density and the percentage of polymerized resin have an essential enhancement role in both E_f and σ_f. In addition, inspection of Table 3 and Figures 10 and 11 indicates that composite specimen # 2 is the best among the bound ones and # 8, 5, and 9 present the best hybrid composites in descending order. Another factor affecting the flexure modulus is the material's compactness (small thickness). Therefore, it could be concluded that as the density and the percentage of resin increase, a small thickness enhancement would be expected in the flexure modulus.

Mechanical parameters' comparisons are presented in logarithmic scale bar charts in Figure 12 for all samples.

Table 3. Flexure Modulus, Flexural Stress, and Flexural Strain of the tested specimens.

Fabricated Specimen No.	Slope (S) (N/mm)	Thickness (δ) (mm)	Density, (kg/m^3)	Polymerized (PVA) Resin to the Total Mass %	Flexure Modulus (MPa), E_f	Flexural Stress (MPa), σ_f	Flexural Strain at Flexural Strength, ε_f
2	3.2	18.0	472.0	34.6	10.5 ± 1.4	0.16 ± 0.01	0.01 ± 0.0005
4	1.4	46.0	149.0	18.7	0.24 ± 0.03	0.12 ± 0.01	0.4 ± 0.018
5	7.9	25.0	260.0	15.0	7.6 ± 0.98	0.37 ± 0.03	0.05 ± 0.002
6	0.45	32.0	225.0	22.0	0.21 ± 0.03	0.06 ± 0.005	0.25 ± 0.011
7	1.7	42.0	174.0	24.0	0.39 ± 0.05	0.10 ± 0.008	0.33 ± 0.015
8	49.0	25.0	329.0	46.0	68.9 ± 8.87	1.52 ± 0.13	0.02 ± 0.0009
9	5.2	29.0	281.0	32.0	4.3 ± 0.55	0.16 ± 0.014	0.02 ± 0.0008

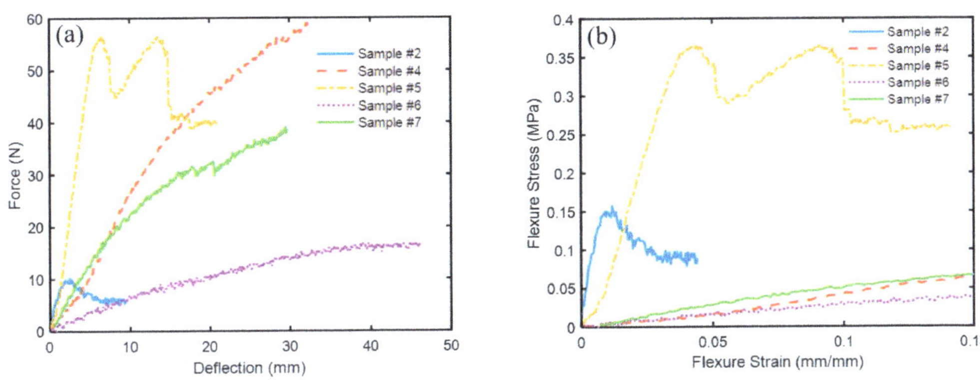

Figure 10. Profiles for the three-point bending test for bound and hybrid composites; (**a**) force–deflection and (**b**) stress–strain.

4.2. Thermal Conductivity Measurements

Figure 13 shows the thermal conductivity coefficient profiles for the samples # 1, 2, 3, 4, 5, 6, and 7. The solid lines through the data represent the linear fitting curves with details given in Table 4.

It should be noted that all values of the thermal conductivity coefficients at the ambient temperature of about 24 °C are less than 0.07 W/(m K), strongly suggesting that the samples can be used as thermal insulation materials. Furthermore, adding resin to the polymer and composite samples increases their thermal conductivity coefficient compared to pure polymer samples. This is clear by comparing the values of k of BTBP (⧫) and its composite BTBC (•) and between DPSFP (×) and DPSFC (o). Furthermore, adding DPSFP to form composite samples such as # 5, 6, or 7 enhanced their thermal conductivity coefficients compared to the bound BTBC sample # 2, keeping in mind the percentage of resin. Based on that, composite sample # 7 represents the best among all composite samples. Figure 14a,b were constructed to clarify the effect of increasing the percentage of resin in the sample. Figure 14a shows that effect, since samples # 7 and # 8 almost have the same polymer mass percentage but the main difference is in the percentage of resin. Sample # 7 has 24% but sample # 8 has 46%. The same applies to samples # 6 and # 9, where sample # 6 has 22% and sample # 9 has 32% resin as shown in Figure 14b.

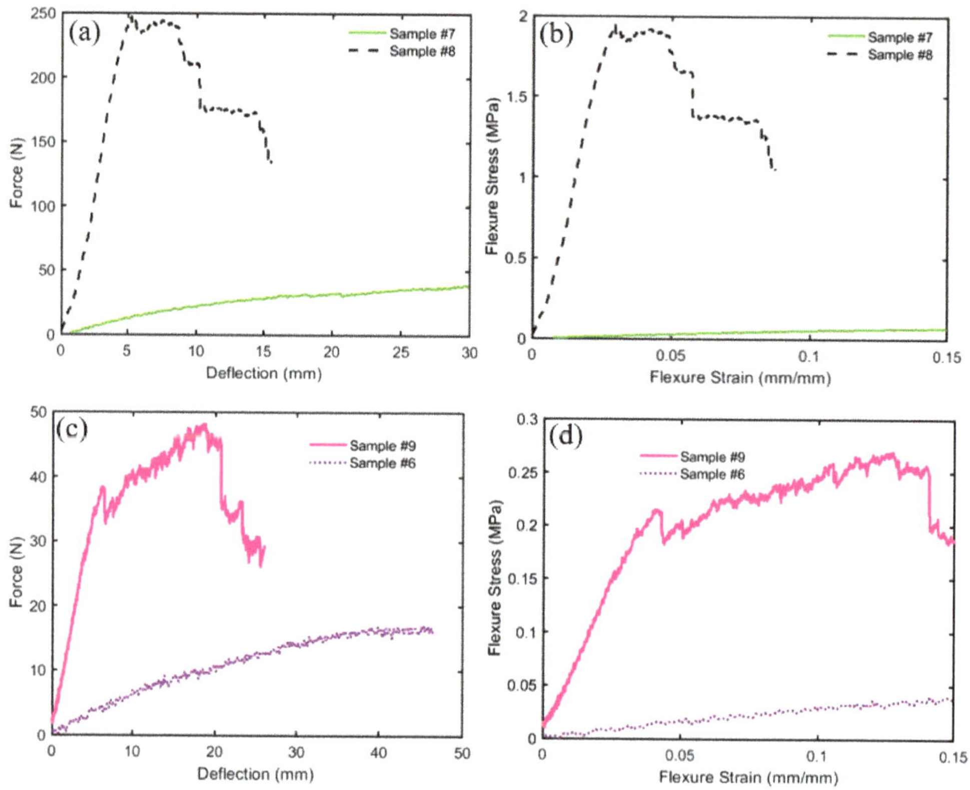

Figure 11. Profiles for the three-point bending test show the effect of increasing the density and the resin in the specimens on force–deflection and flexure stress–strain curves; (**a**,**b**) comparison between samples # 7 and # 8 and (**c**,**d**) comparison between samples # 9 and # 6.

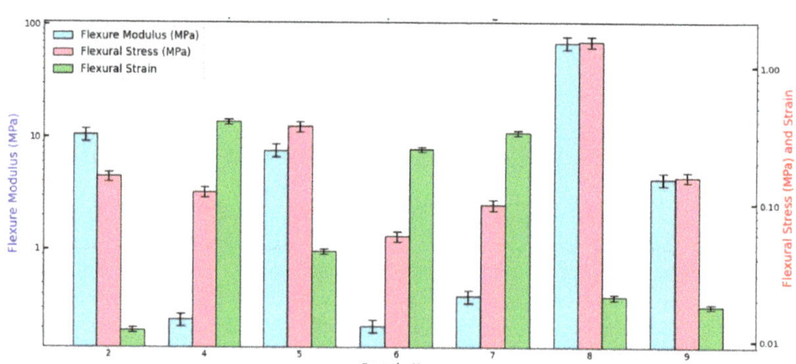

Figure 12. Bar chart presenting Flexure Modulus, Flexural Stress, and Flexural Strain for all samples.

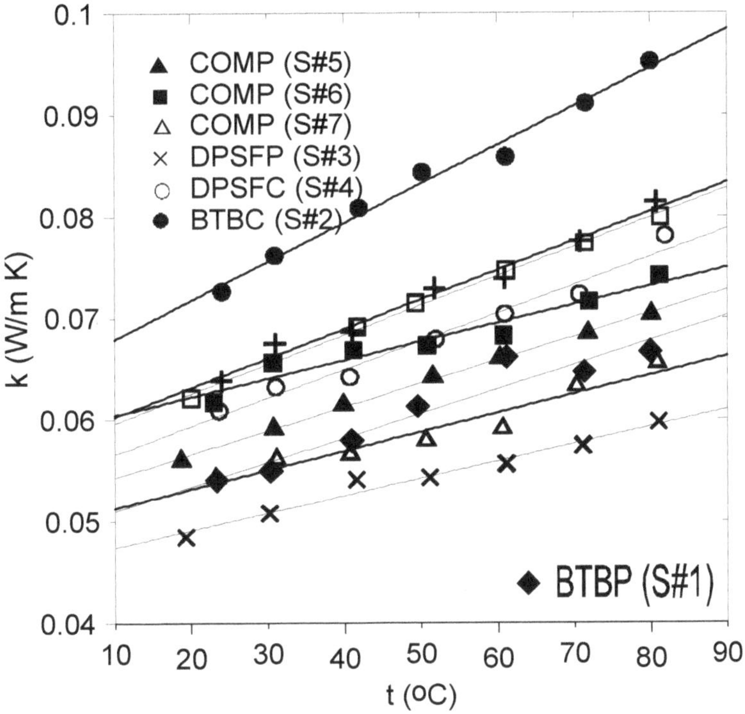

Figure 13. Thermal conductivity coefficient curves for samples # 1–7. Solid lines represent the fitting curves.

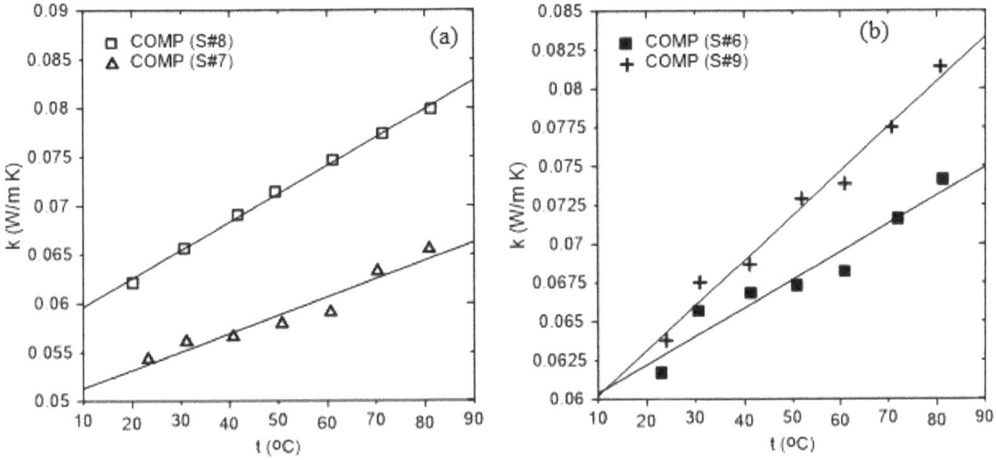

Figure 14. The effect of resin on thermal conductivity coefficient; (**a**) comparison between samples # 8 and # 7 and (**b**) between samples # 6 and # 9.

Table 4. Linear fitting equations for the thermal conductivity coefficient profiles given in Figure 13.

Equation (3)			$k = A + B \times t$		
Sample #	**Symbol**	**A**	**B**	**R^2, %**	**Density, kg/m^3**
1 (BTBP)	♦	0.049	0.0002	92.6	132.0
2 (BTBC)	•	0.064	0.0004	98.7	472.0
3 (DPSFP)	×	0.046	0.0002	97.2	121.0
4 (DPSFC)	o	0.054	0.0003	97.4	149.0
5 (COMP)	▲	0.052	0.0002	99.8	260.0
6 (COMP)	■	0.059	0.0002	93.0	225.0
7 (COMP)	Δ	0.049	0.0002	94.0	174.0
8 (COMP)	□	0.057	0.0003	99.7	329.0
9 (COMP)	+	0.057	0.0003	98.1	281.0

Table 5 compares the thermal conductivity coefficient of this study's developed boards to that of conventional synthetic insulation materials. Even though the thermal conductivity values of the synthetic materials are lower than those of the study's developed boards, these boards are inexpensive, environmentally friendly, and economical compared to the synthetic ones. In addition, these boards have no CO_2 emission, compared to that produced through the manufacturing process of the synthetic insulation materials due to the use of fossil fuels. It should be noted that Asdrubali et al. [35] have classified thermal insulation materials based on their thermal conductivity coefficient as good if it is lower than 0.0566 W/(m K), intermediate for $0.055 < k < 0.091$ W/(m K), and poor for $k > 0.091$.

Table 5. Comparison between the study's developed boards and the conventional synthetic materials.

Materials	Density (kg/m^3)	Thermal Conductivity (W/mK)	References
BTBP (# 1)	132	0.05405–0.0666	This study
BTBC (# 2)	472	0.072636–0.095211	This study
DPSFP (# 3)	121	0.048461–0.059728	This study
DPSFC (# 4)	149	0.060877–0.078064	This study
COMP (# 5)	260	0.05597–0.0703	This study
COMP (# 6)	225	0.0617–0.0741	This study
COMP (# 7)	174	0.0543–0.0655	This study
COMP (# 8)	329	0.062125–0.079914	This study
COMP (# 9)	281	0.063792–0.081400	This study
Recycled polyethylene terephthalate (PET)	30	0.0355	[36]
Recycled (PET) (commercialized)	15–60	0.034–0.039	[35]
Recycled glass fibers (commercialized)	100–165	0.038–0.050	[35]
Polyurethane foam	30–80	0.02–0.027	
Rock wool	40–200	0.033–0.040	[35]
Expanded polystyrene (XPS)	15–35	0.031–0.038	[35]
Extruded polystyrene (EPS)	32–40	0.032–0.037	[35]
Kenaf	30–180	0.034–0.043	[35]
Sheep wool	10–25	0.038–0.054	[35]

4.3. Sound Absorption and Noise Reduction Coefficients

Figure 15 shows the sound absorption coefficients (SACs) for a wide range of frequencies (100–6000 Hz) for bound and hybrid composite samples # 2, 4, 5, 6, and 7. In the range of communication frequency up to 2000 Hz, it is observed that the bound composite sample # 2 of black tea bags has a SAC > 0.5 in the frequency range of 900–2000 Hz with a peak of SAC $\approx$ 0.9 at about 1800 Hz. Furthermore, the hybrid composite samples # 5, 6, and 7 have similarly good SAC. For example, # 5 has SAC > 0.4 for a range of 1000–2000 Hz with a peak of SAC $\approx$ 1 at 2000 Hz, # 6 has SAC > 0.5 for a range of 600–2000 Hz with a peak of SAC $\approx$ 0.85 at about 1200 Hz, and # 7 has SAC > 0.5 at a range of 900–2000 Hz with a peak SAC $\approx$ 0.76 at 1800 Hz. However, the bound composite sample # 4 has a good SAC > 0.4 for a frequency > 3500 Hz and it looks like it does not have a good SAC in the communication range. This could be attributed to the high airflow resistance, which indicates that the pores are close to each other and tend to obstruct the airflow, hence low SAC is expected [24]. It should be noted that the thickness, density, porosity, and percentage of polymerized binders in the board influence the value of SAC. Therefore, the presence of air passages in the boards helps absorb the sound waves and hence increase the SAC, however, if the boards have continuous pores or passages through their thickness, it allows for sound waves to pass and reduce the SAC. The percentage of polymerized binder in the sample controls its structural rigidity and hence the SAC values. The dashed horizontal line in Figure 15 presents a borderline SAC = 0.4 and, when the SAC exceeds this value, the boards are considered good sound absorbers [37]. The average values for a one-third octave of the sound absorption coefficients at frequencies 250, 500, 1000, and 2000 Hz were used to estimate the noise reduction coefficient (NRC). These values were rounded to the nearest 0.05 following [24,38]. Table 6 shows the values of the noise reduction coefficient (NRC) for the samples # 2, 4, 5, 6, and 7. Accordingly, as the NRC exceeds 0.2 or 0.4, the material is considered sound absorbing or has a practical value, respectively [39–41]. Therefore, samples # 2 and # 5 are considered good sound absorbers, and samples # 6 and # 7 are better yet and have practical value.

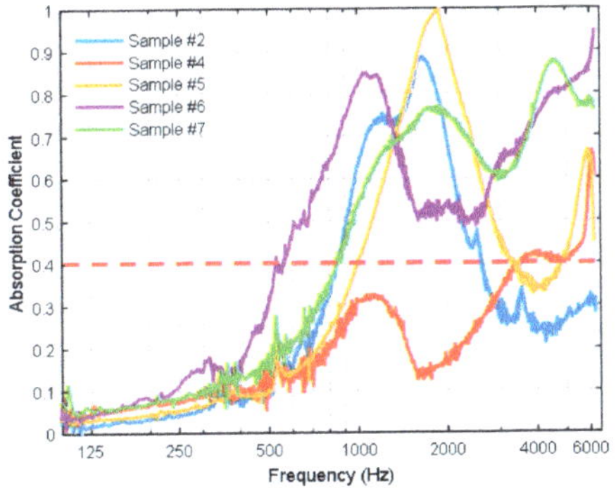

Figure 15. The sound absorption coefficient profiles for samples # 2, 4, 5, 6, and 7 at a frequency range 100–6000 Hz. Dashed line presents a borderline SAC = 0.4.

Figure 16a shows a bar chart of the noise reduction coefficient for bound and hybrid composite samples # 2, 4, 5, 6, and 7. Figure 16b compares the NRC of the samples and the SAC at one-third octave values. Inspection of SACs and the NRCs indicates that the above-developed bound composite sample of BTB (# 2) and its hybrid composites with DPSF, samples # 5, 6, and 7, have a better acoustic characteristic in absorbing the sound, which promotes them as suitable for building applications.

Table 6. Noise reduction coefficient for samples # 2, 4, 5, 6, and 7.

Sample Number	Density, kg/m^3	Frequency (Hz)				NRC
		250	500	1000	2000	
		Sound Absorption Coefficients (SAC)				
2	472	0.0397	0.1162	0.6149	0.7098	0.35
4	149	0.0736	0.1459	0.3029	0.1683	0.15
5	260	0.054	0.0933	0.391	0.9402	0.35
6	225	0.1097	0.2862	0.8149	0.5376	0.45
7	174	0.0818	0.1736	0.5362	0.7534	0.40

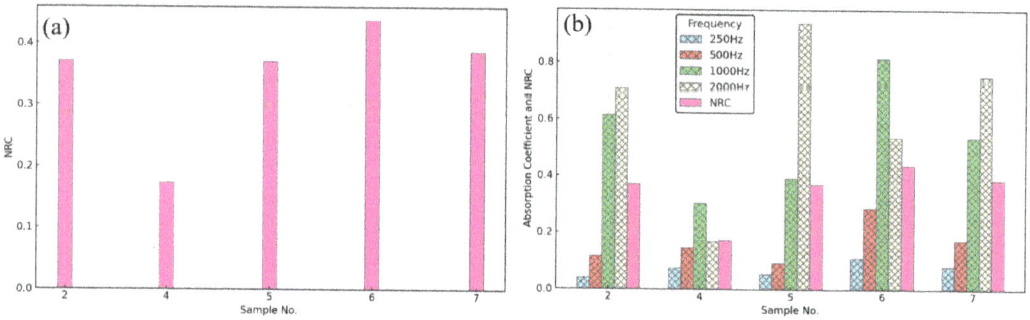

Figure 16. Noise reduction and sound absorption coefficient profiles for samples # 2, 4, 5, 6, and 7; (**a**) NRCs and (**b**) comparison between NRC and SAC at one-third octave values.

Table 7 compares the developed samples' sound absorption and noise reduction coefficient with similar agricultural and synthetic insulation materials using the same technique, which are in the same order of magnitude as similar materials.

Table 7. Comparison of SAC and NRC with some similar materials in the literature.

Materials	Density (kg/m^3)	Thickness of the Board or Fiber (m)	Sound Absorption Coefficient, SAC				NRC	References
			Frequency, Hz					
			250	500	1000	2000		
BTBC (# 2)	472	0.018	0.0397	0.1162	0.6149	0.7098	0.35	This study
DPSFC (# 4)	149	0.046	0.0736	0.1459	0.3029	0.1683	0.15	This study
COMP (# 5)	260	0.025	0.054	0.0933	0.391	0.9402	0.35	This study
COMP (# 6)	225	0.032	0.1097	0.2862	0.8149	0.5376	0.45	This study
COMP (# 7)	174	0.042	0.0818	0.1736	0.5362	0.7534	0.40	This study
Polyurethane foam	95	------------	0.02	0.01	0.11	0.16	0.08	[37]
Kenaf (light)	50	0.06	0.19	0.33	0.68	0.9	0.55	[24]
Wood (fibers)	100	0.04	0.40	0.50	0.65	0.91	0.60	[24]
Coconut	60	0.04/0.06	0.2	0.34	0.67	0.79	0.50	[24]
Cork	100	0.03	0.02	0.10	0.30	0.86	0.30	[24]
Cane (only wooden)	400	0.04	0.06	0.12	0.47	0.43	0.25	[24]
Fleece (100% polyester) fiber	60	0.0035	0.08	0.12	0.19	0.21	0.15	[42]
Queenscord fiber	160	0.0019	0.05	0.14	0.34	0.30	0.20	[42]

Table 7. *Cont.*

Materials	Density (kg/m^3)	Thickness of the Board or Fiber (m)	Sound Absorption Coefficient, SAC				NRC	References
			Frequency, Hz					
			250	500	1000	2000		
Mesh fiber	100	0.0033	0.18	0.02	0.05	0.07	0.10	[42]
Suede fiber	300	0.0006	0.09	0.13	0.24	0.28	0.20	[42]
Wood fiberboard	480	0.018	0.11	0.14	0.21	0.34	0.20	[43]
Palm oil leaves	152	0.010	---	0.05	0.08	0.19	0.10	[44]
Lemongrass	201	0.010	---	0.06	0.15	0.45	0.20	[44]

4.4. Microstructure of the Black Tea Bags

Figure 17a,b show the morphology analysis of a tea bag at two different magnifications (100 and 250) using scanning electron microscopy (SEM). The surface structure shows the external morphology of the tea bag's texture. The texture has many pores as shown in the SEM images, marked as red spots in Figure 17c,d and white spots in Figure 17e,f. The BTB total surface optical porosity percentage was determined by using ImageJ software, version 1.8.0. Digital SEM photomicrographs (Figure 17a,b) were preprocessed and analyzed using this software. The analysis was performed two times using different thresholds, and then the average was calculated. The porosity of a given sample can be described as the percentage ratio of available pore area to the total analyzed area. The 2D porosity range was between 18 and 20% for the surface as shown in Figure 17c,d. Accordingly, the relative density of the BTB was calculated using the following relationship:

$$\text{Relative Density} = 1 - \text{Porosity} \quad (3)$$

where porosity is expressed as a decimal (e.g., 18% porosity is 0.18). Based on this calculation, the relative density of the BTB is between 0.80 and 0.82.

Figure 18 shows the EDX analysis of the tea bag texture at two different spots. The constituents of the tea bag are carbon (66.65% to 71.23%) and oxygen (28.77% to 33.35%) and their percentages are different from one spot to another as shown in the figure.

4.5. Thermal Stability of the Tea Bags and Their Composites

Figure 19a,b show the thermal stability analysis of the components of the tea bag such as the tea bag itself, label, and string. The decomposition of the composite sample # 7 is also shown in the figure for comparison. Thermogravimetric analysis (TGA) and its differential (DTGA) are presented in Figure 19a,b, respectively. Figure 19a indicates that all the tested components, tea bag, label, and string, and composite material of # 7 are thermally stable up to 284.5, 300.7, 287, and 291 °C, respectively, since they lost only about 10% of their mass, which is the water vapor in the components and the sample. These stable temperatures are comparable to those obtained by Alemdar and Sain [45] for untreated wheat straw fiber. The major points of decomposition and degradation of these components and sample # 7 are shown in Table 8. The major degradation temperature for almost all components and the composite sample ranges between 300 and 400 °C as demonstrated in Figure 19b. It should be noted that composite sample # 7 reached a char formation at 800 °C with a 20% mass. These degradation results agreed with the review study by Asim et al. [46] for most natural fibers.

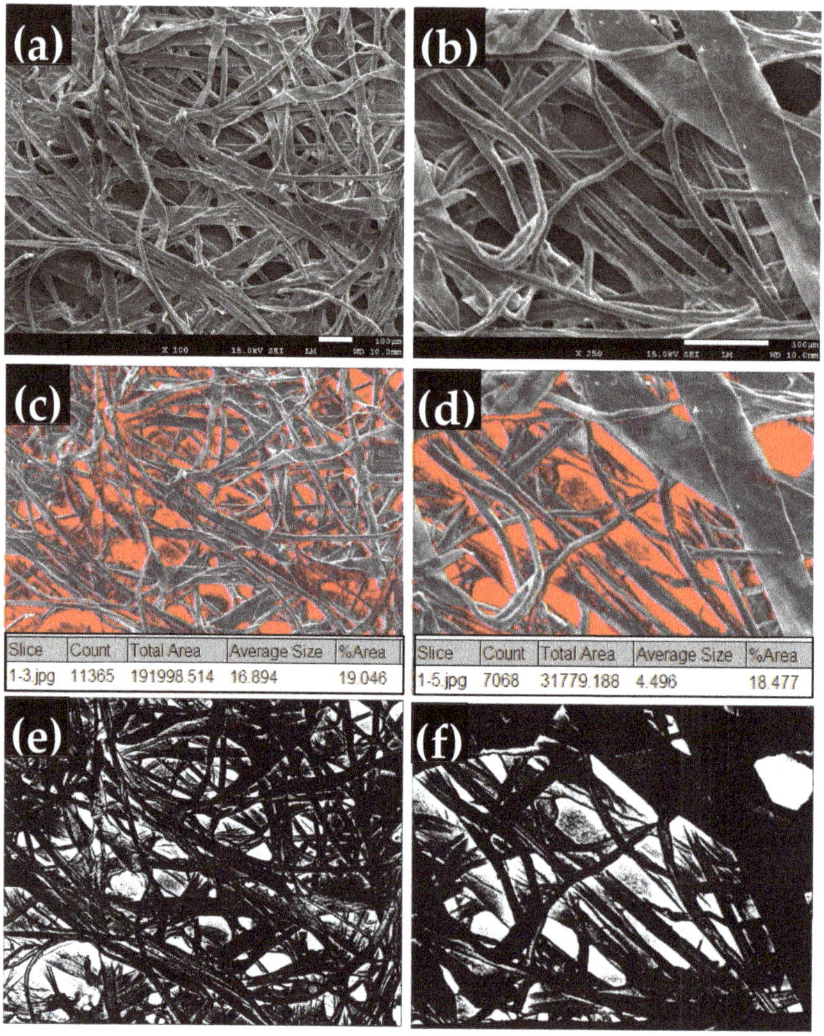

Figure 17. (**a**,**b**) Black tea bag texture shows the fibers and pores at two magnifications; 100× and 250×, (**c**–**f**) the porosity analysis using ImageJ software at the corresponding magnifications.

4.6. Moisture Content of the Tea Bags

Figure 20 shows the steady state profiles of the moisture content for the loose polymer BTB (# 1), bound BTB composite (# 4), bound DPSF composite, and the hybrid composites of both BTB and DPSF (# 5, 6, and 7) following Equation (2). Inspection of the figure indicates that the loose sample (BTB, # 1) absorbed more moisture, about 2.1%, than the bound sample (BTBC, # 2), which absorbed about 1.1%. This could be attributed to the fact that adding resin to the polymer blocked most of the pore spaces and hence reduced the composite's ability to absorb the water. It is also observed that as the percentage of the DPSF increases in the composite samples (# 5, 6, and 7) the moisture content increases due to the increasing number of pores in the composites. Furthermore, as the percentage of resin increases in the composite, this leads to a decrease in the moisture content as shown for similar samples # 7 and # 8 and between samples # 6 and # 9. Accordingly, for all tested polymers and composites, the maximum percentage of moisture content is about 2.1%, which is much lower than the safe moisture content of 16% of similar natural straw fibers as recommended by Bainbridge [47].

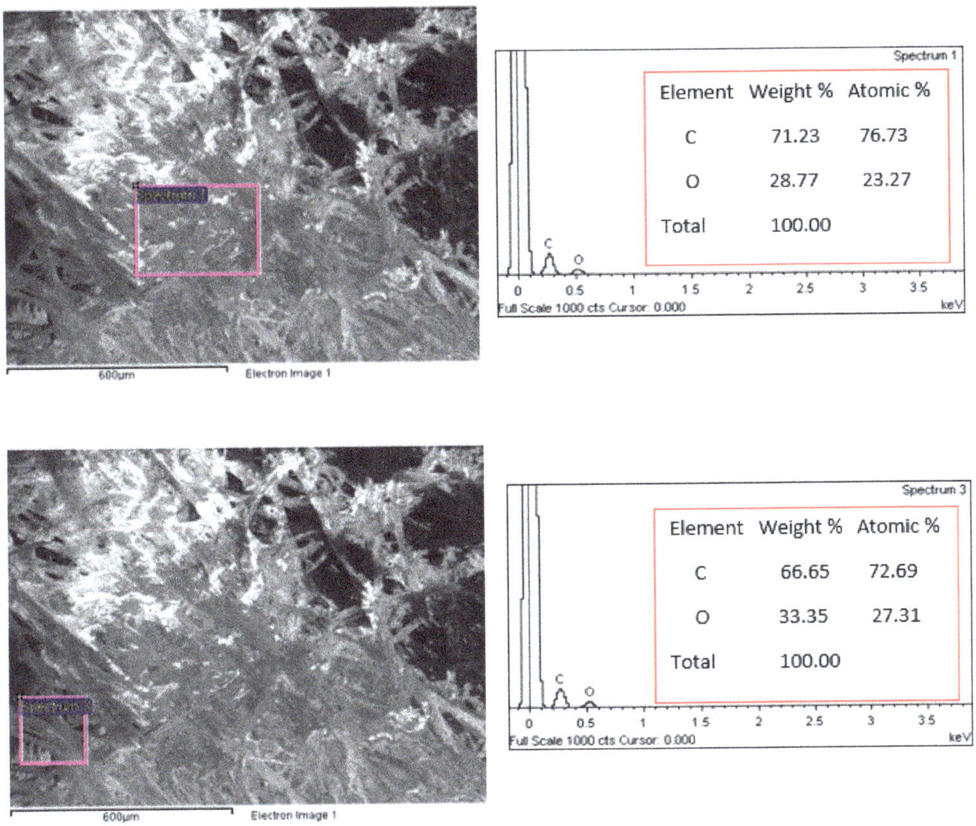

Figure 18. EDX analyses of the tea bag texture at two different spots.

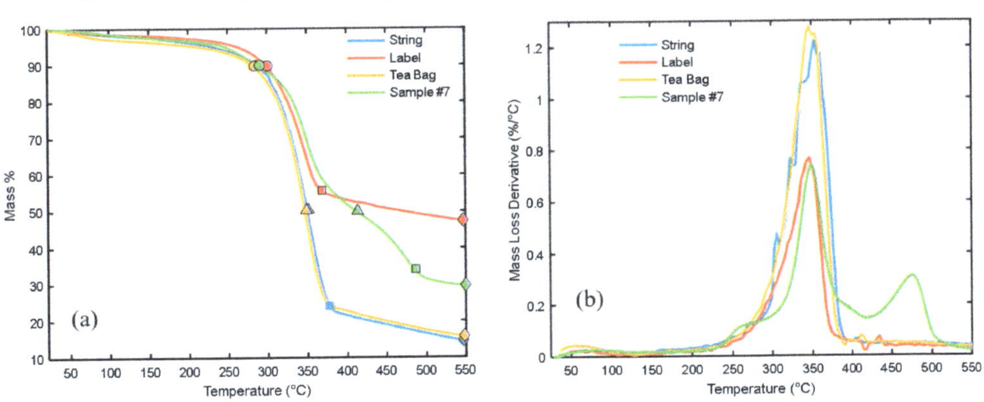

Figure 19. Thermal stability analysis for the tea bag, label, string, and composite sample # 7; (**a**) Thermogravimetric analysis (TGA) and (**b**) Differential thermogravimetric analysis (DTGA). Blue, orange, yellow, and green symbols present string, label, tea bag, and sample 7, respectively.

Table 8. Degradation of tea bag components and hybrid composite sample # 7.

Thermally Stable Zone		**$T_{50\%}$ Degradation**		**End of Major Degradation**		**Char Formation**	
			Tea Bag, 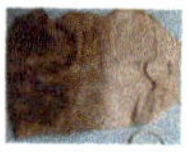				
Mass % 90	Temp. (°C) 284.5	Mass % 50	Temp. (°C) 349.0	Mass % 24.3	Temp. (°C) 379.5	Mass % 16	Temp. (°C) 548.0
			String, 				
Mass % 90	Temp. (°C) 287.0	Mass % 50	Temp. (°C) 351.0	Mass % 24	Temp. (°C) 378.0	Mass % 15	Temp. (°C) 547.0
			Label, 				
Mass % 90	Temp. (°C) 300.7	Mass % 50	Temp. (°C) 473.0	Mass % 56	Temp. (°C) 368.6	Mass % -------	Temp. (°C) -------
			Composite Sample # 7, 				
Mass % 90	Temp. (°C) 291.0	Mass % 50	Temp. (°C) 413.5	Mass % 34	Temp. (°C) 487.0	Mass % 20.4	Temp. (°C) 800.0

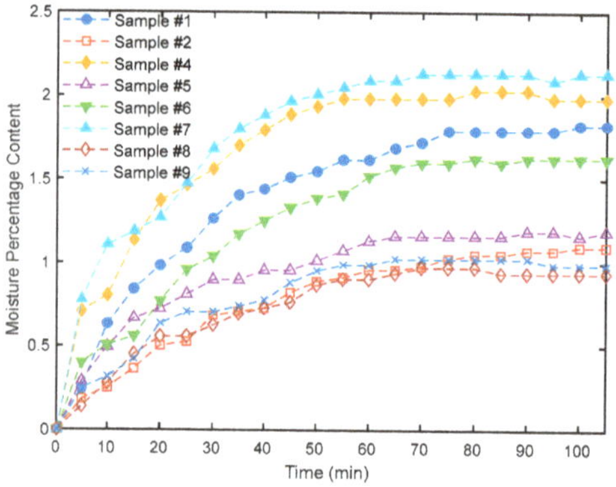

Figure 20. Moisture profiles for loose polymer BTB (sample # 1), bound composite samples (# 2 and # 4), and hybrid composite ones (# 5, 6, 7, 8, and 9).

5. Conclusions

New eco-friendly composite boards were developed on the laboratory scale from waste black tea bags (BTBs) and discarded date palm tree surface fibers (DPSFs) using polyvinyl acetate resin. These composites were tested to be used as thermal insulation and sound absorption materials to compete with synthetic ones developed from petrochemicals. The thermal conductivity coefficient of the composite bound BTBs was only in the range of 0.0726 to 0.0952 W/(m K) for a temperature range of 24.0 °C to 80.0 °C. However, adding DPSF to the BTBs forms hybrid composite materials with enhanced thermal conductivity coefficient for samples # 5, 6, and 7 from 0.056 to 0.070 W/(m K), 0.062 to 0.074 W/(m K), and 0.054 to 0.066 W/(m K) compared to the bound composite BTBC # 2, for the same temperature range, respectively. The noise reduction coefficient is greater than 0.35 for composite samples # 2, 5, 6, and 7 using the average value of a one-third octave. The composite sample boards are thermally stable up to 291.0 °C. Furthermore, the tested composite specimens' flexure modulus, flexural stress, and flexural strain have an average value of 0.21–10.5 MPa, 0.12–0.37 MPa, and 0.01–0.40, respectively. Moreover, the newly developed samples proved to have a low percentage of moisture content below 2.2%. These promising conclusions encourage using these composite materials for thermal insulation and sound absorption in buildings since they are biodegradable, eco-friendly, and natural. Moreover, using such newly developed composites will reduce the negative environmental impacts. In addition to that, these composites offer several advantages, such as low cost, availability of renewable natural resources, and biodegradability over synthetic and petrochemical materials.

Author Contributions: Conceptualization, M.A., K.A.-S., H.A. and A.N.; Methodology, R.A., K.A.-S., A.N., A.A. (Abdullah Alabdullatif) and A.A. (Abdulrahman Almubayrik); Validation, M.A., R.A., K.A.-S. and A.N.; Formal analysis, M.A., R.A., H.A., A.A. (Abdullah Alabdullatif) and A.A. (Abdulrahman Almubayrik); Investigation, M.A., K.A.-S., H.A. and A.N.; Resources, M.A.; Data curation, R.A., K.A.-S., H.A. and A.N.; Writing—original draft, M.A. and H.A.; Writing—review & editing, M.A.; Supervision, M.A.; Funding acquisition, M.A. All authors have read and agreed to the published version of the manuscript.

Funding: The authors would like to appreciate the support and funding from Researchers Supporting Project (number RSPD2024R983), King Saud University, Riyadh, Saudi Arabia.

Institutional Review Board Statement: Not applicable.

Data Availability Statement: The original contributions presented in the study are included in the article, further inquiries can be directed to the corresponding author.

Conflicts of Interest: The authors declare that there are no conflicts of interest.

Nomenclatures

A	Constant
B	Constant
BTBP	Black tea bag polymer (sample # 1)
BTBC	Black tea bag composite (sample # 2)
COMP	Composite sample of BTB, DPSF, and the binder
DPSFP	Date palm surface fiber polymer (sample # 3)
DPSFP	Date palm surface fiber composite (sample # 4)
DTGA	Differential thermogravimetric analysis
k	Thermal conductivity coefficient
NRC	Noise reduction coefficient
SAC	Sound absorption coefficient
S	Slope
t	Temperature (°C)
TGA	Thermogravimetric analysis

References

1. Crespo-López, L.; Coletti, C.; Arizzi, A.; Cultrone, G. Effects of using tea waste as an additive in the production of solid bricks in terms of their porosity, thermal conductivity, strength and durability. *Sustain. Mater. Technol.* **2024**, *39*, e00859. [CrossRef]
2. Projected Production: The Twenty-Third Session of the Intergovernmental Group on Tea. 2018. Available online: https://openknowledge.fao.org/server/api/core/bitstreams/26d2c475-f7e3-43d6-a8a7-48e0462a1723/content (accessed on 20 June 2024).
3. Hussain, S.; Anjali, K.P.; Hassan, S.T.; Dwivedi, P.B. Waste tea as a novel adsorbent: A review. *Appl. Water Sci.* **2018**, *8*, 165. [CrossRef]
4. Negi, T.; Kumar, Y.; Sirohi, R.; Singh, S.; Tarafdar, A.; Pareek, S.; Awasthi, M.K.; Sagar, N.A. Advances in bioconversion of spent tea leaves to value-added products. *Bioresour. Technol.* **2022**, *346*, 126409. [CrossRef]
5. FAOSTAT. *Crop Production, Statistics Division, Food and Agriculture Organization of United Nations*; FAO: Rome, Italy, 2010.
6. McKendry, P. Energy production from biomass (part 1): Overview of biomass. *Bioresour. Technol.* **2002**, *83*, 37–46. [CrossRef] [PubMed]
7. Wazzan, A.A. The Effect of Surface Treatment on the Strength and Adhesion Characteristics of Phoenix dactylifera-L(Date Palm) Fibers. *Int. J. Polym. Mater. Polym. Biomater.* **2006**, *55*, 485–499. [CrossRef]

8. Alotaibi, M.D.; Alshammari, B.A.; Saba, N.; Alothman, O.Y.; Sanjay, M.; Almutairi, Z.; Jawaid, M. Characterization of natural fiber obtained from different parts of date palm tree (*Phoenix dactylifera* L.). *Int. J. Biol. Macromol.* **2019**, *135*, 69–76. [CrossRef]
9. Wazzan, A.A. Effect of fiber orientation on the mechanical properties and fracture characteristics of date palm fiber reinforced composites. *Int. J. Polym. Mater. Polym. Biomater.* **2005**, *54*, 213–225. [CrossRef]
10. Al-Sulaiman, F.A. Mechanical Properties of Date Palm Fiber Reinforced Composites. *Appl. Compos. Mater.* **2002**, *9*, 369–377. [CrossRef]
11. Hussien, A.; Al Zubaidi, R.; Jannat, N.; Ghanim, A.; Maksoud, A.; Al-Shammaa, A. The effects of tea waste additive on the physical and mechanical characteristics of structural unfired clay bricks. *Alex. Eng. J.* **2024**, *101*, 282–294. [CrossRef]
12. Ibrahim, J.E.F.; Tihtih, M.; Şahin, E.I.; Basyooni, M.A.; Kocserha, I. Sustainable zeolitic tuff incorporating tea waste fired ceramic bricks: Development and investigation. *Case Stud. Constr. Mater.* **2023**, *19*, e02238. [CrossRef]
13. Ozturk, S.; Sutcu, M.; Erdogmus, E.; Gencel, O. Influence of tea waste concentration in the physical, mechanical and thermal properties of brick clay mixtures. *Constr. Build. Mater.* **2019**, *217*, 592–599. [CrossRef]
14. Anjum, F.; Naz, M.; Ghaffar, A.; Shukrullah, S.; AbdEl-Salam, N.; Ibrahim, K. Study of thermal and mechanical traits of organic waste incorporated fired clay porous material. *Phys. B Condens. Matter* **2020**, *599*, 412479. [CrossRef]
15. Debnath, B.; Duarah, P.; Purkait, M.K. Microwave-assisted quick synthesis of microcrystalline cellulose from black tea waste (Camellia sinensis) and characterization. *Int. J. Biol. Macromol.* **2023**, *244*, 125354. [CrossRef]
16. Wong, K.; Ahsan, Q.; Putra, A.; Subramonian, S.; Nor, M.J.M. Preliminary study on the sound absorption behavior of spent tea leaves filled with natural rubber latex binder. *J. Teknol. (Sci. Eng.)* **2017**, *79*, 59–64. [CrossRef]
17. Ersoy, S.; Küçük, H. Investigation of industrial tea-leaf-fibre waste material for its sound absorption properties. *Appl. Acoust.* **2009**, *70*, 215–220. [CrossRef]
18. Ali, M.; Al-Suhaibani, Z.; Almuzaiqer, R.; Albahbooh, A.; Al-Salem, K.; Nuhait, A. New Composites Derived from the Natural Fiber Polymers of Discarded Date Palm Surface and Pineapple Leaf Fibers for Thermal Insulation and Sound Absorption. *Polymers* **2024**, *16*, 1002. [CrossRef]
19. Raza, M.; Al Abdallah, H.; Kozal, M.; Al Khaldi, A.; Ammar, T.; Abu-Jdayil, B. Development and characterization of Polystyrene–Date palm surface fibers composites for sustainable heat insulation in construction. *J. Build. Eng.* **2023**, *75*, 106988. [CrossRef]
20. Raza, M.; Al Abdallah, H.; Abdullah, A.; Abu-Jdayil, B. Date Palm Surface Fibers for Green Thermal Insulation. *Buildings* **2022**, *12*, 866. [CrossRef]
21. Ali, M.E.; Alabdulkarem, A. On thermal characteristics and microstructure of a new insulation material extracted from date palm trees surface fibers. *Constr. Build. Mater.* **2017**, *138*, 276–284. [CrossRef]
22. Alabdulkarem, A.; Ali, M.; Iannace, G.; Sadek, S.; Almuzaiqer, R. Thermal analysis, microstructure and acoustic characteristics of some hybrid natural insulating materials. *Constr. Build. Mater.* **2018**, *187*, 185–196. [CrossRef]
23. Fouladi, M.H.; Nassir, M.H.; Ghassem, M.; Shamel, M.; Peng, S.Y.; Wen, S.Y.; Xin, P.Z.; Nor, M.J.M. Utilizing Malaysian natural fibers as sound absorber. In *Modeling and Measurement Methods for Acoustic Waves and for Acoustic Microdevices*; BoD—Books on Demand: Norderstedts, Germany, 2013; pp. 161–170.
24. Berardi, U.; Iannace, G. Acoustic characterization of natural fibers for sound absorption applications. *Build. Environ.* **2015**, *94*, 840–852. [CrossRef]
25. Available online: https://www.gulfindustrialgroup.com/saaf/wp-content/uploads/2016/08/WOOD-ADHESIVE-78-1040-MSDS.pdf (accessed on 10 June 2024).
26. *ASTM D790-03*; Standard Test Methods for Flexural Properties of Unreinforced and Reinforced Plastics and Electrical Insulating Materials. ASTM International: West Conshohocken, PA, USA, 2003.
27. *ASTM-C518*; Standard Test Method for Steady-State Thermal Transmission Properties by Means of the Heat Flow Meter Apparatus (C 518). American Society of Testing and Materials (ASTM): West Conshohocken, PA, USA, 2010; pp. 152–166.
28. Ali, M.; Almuzaiqer, R.; Al-Salem, K.; Alabdulkarem, A.; Nuhait, A. New novel thermal insulation and sound-absorbing materials from discarded facemasks of COVID-19 pandemic. *Sci. Rep.* **2021**, *11*, 23240. [CrossRef] [PubMed]
29. *ISO 10534-1*; Determination of Sound Absorption Coefficient and Impedance in Impedance Tubes—Part 1: Method Using Standing Wave Ratio. ISO: Geneva, Switzerland, 1996.
30. *ISO 10534-2*; Determination of Sound Absorption Coefficient and Impedance in Impedance Tubes—Part 2: Transfer-Function Method. ISO: Geneva, Switzerland, 1998.
31. *ASTM D2974—07A ASTM D2974-07*; Standard Test Methods for Moisture, Ash, and Organic Matter of Peat and Other Organic Soils. ASTM International: West Conshohocken, PA, USA, 2007.
32. RILEM-TC. Test for the determination of modulus of rupture and limit of proportionality of thin fibre reinforced cement sections. In *RILEM Recommendations for the Testing and Use of Constructions Materials*; RILEM, Ed.; E & F SPON: London, UK, 1984; pp. 161–163, ISBN 2351580117.
33. Nguyen, D.M.; Grillet, A.-C.; Diep, T.M.H.; Bui, Q.B.; Woloszyn, M. Influence of thermo-pressing conditions on insulation materials from bamboo fibers and proteins based bone glue. *Ind. Crops Prod.* **2018**, *111*, 834–845. [CrossRef]
34. Dukhan, N.; Rayess, N.; Hadley, J. Characterization of aluminum foam–polypropylene interpenetrating phase composites: Flexural test results. *Mech. Mater.* **2010**, *42*, 134–141. [CrossRef]

35. Asdrubali, F.; D'Alessandro, F.; Schiavoni, S. A review of unconventional sustainable building insulation materials. *Sustain. Mater. Technol.* **2015**, *4*, 1–17. [CrossRef]
36. Intini, F.; Kühtz, S. Recycling in buildings: An LCA case study of a thermal insulation panel made of polyester fiber, recycled from post-consumer PET bottles. *Int. J. Life Cycle Assess.* **2011**, *16*, 306–315. [CrossRef]
37. Martwong, E.; Yingshataporn-A-Nan, T.; Minanandana, T.; Puksuwan, K.; Junthip, J.; Sukhawipat, N. Sound absorption and thermal insulation materials from waste palm oil for housing application: Green polyurethane/water hyacinth fiber sheet composite. *Constr. Build. Mater.* **2024**, *438*, 137007. [CrossRef]
38. Danihelová, A.; Němec, M.; Gergeľ, T.; Gejdoš, M.; Gordanová, J.; Sčensný, P. Usage of Recycled Technical Textiles as Thermal Insulation and an Acoustic Absorber. *Sustainability* **2019**, *11*, 2968. [CrossRef]
39. Sun, Y.; Xu, Y.; Li, W.; Yue, X. Functional modification of softwood fiber and its application in natural fiber-based sound-absorbing composite. *Ind. Crops Prod.* **2024**, *218*, 119044. [CrossRef]
40. Sharma, S.K.; Shukla, S.R.; Sethy, A.K. Acoustical behaviour of natural fibres-based composite boards as sound-absorbing materials. *J. Indian Acad. Wood Sci.* **2020**, *17*, 66–72. [CrossRef]
41. Zhang, C.; Li, H.; Gong, J.; Chen, J.; Li, Z.; Li, Q.; Cheng, M.; Li, X.; Zhang, J. The review of fiber-based sound-absorbing structures. *Text. Res. J.* **2023**, *93*, 434–449. [CrossRef]
42. Nandanwar, A.; Kiran, M.C.; Varadarajulu, K.C. Influence of Density on Sound Absorption Coefficient of Fibre Board. *Open J. Acoust.* **2017**, *7*, 73876. [CrossRef]
43. Na, Y.; Lancaster, J.; Casali, J.; Cho, G. Sound Absorption Coefficients of Micro-fiber Fabrics by Reverberation Room Method. *Text. Res. J.* **2007**, *77*, 330–335. [CrossRef]
44. Shahid NS, M.; Ahmad, M.A.; Tahir, F.L. Sound Absorption Coefficient of Different Green Materials Polymer on Noise Redction. *Int. J. Innov. Tech. Explor. Eng* **2020**, *9*, 2773–2777. [CrossRef]
45. Alemdar, A.; Sain, M. Biocomposites from wheat straw nanofibers: Morphology, thermal and mechanical properties. *Compos. Sci. Technol.* **2008**, *68*, 557–565. [CrossRef]
46. Asim, M.; Paridah, M.T.; Chandrasekar, M.; Shahroze, R.M.; Jawaid, M.; Nasir, M.; Siakeng, R. Thermal stability of natural fibers and their polymer composites. *Iran. Polym. J.* **2020**, *29*, 625–648. [CrossRef]
47. Bainbridge, D.A. High performance low cost buildings of straw. *Agric. Ecosyst. Environ.* **1986**, *16*, 281–284. [CrossRef]

polymers

MDPI

Article

Using Thin Ultra-High-Molecular-Weight Polyethylene Coatings to Reduce Friction in Frost-Resistant Rubbers

Elena Torskaya [1,*], Ivan Shkalei [1], Fedor Stepanov [1], Yulia Makhovskaya [1], Afanasy Dyakonov [2] and Natalia Petrova [2]

[1] Ishlinsky Institute for Problems in Mechanics RAS, 119526 Moscow, Russia; ioann_shiva@list.ru (I.S.); stepanov_ipm@mail.ru (F.S.); makhovskaya@mail.ru (Y.M.)
[2] Department of Chemistry, North-Eastern Federal University, 677000 Yakutsk, Russia; afonya71185@mail.ru (A.D.); pnn2002@mail.ru (N.P.)
* Correspondence: torskaya@mail.ru

Abstract: Frost-resistant rubbers retain their highly elastic properties over a wide temperature range. They are used in various friction units (e.g., seals), but their high friction coefficient and low wear resistance lead to the need for frequent replacement. In this paper, we propose applying thin (several hundred microns) UHMWPE coatings to formed rubber rings. The application technology depends on the required coating thickness. Friction tests of the coatings and pure UHMWPE were performed using the ball-on-disk (unidirectional sliding) scheme for various loads and velocities. In the experiments, the friction coefficients and temperatures near the contact area were determined. Friction tracks were studied using microscopy methods. The sliding contact of the ball and the two-layer material was modeled to obtain the dependences of the deformation component of friction on the sliding velocity for coatings of different thicknesses. UHMWPE is sensitive to frictional heating, so the thermal problem of determining the temperature in the contact area was also solved. It is shown that the minimum friction coefficient occurs for coatings with a thickness of 600 μm. At the same time, in the case of the 300 μm coating, the surface of the friction track is practically no different from the initial one. Thus, the studied combination of polymers provides antifrictional properties and wear resistance to the surface layer while maintaining the damping properties of rubber.

Keywords: UHMWPE coating; rubber; friction; contact problem; frictional heating

1. Introduction

Polymer composite materials (PCMs) are widely used in vehicles, quarry equipment, and process equipment since they have sufficiently high mechanical strength and processability, low density, corrosion resistance, and, in some cases, depending on the chemical nature of the selected polymers, antifriction properties, resistance in various environments, and frost resistance [1]. PCMs can be characterized as materials consisting of two or more phases separated by an interphase surface. To obtain them, polymers with different properties are used, including those belonging to different classes, such as thermoplastic and elastomer, in order to obtain a new set of properties [2,3]. Polymer phases are distributed differently in dispersed-filled, fibrous, and layered PCMs. A thin layer of polymeric material, which is applied to the surface of another material (substrate), can be considered as a coating [4,5]. Coatings have a variety of functional purposes, such as imparting high wear and abrasion resistance, and they have dielectric or electrically conductive properties while maintaining the properties of the volumetric matrix. This allows them to be used in sensors, bearings, and other rubber products (RPs), operating under complex loading conditions in extreme operating conditions [6,7]. In [8], the tribological characteristics of tantalum coatings sprayed onto a rubber substrate were studied. The use of such coatings leads to a significant improvement in the tribological characteristics of rubber (the friction coefficient decreases by an average of 1.7 times, and abrasive wear decreases by 1.8 times). The authors of [9] increased the wear resistance of elastomers by applying composite antifriction coatings. In [10], it is proposed to use the surface modification of silicone rubber by an electron beam and also aging to improve the service life of insulators. In [11], a titanium dioxide layer was successfully formed on a natural rubber substrate by the liquid-phase deposition method, which is expected to be used as an anti-aging layer for natural rubber.

Of considerable interest in the creation of PCMs is ultra-high-molecular-weight polyethylene (UHMWPE), which has a unique set of properties: high strength, wear resistance due to a low coefficient of friction, abrasion resistance, and high impact toughness, which is maintained at extremely low temperatures. Combining it with elastomeric materials that have high elasticity and damping properties gives interesting results [12–14].

As has been shown in a number of studies [15–17], an important mechanism of frictional interaction of UHMWPE with most materials is adhesion, which is caused by the high surface energy of the polymer.

The surface energy, in turn, depends significantly on temperature, so it is important to note the dependence of the tribological characteristics of UHMWPE on it. With an increasing temperature, the friction coefficient of UHMWPE increases [18]. Despite the fact that UHMWPE's melting temperature is about 140 °C [15], and it is thermally stable up to 435 °C [19] (result of a thermal gravimetric analysis), during friction, a relatively small increase in temperature changes the frictional contact conditions of the material. In [20], it was shown that when UHMWPE slides against CoCr, an increase in the ambient temperature from 20 to 37 °C leads to a decrease in the friction coefficient and wear. On the contrary, in ballistic experiments on impact resistance, an increase in temperature to 80 °C does not have a significant effect on the deformation and destruction of UHMWPE laminates [21]. In addition, we should not forget about frictional heating. The contact temperature increases with increasing contact pressure and sliding velocity, which is shown in [22] for the friction of UHMWPE paired with ceramics. It is also noted that the gradient of temperature increase is greater at the beginning and gradually decreases over time. Thus, the effect of frictional heating on the tribological properties of the polymer must be taken into account.

Another friction mechanism important for the composite is energy dissipation due to the imperfect elasticity of the rubber. If the UHMWPE coating is thin, the deformation component can make a significant contribution to the total friction force. There are many experimental studies of rubber friction, particularly its deformation component. Pioneering papers [23–25] are those on rolling and sliding with lubrication. Many researchers have also developed models for studying hysteretic friction. In relation to the subject of this study, the most interesting subjects are spatial contact problems in quasi-static formulation for a slider and viscoelastic half-space [26–29]. The solutions are usually based on the results of sliding the concentrated force along the boundary of the viscoelastic half-space [30]. Problems were investigated in which the material model included a spectrum of relaxation times [26], which is more realistic for polymers. The influence of adhesive friction forces arising during sliding was investigated in [27,28]; it was found that for low compressible materials, the effect of friction on the contact problem solution is negligible. Inclusions in viscoelastic materials have an effect on energy dissipation in steady sliding contact [29]. The presence of a rigid coating on a viscoelastic body makes it necessary to complicate the model, which was carried out in [31].

The aim of this work is to study the features of dry friction of frost-resistant rubber with a UHMWPE coating using friction tests and modeling.

2. Materials and Experimental Methods

2.1. Materials

Two-layer composite material consists of ultra-high-molecular-weight polyethylene (GUR-4022, Ticona, Kelsterbach, Germany) as a coating and rubber as a substrate; nitrile butadiene rubber (a synthetic rubber derived from acrylonitrile and butadiene, NBR) obtained from SIBUR, Russia was used as the base. The composition contained all the necessary ingredients of rubber mixtures: rubber, filler (carbon black P-803), dispersant, and accelerator and activator of vulcanization. Sulfur was used as a vulcanizing agent.

The ingredients were mixed in a closed-type Plastograph EC Plus rubber mixer (Brabender, Duisburg, Germany). The initial temperature of the mixing rollers was 40 °C, the rotation speed was 25 rpm, the force was 25 N, and the total mixing time was 20 min. Vulcanization of the samples was performed on a GT-7014-H10C thermohydraulic press (Gotech, Taichung, Taiwan).

The production of molded rubber products by the compression method includes molding the rubber mixture from prefabricated blanks and their vulcanization at elevated temperature and pressure. The process temperature range is 150–180 °C. One of the methods for processing UHMWPE is hot pressing at an elevated temperature, and the temperature ranges for processing rubber and UHMWPE partially coincide, which allows samples to be obtained in one mold. At the vulcanization temperatures of the rubber mixture, UHMWPE powder begins to melt and create a high-quality coating on the surface of the elastomer. The coating application technology is described in sufficient detail in [13]. Ring samples with UHMWPE coating (600 μm and 300 μm) were obtained. The technology varied when manufacturing samples with different coating thicknesses. In the deposition of a 600 μm coating, UHMWPE powder was first molded (10 MPa for 5 min), combined with the rubber mixture, and held at 155 °C at 10 MPa for 20 min in a mold to vulcanize the elastomer and to press the thermoplastic, and then it was cooled in the same mold. For the thinner coating, the rubber mixture was first pre-vulcanized (155 °C × 10 MPa for 5 min), and then UHMWPE powder was applied (dry impregnation) and combined hot pressing was carried out with subsequent cooling in the mold. The total vulcanization duration was 20 min, including 5 min of pre-vulcanization of the rubber mixture and 15 min of post-vulcanization.

2.2. Methods for Determining Tensile Strength and Elongation

The tensile strength and relative elongation of the elastomer and UHMWPE samples were determined on an Autograph testing machine (Shimadzu, Kyoto, Japan) at room temperature in accordance with ISO 37:2024 [32]. Aggression resistance was determined using the test method for resistance (in unstressed state) to hydrocarbon environments by mass change according to ISO 1817:2024 [33]. AMG-10 oil was selected as the medium. The tests were performed at room temperature for 72 h. Differential scanning calorimetry (DSC, DSC 204 F1 Phoenix (Netzsch, Selb, Germany)) was used to determine the phase transition temperatures in the materials; the glass transition temperature was recorded.

For the successful operation of two-layer materials, it is necessary to ensure high bond strength between the layers. This was determined during their delamination. The idea of the method is to determine the force required to separate the UHMWPE layer from the rubber substrate. This was carried out using a testing machine for peeling off adhesive joints (Autograph (Shimadzu, Japan)), which allows the applied load (tensile force) to be perpendicular to the solid polymer substrate. The samples (double-sided blades) were prepared according to ISO 23529:2016 [34]. The crosshead speed was 500 mm/min. The experiments were performed at room temperature.

2.3. Friction Tests

Tribotester (MFT-5000, Rtec, San Jose, CA, USA) was used in the mode of unidirectional sliding. A steel ball (AISI 52100, diameter 10 mm) was loaded depending on the material: 10 N for bulk UHMWPE and 20 N for relatively thin and thick coatings. A lower load for the bulk material was chosen to minimize plastic deformations during friction [35]. The track radius was 23 mm. In an hour, the linear sliding velocity varied (all other parameters were equal) by increasing from 0.005 m/s to 1 m/s. This method was based on previous experiments with viscoelastic materials: the repeatability of the results is better when the contact pair does not open during the test [36,37]. At each velocity, the duration was 10 min. The dependence of the friction coefficient (COF) on time was recorded, as well as the temperature, which was measured at a distance from the contact zone by the radius of the ball and the thickness of the wall of its holder. The resulting values of the friction coefficient and temperature depending on the linear sliding velocity were obtained by averaging over several tests (not less than three).

It should be noted that the testing method is based on ASTM G99-23 [38], which was developed for testing coatings. The coating load is doubled in comparison to the standard. At the same time, the testing time is significantly shorter than that in ASTM G99-23. Wear was not assessed; only the ability of the composite to withstand the load without plastic deformations at the macro level and without cracks was tested. The absence of cracks and beads along the edges of the track, characteristic of plastic displacement of the material, was checked using optical microscopy methods.

3. Models

For better insights into the experimental results, a numerical model of sliding contact was developed. Modeling consists of two independent problems. The first one is the contact problem for an elastic sphere sliding on a viscoelastic half-space covered with a relatively hard coating. The second problem considers the steady process of heating occurring due to friction between the two bodies. The input and output parameters of the model are summarized in Table 1, and the mathematical formulations of the problems are shown below.

Table 1. The parameters used in modeling.

Input	Q, N	Vertical load	10; 20
	R, m	Radius of ball	0.005
	$E^{(0)}$, MPa	Young modulus of ball	210,000
	$\nu^{(0)}$	Poisson ratio of ball	0.3
	$E^{(1)}$, MPa	Young modulus of coating	1000
	$\nu^{(1)}$	Poisson ratio of coating	0.35
	$E^{(2)}$, MPa	Young modulus of substrate	13.8
	$\nu^{(2)}$	Poisson ratio of substrate	0.45
	μ_i, s	Spectrum of retardation times	0.13
	$1/k_i$, s	Spectrum of relaxation times	1/40.14
	H, m	Coating thickness	0.0003; 0.0006
	$\lambda^{(0)}$, W/(m·K)	Coefficient of thermal conductivity of indenter	50
	$\lambda^{(1)}$, W/(m·K)	Coefficient of thermal conductivity of coating	0.4
	$\lambda^{(2)}$, W/(m·K)	Coefficient of thermal conductivity of substrate	0.15
	μ	Friction coefficient	from the tests
	V, m/s	Sliding velocity	from the tests
Output	$P(x,y)$, Pa	Contact pressure distribution	
	D, m	Approach of indentor	
	$\mu^{(d)}$	Deformation component of friction coefficient	
	$\mu^{(a)}$	Adhesion component of friction coefficient	
	$T(x,y,z)$, °C	Temperature	

3.1. Contact Problem

We consider an elastic sphere (indenter) with a radius R sliding with a constant velocity V over the viscoelastic half-space covered with a relatively hard coating of thickness H. The sphere is loaded with vertical force Q and an unknown tangential force Q_t. The elastic modulus and Poisson ratio of the considered materials are $E^{(i)}$ and $\nu^{(i)}$, correspondingly, where index $i = 0$ is declared for the indenter, $i = 1$ for the coating, and $i = 2$ for the half-space. It is assumed that $E_1 >> E_2$.

The viscoelastic behavior of the half-space material can be expressed by substituting the elastic modulus $1/E_2$ in the Hook's law with the Volterra integral operator $\hat{A}$:

$$\hat{A}(\sigma(t)) = \frac{1}{E_2}\left[\sigma(t) + \int_{-\infty}^{t} \sigma(\tau)K(t-\tau)\,d\tau\right],\ K(t) = \sum_{i=1}^{N} k_i e^{-\frac{t}{\mu_i}} \quad (1)$$

where $\sigma(t)$ is a time-dependent function, and $1/k_i$ and μ_i are the relaxation and retardation time spectrums, correspondingly.

A moving Cartesian coordinate system is used in the solution. Its XY plane corresponds with the undeformed top surface of the layer, the OZ-axis is the indenter axis of symmetry, and the OX-axis is co-directional to the vector of sliding.

The boundary conditions at the top surface of the layer are as follows (z = 0):

$$\begin{gathered} w(x,y) = f(x,y) + D,\ (x,y) \in \Omega \\ \sigma_z = 0,\ (x,y) \notin \Omega \\ \tau_{xz} = 0,\ \tau_{yz} = 0 \end{gathered} \tag{2}$$

Here, w is the normal displacement of the top surface of the layer; f is the function defining the shape of the indenter; σ_z, τ_{xz}, and τ_{yz} are the normal and tangential stresses; Ω is the area of contact; and D is the slider penetration. We do not take into account tangential stresses in the contact for simplification, considering that they insignificantly change the normal pressure distribution.

The conditions at the layer-substrate interface ($z = H$) are as follows:

$$w^{(1)} = w^{(2)},\ u_x^{(1)} = u_x^{(2)},\ u_y^{(1)} = u_y^{(2)} \tag{3}$$

Here, u_x and u_y are tangential displacements.

The equilibrium condition in the OZ direction is considered in the following form:

$$Q = \iint\limits_{\Omega} P(x,y)dxdy \tag{4}$$

The contact problem is solved using the Boundary Element Method. A rectangular mesh is built in an area exceeding the contact zone. Using a piecewise constant approximation, a linear system of equations is constructed based on the boundary conditions (2) and the equilibrium condition (4). The influence coefficients related to the normal displacement of the coated half-space are obtained using double integral Fourier transform (FT):

$$\begin{gathered} \kappa_i^j = -\frac{1}{G^{(1)}} \int\limits_0^{\pi/2} \int\limits_0^{\infty} \Delta(\lambda,\varphi,\gamma,\chi) \cdot \cos(y_{ij}\gamma\sin(\varphi)) \cdot \\ \left[\cos(x_{ij}\gamma\cos(\varphi)) + \sum_l^N k_l\omega_l \frac{\omega_l V\gamma\cos(\varphi)\sin(x_{ij}\gamma\cos(\varphi)) + \cos(x_{ij}\gamma\cos(\varphi))}{1 + (V\omega_l\gamma\cos(\varphi))^2} \right] d\varphi d\gamma \end{gathered} \tag{5}$$

φ and γ are the coordinates in the space of FT; $\chi = E^{(1)}(1 + \nu^{(2)})/E^{(2)}(1 + \nu^{(1)})$; Δ is a very cumbersome expression derived in [39] and linearly depends on pressure in the FT dimension. The derivation of the influence coefficients and solution of the mentioned linear system of equations is explicitly discussed in [31]. As a result, the contact pressure distribution P_i and the penetration D are calculated. It is possible to calculate the deformation component of the friction force.

$$\mu^{(d)} = \frac{\sum\limits_{i=1}^{n} P_i \frac{\partial f(x_i,y_i)}{\partial x}}{Q} \tag{6}$$

3.2. Thermal Problem

When the contact problem is solved, it is possible to calculate the heating occurring due to friction both in the ball and in the polymer sample. The heating process is considered to be stationary. This leads us to the following equations for the coating (i = 1) and the half-space (i = 2):

$$\Delta T^{(i)} = 0,\ i = 1,2 \tag{7}$$

We will apply the DC-FFT algorithm [40] using the same mesh as in the contact problem in Section 3.1 to calculate the temperature. First, a frequency response function needs to be derived in the space of double FT.

Each i element of the mesh within the contact area Ω is subjected to a constant friction force moving with velocity V and therefore generating the specific power of frictional heat release:

$$q_i = \mu V P_i \tag{8}$$

Let us consider that $\alpha < 1$ is the part of the heat flux that is absorbed by the specimen and $1 - \alpha$ is absorbed by the ball. At this point, α is unknown and is to be found later.

For a rectangular element on the surface with sides $2a$ and $2b$, the heat flux is expressed in the following form ($z = 0$):

$$\left.\frac{\partial T^{(1)}}{\partial z}\right|_{z=0} = \begin{cases} -q(x,y)/\lambda^{(1)} & |x| < a \wedge |y| < b \\ 0 & |x| > a \vee |y| > b \end{cases} \tag{9}$$

At the interface ($z = H$),

$$\lambda^{(1)}\left.\frac{\partial T^{(1)}}{\partial z}\right|_{z=H} = \lambda^{(2)}\left.\frac{\partial T^{(2)}}{\partial z}\right|_{z=H},\ T^{(1)}(x,y,H) = T^{(2)}(x,y,H) \tag{10}$$

λ_i, $i = 1, 2$ is the coefficient of thermal conductivity.

By applying double integral Fourier transform to the temperature T, Equation (7) can be reduced as follows:

$$\frac{\partial^2 \widetilde{T}^{(i)}}{\partial z^2} - \gamma^2 \widetilde{T}^{(i)} = 0\ (\gamma^2 = \varphi^2 + \beta^2) \tag{11}$$

where $\widetilde{T}^{(i)}(\varphi,\beta,z)$—the transformed temperature; φ and β—coordinates in FT. The boundary conditions (9) and (10) after the transformation are the following:

$$\begin{array}{c} \left.\frac{\partial \widetilde{T}^{(1)}}{\partial z}\right|_{z=0} = -\frac{\widetilde{q}}{\lambda^{(1)}},\ \left.\frac{\partial \widetilde{T}^{(1)}}{\partial z}\right|_{z=H} = \left.\frac{\partial \widetilde{T}^{(2)}}{\partial z}\right|_{z=H}, \\ \widetilde{T}^{(1)}(\varphi,\beta,H) = \widetilde{T}^{(2)}(\varphi,\beta,H),\ \widetilde{q} = \frac{q}{\pi^2}\frac{\sin(a\varphi)\sin(b\beta)}{\varphi\beta} \end{array} \tag{12}$$

The general solution of Equation (11) is represented as follows [41]:

$$\widetilde{T}^{(i)}(\gamma,z) = A^{(i)}(\gamma)e^{-\gamma z} + B^{(i)}(\gamma)e^{\gamma z} \tag{13}$$

Using the boundary conditions (9) and (10), we obtain $A^{(i)}$ and $B^{(i)}$:

$$\begin{array}{ll} A^{(1)} = \widetilde{q}\frac{1}{\gamma(1+\omega e^{-2\gamma H})}, & B^{(1)} = \widetilde{q}\frac{-\omega e^{-2\gamma H}}{\beta(1+\omega e^{-2\gamma H})}, \\ A^{(2)} = \widetilde{q}\frac{1+\omega}{\beta(1+\omega e^{-2\gamma H})}, & B^{(2)} = 0,\ \omega = \frac{\lambda^{(2)}-\lambda^{(1)}}{\lambda^{(2)}+\lambda^{(1)}} \end{array} \tag{14}$$

Using expression (13) together with (14), we obtain the frequency response function and then use it in the DC-FFT algorithm [40] for calculating the temperature distribution at any depth z due to the applied αq_i. In order to find the unknown coefficient α, we consider only one loaded element of the mesh. Knowing the coefficients of thermal conductivity of the contacting bodies, one can calculate the surface temperature in the element for the indenter T_I and the sample T_S depending on α, which is then obtained numerically from the following nonlinear equation:

$$T_I(\alpha) - T_S(\alpha) = 0,\ |\alpha| < 1 \tag{15}$$

All input and output parameters used in modelling are presented in Table 1.

The properties of UHMWPE were obtained from the nanoindentation test (elastic properties) and from [35,42] thermal conductivity. Rubber characterization was carried out by nanoDMA testing (see Appendix A).

4. Results

When selecting the materials to be combined, we proceeded from the need to ensure sufficient strength and elasticity of the two-layer material, its resistance to a hydrocarbon environment, and the possibility of its use at sufficiently low temperatures in Arctic conditions. UHMWPE and NBR fully comply with the selected criteria. This is confirmed by the data in Table 2, which show the properties of rubbers based on NBR and UHMWPE samples. When forming a two-layer material, the rate of structuring of the elastomer layer and the possibility of crosslinking at the rubber–thermoplastic phase interface are of great importance to ensure the integrity and monolithicity of the resulting products. Special attention was paid to the choice of vulcanizing system. The sulfur vulcanizing system is most often used for structuring rubbers. Varying the type of vulcanization accelerator allows for the rate and degree of elastomer vulcanization to be controlled and allows chemical crosslinking of NBR and UHMWPE. The following medium-activity accelerators were tested: mercaptobenzthiazole (MBT), which belongs to the group of thiazole-type accelerators; the amine-type accelerator diphenylguanidine (DPG); and an ultra-accelerator that provides an induction period—sulfenamide C—as well as their combinations with each other. For further studies, two systems were selected: sulfenamide C in combination with sulfur, zinc oxide, and a dispersant (sample 1) and a mixture of MBT + DPG with the same vulcanizing agent and activator (sample 2), since they provided the highest strength, the lowest degree of swelling, and sufficiently high frost resistance of the elastomer, as well as the highest adhesion between layers. The delamination force for the two-layer material obtained using sulfenamide C is 7.4 N/mm when using the MBT + DPG composition—7.3 N/mm—which is significantly higher than that for the other tested vulcanizing systems. In this case, the destruction is cohesion, i.e., in the process of delamination of the layers, destruction occurs along the elastomer, which indicates the formation of a high-quality and reliable connection between UHMWPE and the elastomer (the strength exceeds the cohesive strength of NBR). The glass transition temperature for the elastomer layer is in the range of −44.3–−45.4 °C; this indicator for UHMWPE could not be determined by the same method. According to published data [15], it is known that the glass transition temperature of UHMWPE is −120 °C.

Table 2. Properties of rubber based on NBR and UHMWPE samples.

Properties	Tensile Strength, MPa,	Relative Elongation, %	The Degree of Swelling in Oil (AMG-10), %	Tg, °C
NBR, (sample 1)	11.7 ± 0.95	273 ± 21	15.5 ± 0.5	−44.3
NBR, (sample 2)	12.3 ± 0.8	238 ± 18	14.6 ± 0.4	−45.4
UHMWPE	37.4 ± 1.0	324 ± 24	0	−120 [15]

A sulfur vulcanizing system with accelerators MBT + DPG (sample 2 in Table 2) was selected for the next step, which is tribological testing. When using this vulcanizing system, the rubber has higher strength values, slightly lower swelling values, and lower Tg values compared to another system (sample 1 in Table 2), i.e., it is more suitable for Arctic operating conditions.

On a noncontact optical profilometer S neox 3D (Sensofar-Tech, Barcelona, Spain), using a 20× objective, optical images of cross sections were obtained from which the actual coating thickness was determined (Figure 1).

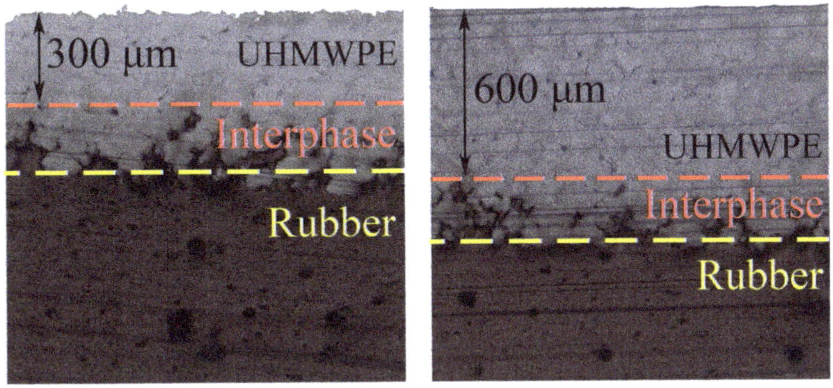

Figure 1. Optical images of cross sections of samples with thin and thick coatings.

The friction tracks formed during the full cycle of friction tests are shown in Figure 2. The friction track on the coating surface is wider than that on pure UHMWPE, which is explained by the significantly higher integral compliance of the composite. The surface structure of bulk UHMWPE and the coating differ due to differences in the formation technology. With the exception of abrasive plowing traces, the changes in the UHMWPE surface during friction are more noticeable for bulk UHMWPE. This is due to higher contact pressures. For thinner coatings, there are fewer abrasive traces, and the friction track boundaries are difficult to identify.

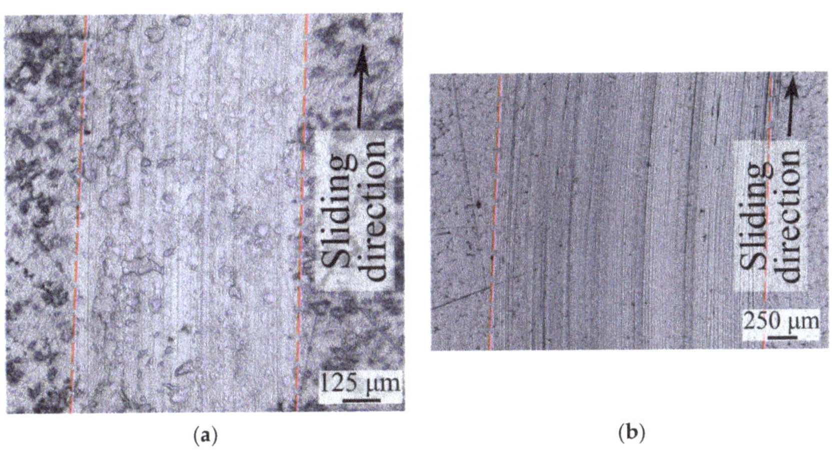

Figure 2. Optical images of friction tracks for bulk UHMWPE (**a**) and for composite with 600 μm UHMWPE coating (**b**).

The temperature and friction coefficient depend on the sliding velocity (Figure 3). It should be noted that the friction force in the case of bulk UHMWPE does not include the deformation component. Friction in a similar rubber without coating was studied earlier [37]; the friction coefficient was fixed in the range of 0.4–1.1 depending on the load and velocity parameters. The temperature was measured using a thermocouple (type K, tolerance class ±1.5 °C), located at some distance from the contact patch; therefore, it only allows for a qualitative assessment of the effect of velocity on heat flow going into the ball and caused by frictional heating.

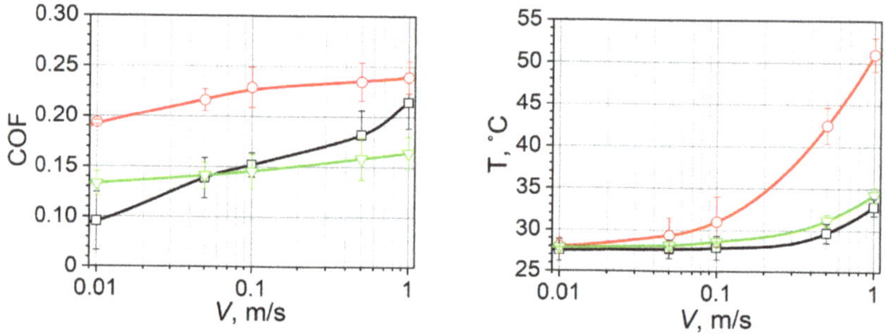

(a) (b)

Figure 3. The dependence of the friction coefficient (COF) (**a**) and temperature (T) (**b**) on the sliding velocity V for bulk UHMWPE (black line, square), the 300 μm coating (red line, circle), and the 600 μm coating (green line, triangle).

To analyze the friction mechanisms, it is necessary to consider the results of modeling the sliding contact. Figure 4 shows that the contact area size and the maximum contact pressure depend on the velocity for two coating thicknesses. The floating effect, characteristic of sliding over the surface of viscoelastic bodies, is expressed through a decrease in the contact area and an increase in the contact pressure. With an increase in velocity, these parameters tend to be constant due to the instantaneous modulus of elasticity of the viscoelastic material. The thicker the coating, the lower the integral compliance of the coating–substrate system, and the higher the maximum pressure. These parameters are directly related to the adhesive component of the friction force. The larger the contact area, the greater the adhesion [43]; therefore, for a thinner coating, adhesive friction should be more observable.

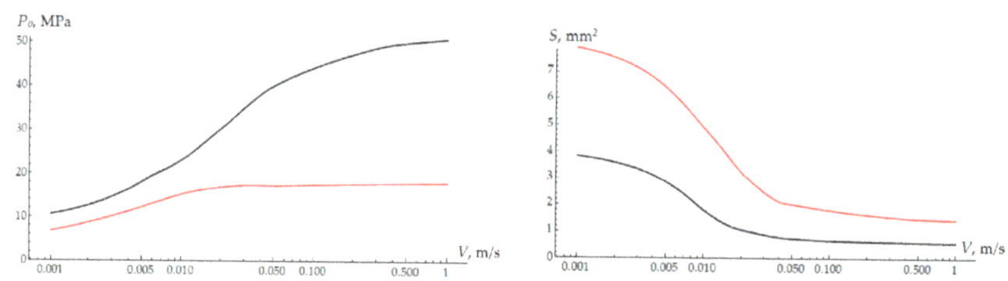

(a) (b)

Figure 4. The maximum of the contact pressure P_0 (**a**) and area of contact S (**b**) depending on the sliding velocity. H = 300 μm—red curve; H = 600 μm—black curve.

The friction coefficient due to hysteresis losses in rubber also depends on velocity and the coating thickness (Figure 5a). As shown earlier [27,44], for weakly compressive materials, the effect of adhesive friction on deformation friction is negligible. As expected, hysteresis losses are greater for a thinner coating, especially at velocities at which the viscoelastic properties of the substrate are maximal. Subtracting the deformation component of the friction force from the total one recorded in the experiment allows us to analyze the adhesive component (Figure 5b). For samples with coatings at low velocities, the effect of frictional heating is negligible, while the effect of floating (reductions in the contact area and adhesion) is quite obvious. With an increase in velocity, heating leads to an increase in the surface energy of the polymer and, consequently, to an increase in adhesion.

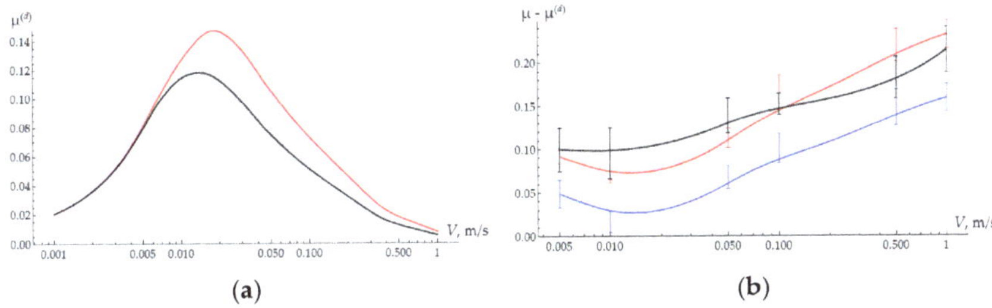

Figure 5. (**a**) The deformation component of the friction force depending on the sliding velocity. H = 300 μm—red curve; H = 600 μm—black curve. (**b**) The difference between the friction coefficient obtained in the experiment and the deformation component of the friction coefficient obtained from modeling. H = 300 μm—red curve; H = 600 μm—blue curve; clear UHMWPE—black curve.

As the results of the thermal problem solution show, the average temperature in the contact area is approximately the same for all three types of samples under consideration (Figure 6a). Due to the smaller contact area, the temperature of bulk UHMWPE is slightly higher, but in general, this explains the approximately identical behavior of the curves in Figure 5b. The difference in the curves obtained for the samples with coatings is explained by the difference in the contact area. Plastic deformation at the micro level also contributes to the total friction force, and it is greater the higher the contact pressure values. The difference in the surface topography of the coating and pure UHMWPE also affects the adhesion of the polymer. The curves of the heat flow into the ball as a function of velocity are shown in Figure 6b. The relative position of the curves coincides with the results in Figure 3b, although the experiments for the thick coating and bulk UHMWPE show similar results. The lowest flow in contact with bulk UHMWPE is explained by the half-lower load.

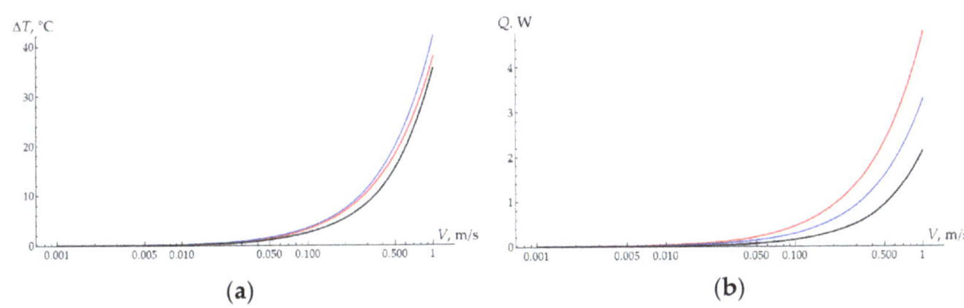

Figure 6. (**a**) The mean increase in the temperature within the contact area. H = 300 μm—red curve; H = 600 μm—blue curve; clear UHMWPE—black curve. (**b**) The heat flux depending on the sliding velocity. H = 300 μm—red curve; H = 600 μm—blue curve; clear UHMWPE—black curve.

5. Conclusions

In conclusion, the combination of two polymers of different types allows us to obtain a damping material with a surface modification providing wear resistance and antifriction properties. The friction coefficient for the composite changes from 0.1 to 0.22, which is essentially smaller than the rubber COF (0.4–1.1). The disadvantage of UHMWPE is a low yield point, which, in combination with a high (for polymers) Young's modulus, leads to the development of plastic deformation. The combination of thin layers of UHMWPE with rubber leads to decreases in contact and internal stresses, which allows us to avoid the demonstration of plasticity. The experimental results (dry friction) and modeling show that the friction coefficient depends on the coating thickness and sliding velocity. At the same time, the thicker the coating, the lower the damping capacity of the material. This means that for specific friction units and loading conditions, there is an optimal coating thickness, providing a combination of damping and antifriction properties.

Author Contributions: Conceptualization, E.T. and N.P.; Formal Analysis, E.T. and N.P.; Investigation, I.S., Y.M. and A.D.; Methodology, I.S. and A.D.; Software, F.S.; Supervision, E.T.; Visualization, F.S.; Writing—Original Draft, E.T., F.S. and Y.M.; Writing—Review and Editing, I.S. and N.P. All authors have read and agreed to the published version of the manuscript.

Funding: This research received financial support from the Russian Science Foundation (project No. 23-19-00484).

Institutional Review Board Statement: Not applicable.

Data Availability Statement: The data are contained within the article.

Conflicts of Interest: The authors declare no conflicts of interest. The funders had no role in the design of the study; in the collection, analyses, or interpretation of data; in the writing of the manuscript; or in the decision to publish the results.

Appendix A

A scheme of contact during a typical nanoDMA test is presented in Figure A1. A rigid ball (1) is in contact with a rubber specimen (2) under the controlled normal force oscillating harmonically with time t:

$$P(t) = P_0 + P_s \sin \omega t \tag{A1}$$

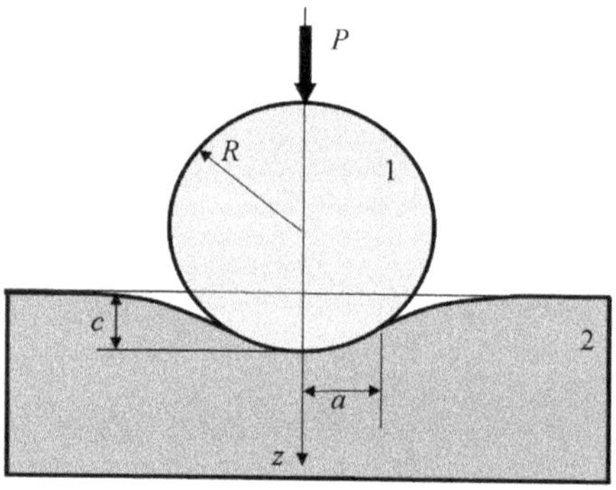

Figure A1. Scheme of contact in nanoDMA testing.

It is assumed that in the regime of steady oscillations, the harmonic load (A1) makes the indentation depth c of the ball into the specimen vary harmonically with the same frequency ω but with a phase shift δ:

$$c(t) = c_0 + c_s \sin(\omega t - \delta) \tag{A2}$$

The contact radius a is a function of time as well, so the contact condition for the normal displacement $w(r,t)$ of the specimen surface ($z = 0$) has the following form:

$$w(r,t) = c(t) - \frac{r^2}{2R}, \ r \le a(t) \tag{A3}$$

where r is the radial coordinate of the cylindrical system of coordinates (r, z, θ), while the problem is axisymmetric with all functions independent of the angle θ.

The specimen is modeled by a linear viscoelastic half-space with the relaxation kernel presented as a sum of exponential terms, which, for a uniaxial loading, can be presented as follows [45]:

$$\sigma(t) = E_1\left[\varepsilon(t) - \int\limits_{-\infty}^{t} \varepsilon(\tau)\Gamma(t-\tau)\, d\tau\right], \ \Gamma(t) = \sum_{i=1}^{N} B_i e^{-\frac{t}{T_i}} \tag{A4}$$

where E_1 is the instantaneous modulus of elasticity and $\Gamma(t)$ is the exponential relaxation kernel. The reverse relationship through the creep kernel $K(t)$ can be presented as follows:

$$\varepsilon(t) = \frac{1}{E_1}\left[\sigma(t) + \int\limits_{-\infty}^{t} \sigma(\tau)K(t-\tau)\, d\tau\right], \ K(t) = \sum_{i=1}^{N} k_i e^{-\frac{t}{\mu_i}} \tag{A5}$$

Here k_i and μ_i are the parameters in the creep kernel ($i = 1\ldots N$).

For the axisymmetric loading of the viscoelastic half-space with a constant Poisson's ratio ν, from (A4) and the viscoelastic correspondence principle [45,46], it follows that the relation between the normal contact pressure p and displacement w has the following form:

$$w(r,t) = \frac{4}{\pi E_1^*}\int\limits_{0}^{\infty} \mathbf{K}\left(\frac{2\sqrt{rr'}}{r+r'}\right)\left(p(r',t) + \int\limits_{-\infty}^{t} p(r',\tau)\sum_{i=1}^{N} K_i e^{-\frac{t-\tau}{\mu_i}}\, d\tau\right)\frac{r'dr'}{r+r'}, \tag{A6}$$

where $E_1^* = E_1/(1-\nu^2)$ is the instantaneous reduced modulus of elasticity. The equilibrium condition has the following form:

$$P(t) = 2\pi\int\limits_{0}^{\infty} rp(r,t)dr \tag{A7}$$

The inverse contact problem needs to be solved, which involves the determination of the instantaneous modulus E_1^* and the parameters (B_i and T_i) of the relaxation kernel and those (k_i and μ_i) of the creep kernel from the prescribed functions of harmonic loading (A1) and response (A2).

Following [47,48], we introduce the following functions:

$$q(x,t) = -2\int\limits_{x}^{\infty} \frac{rp(r,t)}{\sqrt{r^2-x^2}}dr, \ u(x,t) = \frac{\partial}{\partial x}\int\limits_{0}^{x} \frac{rw(r,t)}{\sqrt{x^2-r^2}}dr \tag{A8}$$

Relation (A8) specifies an Abel-type transformation which reduces Equation (A6) to the equivalent form [49,50]:

$$q(x,t) = E_1^*\left[u(x,t) - \int\limits_{-\infty}^{t} u(x,\tau)\sum_{i=1}^{N} B_i e^{-\frac{t-\tau}{T_i}}\, d\tau\right] \tag{A9}$$

whereas the contact condition (A3) accounting for (A2) takes the following form:

$$u(x,t) = c_0 + c_s\sin(\omega t - \delta) - \frac{x^2}{R}, \ x \le a(t) \tag{A10}$$

By substituting the contact condition (A10) into Equation (A9) and dropping a decaying exponential term, one can obtain the following expression for the function q:

$$q(x,t) = E^*\left(c_0 - \frac{x^2}{R}\right) + E'c_s\sin(\omega t - \delta) + E''c_s\cos(\omega t - \delta) \tag{A11}$$

where E^* is the relaxed modulus of elasticity, while E' and E'' are the storage and loss moduli, respectively, which are specified by the following expressions:

$$E^* = \frac{E_1^*}{\alpha},\ E' = \sum_{i=1}^{N} E_i\frac{\alpha T_i^2\omega^2 + 1}{T_i^2\omega^2 + 1},\ E'' = \sum_{i=1}^{N} E_i\frac{\omega T_i(\alpha - 1)}{T_i^2\omega^2 + 1} \tag{A12}$$

where the values α and E_i ($i = 1\ldots N$) are introduced so that the relations are satisfied as follows:

$$B_i = \frac{(\alpha - 1)E_i}{E_1^* T_i},\ i = 1\ldots N,\ E^* = \sum_{i=1}^{N} E_i \tag{A13}$$

The obtained expression (A11) for the function q satisfies Equation (A9) only for the points of contact area which always stay in contact with the ball, i.e., for $x \leq a_{\min}$, where $a_{\min} \leq a(t) \leq a_{\max}$ is the range of variation of the contact radius during the cyclic indentation. Under the condition$c_s \ll c_0$, we can assume that expression (A11) is approximately satisfied for all points of the contact area. Then, from the continuity condition $q(a,t) = 0$, we obtain the approximate expression for the contact radius evolution:

$$a(t) = \sqrt{Rc_0}\left(1 + \frac{c_s}{c_0}H(t)\right)^{1/2}, \tag{A14}$$

where

$$H(t) = \frac{1}{E^*}\left[E'\sin(\omega t - \delta) + E''\cos(\omega t - \delta)\right] \tag{A15}$$

The contact pressure *p(r,t)* is determined from the inverse transformation to (A8) [50]:

$$p(r,t) = -\frac{1}{\pi}\int_{r}^{a(t)} \frac{1}{\sqrt{x^2 - r^2}}\frac{\partial q(x,t)}{\partial x}dr \tag{A16}$$

By substituting (A11) with (A14), by putting (A15) into (A16), taking the integral, and substituting the result into the equilibrium condition (A7), we obtain the following for the load:

$$P(t) = \frac{4E^*}{3R}\sqrt{Rc_0^3}\left(1 + \frac{c_s}{c_0}H(t)\right)^{3/2} \approx \frac{4E^*}{3R}\sqrt{Rc_0^3}\left(1 + \frac{3}{2}\frac{c_s}{c_0}H(t)\right) \tag{A17}$$

Experimental setup. Measurements were carried out on the NanoScan-4D hardness tester (TISNUM, Moscow, Russia). The design of the device makes it possible to set the load with an accuracy of about 100 μN in the range of up to 3.5 N. The displacement is measured with an accuracy of about 3 nm in the range of up to 200 μm.

The NBR specimen was tested at room temperature (T = 22 ± 3 °C). To avoid the Patrikeev–Mullins effect, the specimen surface was preliminarily indented with a maximum frequency of 60 Hz for 60 s, and then the surface was left alone for 20 s, and after that, the surface was indented again at the same place with a chosen frequency. The relaxation of each specimen before the next measurement lasted 40 s, and the oscillations lasted 60 s.

An Al_2O_3 ball with a radius R of 2 mm was used as an indenter. The load was set in accordance with the harmonic law (A1), where P_0 = 100 mN, P_s = 10 mN, and $\omega = 2\pi f$. A typical time diagram of a nanoDMA test with the controlled load P and registered depth c variating in time is shown in Figure A2.

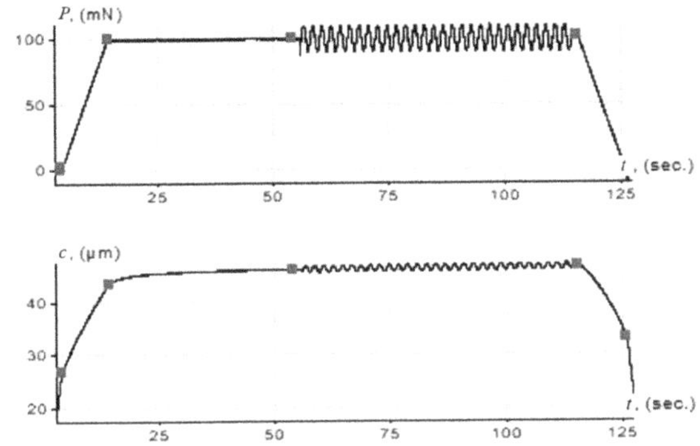

Figure A2. A typical diagram of a single nanoDMA test, including the regions of loading and unloading.

Tests were performed at six frequencies f_j ($j = 1 \ldots n,\ n = 6$). For each frequency, the corresponding mean depth c_{0j} and amplitude depth c_{sj} of the ball indentation were registered, as well as the phase difference δ_j between the load and depth oscillations. The results of the measurements are presented in Table A1.

Table A1. The results of the nanoDMA test for an NBR specimen.

j	f_j, Hz	c_{0j}, nm	c_{sj}, nm	δ_j, rad
1	0.1	87,493	7033	0.138
2	0.2	85,832	6993	0.126
3	0.5	84,761	6317	0.044
4	1	83,525	6018	0.038
5	6	83,607	5333	0.057
6	30	83,607	6700	0.161

Numerical algorithm. To obtain the expressions for the numerical processing of the testing results, the approximate solution for the load P on the right-hand side of (A17) is compared with the exact form (A1). In these relations, by setting the terms containing $\cos \omega t$ equal to zero and equating the terms containing $\sin \omega t$ and the free terms, we obtain the expressions for the storage E'_j and loss E''_j moduli and a relaxed elastic modulus for a j-th frequency:

$$E'_j = \frac{P_s \cos \delta_j}{2c_{sj}\sqrt{Rc_{0j}}},\ E''_j = \frac{P_s \sin \delta_j}{2c_{sj}\sqrt{Rc_{0j}}},\ E^*_j = \frac{3P_0}{4c_{0j}^{3/2}R^{1/2}} \tag{A18}$$

The first two relations of (A18) coincide with the formulas for the storage and loss moduli obtained in (A18), which were obtained in [51], while the third relation coincides with the Hertz formula [47]. Taking into account Equation (A12), from (A18), we obtain the following system of equations ($j = 1 \ldots n$):

$$\sum_{i=1}^{N} E_i \frac{\alpha T_i^2 \omega_j^2 + 1}{T_i^2 \omega_j^2 + 1} = E'_j,\ \sum_{i=1}^{N} E_i \frac{\omega_j T_i(\alpha - 1)}{T_i^2 \omega_j^2 + 1} = E''_j,\ \sum_{i=1}^{N} E_i = E^* \tag{A19}$$

where $\omega_j = 2\pi f_j$. System (A19) is to be solved for the unknown constants (E_i, T_i, and α) of the relaxation spectrum ($i = 1 \ldots N$). To make solving this nonlinear system easier, one can set values of T_i, α, and E^* and then solve the obtained linear system for a set of E_i [51].

To set the values of T_i, α, and E^*, we use the solution obtained in the case of $N = 1$. In this case, the system of equations (A19) with (A18) is solved analytically to obtain the values T_j and α_j for each j-th frequency:

$$T_j = \frac{1}{\omega_j}\sqrt{\frac{P_s^2 - F_j^2}{F^2\alpha_j^2 - P_s^2}} \tag{A20}$$

$$\alpha_j = \frac{P_s}{F_j}\frac{P_sF_j\sin^2\delta_j + (P_s^2 - F_j^2)\cos\delta_j}{P_s^2\cos^2\delta_j - F_j^2} \tag{A21}$$

where $F_j = 2E_j^*\sqrt{Rc_{0j}}c_{sj}$. We set the values of α and E^* as averaged over $j = 1\ldots n$:

$$\alpha = \frac{1}{n}\sum_{j=1}^{n}\alpha_j,\ E^* = \frac{1}{n}\sum_{j=1}^{n}E_j^*, \tag{A22}$$

while the relaxation times T_j are evaluated by (A20). We set the number of relaxation times in the spectrum ($N = 2$) and take the smallest and the largest values of those calculated by (A20) as T_j.

Then, Equation (A19) specifies an overdetermined linear system ($n = 6$; $N = 2$) for a set of moduli E_i which is solved by the least squares method. After this, the instantaneous elastic modulus is calculated as $E_1^* = \alpha E^*$, while the parameters B_i of the relaxation kernel $\Gamma(t)$ (A4) are calculated from (A13).

After calculating the parameters B_i and T_i of the relaxation kernel $\Gamma(t)$ (A4), one can determine the parameters k_i and μ_i of the creep kernel $K(t)$ (A5). It is shown that the parameters of the exponential creep and relaxation kernels are related as follows [45]:

$$1 + \sum_{i=1}^{N}\frac{k_i}{\mu_i^{-1} - T_s^{-1}} = 0,\ 1 + \sum_{s=1}^{N}\frac{B_s}{\mu_i^{-1} - T_s^{-1}} = 0,\ i,s = 1\ldots N \tag{A23}$$

The equations shown in (A23) are an algebraic system to be solved for k_i and μ_i. For $N = 2$, this system is reduced to a quadratic equation and is solved analytically. The parameters of the relaxation B_i, T_i, and creep k_i and μ_i kernels, as well as the instantaneous modulus E_1^* and ratio α of retardation to relaxation times, which are calculated based on the results from Table 1, are presented in Table A2.

Table A2. The parameters of the relaxation and creep kernels for an NBR specimen.

i	T_i, s	B_i	μ_i, s	k_i	E_1^*, MPa	α
1	4.77	0.14	0.030	5.54	2.76	5.03
2	0.025	5.59	20.67	0.18		

References

1. Feng, W.; Patel, S.H.; Young, M.; Zunino, J.L.; Xanthos, M. Smart Polymeric Coatings—Recent Advances. *Adv. Polym. Technol.* **2007**, *26*, 1–13. [CrossRef]
2. Martínez-Martínez, D.; Tiss, B.; Glanzmann, L.N.; Wolthuizen, D.J.; Cunha, L.; Mansilla, C.; De Hosson, J.T.M. Protective Films on Complex Substrates of Thermoplastic and Cellular Elastomers: Prospective Applications to Rubber, Nylon and Cork. *Surf. Coat. Technol.* **2022**, *442*, 128405. [CrossRef]
3. Zhou, B.; Liu, Z.; Xu, B.; Rogachev, A.V.; Yarmolenko, M.A.; Rogachev, A.A. Modification of Cu-PE-PTFE Composite Coatings on Rubber Surface by Low-energy Electron Beam Dispersion with Glow Discharge. *Polym. Eng. Sci.* **2018**, *58*, 103–111. [CrossRef]
4. Liu, Y.; Bian, D.; Wang, J.; Zhao, Y. Influence of Pore-Forming Agent on Wear Resistance of Composite Coating. *Surf. Eng.* **2022**, *38*, 208–215. [CrossRef]
5. Aliyu, I.K.; Samad, M.A.; Al-Qutub, A. Tribological Characterization of a Bearing Coated with UHMWPE/GNPs Nanocomposite Coating. *Surf. Eng.* **2021**, *37*, 60–69. [CrossRef]
6. Smith, J.R.; Lamprou, D.A. Polymer Coatings for Biomedical Applications: A Review. *Trans. IMF* **2014**, *92*, 9–19. [CrossRef]

7. Abdul Samad, M. Recent Advances in UHMWPE/UHMWPE Nanocomposite/UHMWPE Hybrid Nanocomposite Polymer Coatings for Tribological Applications: A Comprehensive Review. *Polymers* **2021**, *13*, 608. [CrossRef]
8. Surikov, V.I.; Nikolaev, I.V.; Polonyankin, D.A.; Tselykh, E.P.; Rogachev, E.A.; Surikov, V.I. Morphology, Composition and Tribological Properties of Tantalum Coatings Deposited onto a Rubber Substrate. *J. Phys. Conf. Ser.* **2017**, *858*, 012034. [CrossRef]
9. Tikhomirov, L.A.; Tarasenko, V.A.; Kostina, T.Y.; Dorofeeva, L.V. The Effect of Molybdenum Disulphide on the Tribotechnical Characteristics of Polyamide Coatings on Nitrile Rubber Compounds. *Int. Polym. Sci. Technol.* **2015**, *42*, 31–34. [CrossRef]
10. Ning, K.; Lu, J.; Xie, P.; Hu, J.; Huang, J.; Sheng, K. Study on Surface Modification of Silicone Rubber for Composite Insulator by Electron Beam Irradiation. *Nucl. Instrum. Methods Phys. Res. Sect. B Beam Interact. Mater. At.* **2021**, *499*, 7–16. [CrossRef]
11. Jin, M.; Zhang, X.; Emeline, A.V.; Numata, T.; Murakami, T.; Fujishima, A. Surface Modification of Natural Rubber by TiO_2 Film. *Surf. Coat. Technol.* **2008**, *202*, 1364–1370. [CrossRef]
12. Dyakonov, A.A.; Shadrinov, N.V.; Sokolova, M.D.; Fedorov, A.L.; Sleptsova, S.A.; Okhlopkova, A.A. Research of the Effect of Diphenylguanidine on the Adhesive Interaction of Elastomers with Ultrahigh Molecular Weight Polyethylene. *J. Sib. Fed. University. Eng. Technol.* **2019**, *12*, 476–487. [CrossRef]
13. Dyakonov, A.A.; Vasilev, A.P.; Danilova, S.N.; Okhlopkova, A.A.; Tarasova, P.N.; Lazareva, N.N.; Ushkanov, A.A.; Tuisov, A.G.; Kychkin, A.K.; Vinokurov, P.V. Two-Layer Rubber-Based Composite Material and UHMWPE with High Wear Resistance. *Materials* **2022**, *15*, 4678. [CrossRef]
14. Shadrinov, N.V.; Sokolova, M.D.; Dyakonov, A.A. Mechanical Properties and Deformation Features of Nitrile Butadiene Rubber with Protective Coating from Ultra High-Molecular Weight Polyethylene. *IOP Conf. Ser. Mater. Sci. Eng.* **2021**, *1079*, 042015. [CrossRef]
15. Petrica, M.; Duscher, B.; Koch, T.; Archodoulaki, V.-M. Studies on Tribological Behavior of PEEK and PE-UHMW. In Proceedings of the Regional Conference Graz 2015–Polymer Processing Society, Graz, Austria, 21–25 September 2015; p. 070001.
16. Han, Y.; Chen, J. Experimental Investigation on Tribological Properties of UHMWPE with the Addition of Basalt Fiber. *Adv. Compos. Lett.* **2019**, *28*. [CrossRef]
17. Wang, Y.; Yin, Z.; Li, H.; Gao, G.; Zhang, X. Friction and Wear Characteristics of Ultrahigh Molecular Weight Polyethylene (UHMWPE) Composites Containing Glass Fibers and Carbon Fibers under Dry and Water-Lubricated Conditions. *Wear* **2017**, *380–381*, 42–51. [CrossRef]
18. Wang, Q.; Wang, Y.; Wang, H.; Fan, N.; Yan, F. Fretting Wear Behavior of UHMWPE Under Different Temperature Conditions. *J. Macromol. Sci. Part B* **2017**, *56*, 493–504. [CrossRef]
19. Alam, F.; Choosri, M.; Gupta, T.K.; Varadarajan, K.M.; Choi, D.; Kumar, S. Electrical, Mechanical and Thermal Properties of Graphene Nanoplatelets Reinforced UHMWPE Nanocomposites. *Mater. Sci. Eng. B* **2019**, *241*, 82–91. [CrossRef]
20. Saikko, V.; Morad, O.; Viitala, R. Effect of Temperature on UHMWPE and VEXLPE Friction and Wear against CoCr in Noncyclic Tests. *Wear* **2022**, *490–491*, 204190. [CrossRef]
21. Cao, M.; Chen, L.; Xu, R.; Fang, Q. Effect of the Temperature on Ballistic Performance of UHMWPE Laminate with Limited Thickness. *Compos. Struct.* **2021**, *277*, 114638. [CrossRef]
22. Imado, K.; Miura, A.; Nagatoshi, M.; Kido, Y.; Miyagawa, H.; Higaki, H. A Study of Contact Temperature Due to Frictional Heating of UHMWPE. *Tribol. Lett.* **2004**, *16*, 265–273. [CrossRef]
23. Tabor, D. The Mechanism of Rolling Friction II. The Elastic Range. *Proc. R. Soc. Lond. A* **1955**, *229*, 198–220. [CrossRef]
24. Greenwood, J.A.; Tabor, D. The Friction of Hard Sliders on Lubricated Rubber: The Importance of Deformation Losses. *Proc. Phys. Soc.* **1958**, *71*, 989–1001. [CrossRef]
25. Grosch, K.A. The Relation between the Friction and Visco-Elastic Properties of Rubber. *Proc. R. Soc. Lond. A* **1963**, *274*, 21–39. [CrossRef]
26. Aleksandrov, V.M.; Goryacheva, I.G.; Torskaya, E.V. Sliding Contact of a Smooth Indenter and a Viscoelastic Half-Space (3D Problem). *Dokl. Phys.* **2010**, *55*, 77–80. [CrossRef]
27. Goryacheva, I.G.; Stepanov, F.I.; Torskaya, E.V. Sliding of a Smooth Indentor over a Viscoelastic Half-Space When There Is Friction. *J. Appl. Math. Mech.* **2015**, *79*, 596–603. [CrossRef]
28. Goryacheva, I.; Stepanov, F.; Torskaya, E. Effect of Friction in Sliding Contact of a Sphere Over a Viscoelastic Half-Space. In *Mathematical Modeling and Optimization of Complex Structures*; Computational Methods in Applied Sciences, Neittaanmäki, P., Repin, S., Tuovinen, T., Eds.; Springer International Publishing: Cham, Switzerland, 2016; Volume 40, pp. 93–103, ISBN 978-3-319-23563-9.
29. Koumi, K.E.; Chaise, T.; Nelias, D. Rolling Contact of a Rigid Sphere/Sliding of a Spherical Indenter upon a Viscoelastic Half-Space Containing an Ellipsoidal Inhomogeneity. *J. Mech. Phys. Solids* **2015**, *80*, 1–25. [CrossRef]
30. Aleksandrov, V.M.; Goryacheva, I.G. Motion with a Constant Velocity of a Distributed Load over a Viscoelastic Half-Space. In Proceedings of the 5th Russian Conference "Mixed Problems of Deformable Body Mechanics"; Saratov University: Saratov, Russia, 2005; pp. 23–25.
31. Torskaya, E.V.; Stepanov, F.I. Effect of Surface Layers in Sliding Contact of Viscoelastic Solids (3-D Model of Material). *Front. Mech. Eng.* **2019**, *5*, 26. [CrossRef]
32. *ISO 37:2024*; Rubber, Vulcanized or Thermoplastic—Determination of Tensile Stress-Strain Properties. ISO: Geneva, Switzerland, 2024.
33. *ISO 1817:2024*; Rubber, Vulcanized or Thermoplastic—Determination of the Effect of Liquids. ISO: Geneva, Switzerland, 2024.

34. *ISO 23529:2016*; Rubber—General Procedures for Preparing and Conditioning Test Pieces for Physical Test Methods. ISO: Geneva, Switzerland, 2016.
35. Wypych, G. UHMWPE Ultrahigh Molecular Weight Polyethylene. In *Handbook of Polymers*; Elsevier: Berlin/Heidelberg, Germany, 2016; pp. 693–697, ISBN 978-1-895198-92-8.
36. Morozov, A.V.; Murav'eva, T.I.; Petrova, N.N.; Portnyagina, V.V.; Ammosova, V.N.; Zagorskii, D.L. Investigation of the Tribological and Adhesive Properties of Cold-Resistant Rubbers. *Int. Polym. Sci. Technol.* **2016**, *43*, 27–32. [CrossRef]
37. Bukovskiy, P.O.; Morozov, A.V.; Petrova, N.N.; Timofeeva, E.V. Study on the Influence of Activated Carbon Nanotubes on the Tribological Properties of Frost-Resistant Rubber. *Mech. Solids* **2019**, *54*, 1250–1255. [CrossRef]
38. *ASTM G99-23*; Standard Test Method for Wear and Friction Testing with a Pin-on-Disk or Ball-on-Disk Apparatus. ASTM International: West Conshohocken, PA, USA, 2023.
39. Nikishin, V.S.; Shapiro, G.S. *Space Problems of Elasticity Theory for Multilayered Media*; VTs AN SSSR: Moscow, Russia, 1970.
40. Liu, S.; Wang, Q.; Liu, G. A Versatile Method of Discrete Convolution and FFT (DC-FFT) for Contact Analyses. *Wear* **2000**, *243*, 101–111. [CrossRef]
41. Evtushenko, A.A.; Ivanik, E.G.; Evtushenko, E.A. Approximate Method for Determining the Maximum Temperature during Quasistationary Heating of a Piecewise-Homogeneous Half-Space. *J. Appl. Mech. Technol. Phys.* **2005**, *46*, 375–385. [CrossRef]
42. Macuvele, D.L.P.; Nones, J.; Matsinhe, J.V.; Lima, M.M.; Soares, C.; Fiori, M.A.; Riella, H.G. Advances in Ultra High Molecular Weight Polyethylene/Hydroxyapatite Composites for Biomedical Applications: A Brief Review. *Mater. Sci. Eng. C* **2017**, *76*, 1248–1262. [CrossRef]
43. Goryacheva, I.G.; Makhovskaya, Y. *Discrete Contact Mechanics with Applications in Tribology*; Elsevier: Berlin/Heidelberg, Germany, 2022; ISBN 978-0-12-821799-3.
44. Torskaya, E.V.; Stepanov, F.I. Effect of Friction in Sliding Contact of Layered Viscoelastic Solids. In *Advanced Problem in Mechanics II*; Lecture Notes in Mechanical Engineering, Indeitsev, D.A., Krivtsov, A.M., Eds.; Springer International Publishing: Cham, Switzerland, 2022; pp. 320–330, ISBN 978-3-030-92143-9.
45. Rabotnov, Y.N. *Creep of Structural Elements*; Nauka Publ.: Moscow, Russia, 1966.
46. Christensen, R.M. *Theory of Viscoelasticity: An Introduction*; Academic Press: New York, NY, USA, 1971.
47. Johnson, K.L. *Contact Mechanics*, 1st ed.; Cambridge University Press: Cambridge, UK, 1985; ISBN 978-0-521-25576-9.
48. Sneddon, I. *Fourier Transforms*; McGraw Hill: New York, NY, USA, 1951.
49. Haiat, G.; Phan Huy, M.C.; Barthel, E. The Adhesive Contact of Viscoelastic Spheres. *J. Mech. Phys. Solids* **2003**, *51*, 69–99. [CrossRef]
50. Argatov, I.I.; Popov, V.L. Rebound Indentation Problem for a Viscoelastic Half-space and Axisymmetric Indenter—Solution by the Method of Dimensionality Reduction. *Z. Angew. Math. Mech.* **2016**, *96*, 956–967. [CrossRef]
51. Baumgaertel, M.; Winter, H.H. Determination of Discrete Relaxation and Retardation Time Spectra from Dynamic Mechanical Data. *Rheol. Acta* **1989**, *28*, 511–519. [CrossRef]

polymers

MDPI

Article

Coniferous Bark as Filler for Polylactic Acid-Based Biocomposites

Wojciech Jasiński [1,*], Radosław Auriga [1], Seng Hua Lee [2], Łukasz Adamik [1,3] and Piotr Borysiuk [1,*]

[1] Institute of Wood Sciences and Furniture, Warsaw University of Life Sciences-SGGW, 159 Nowoursynowska St., 02-776 Warsaw, Poland; radoslaw_auriga@sggw.edu.pl (R.A.); lukasz.adamik@nowystyl.com (Ł.A.)
[2] Department of Wood Industry, Faculty of Applied Sciences, Universiti Teknologi MARA, Cawangan Pahang Kampus Jengka, Bandar Tun Razak 26400, Malaysia; leesenghua@uitm.edu.my
[3] Nowy Styl sp. z o.o., ul. Pużaka 49, 38-400 Krosno, Poland
* Correspondence: 9.wojciech.jasinski@gmail.com (W.J.); piotr_borysiuk@sggw.edu.pl (P.B.)

Abstract: This study explores the possibilities of utilisation of coniferous bark as a filler in wood–polymer composites (WPCs), its impact on properties such as the modulus of rupture (MOR), modulus of elasticity (MOE), thickness swelling (TS) and water absorption (WA) after 2 h and 24 h of immersion in water and the significance of this impact compared to other factors. Six variants of bark–polylactic acid (PLA) WPCs were manufactured, differentiated by their filler content and filler particle size. As a comparison, analogous composites filled with coniferous sawdust were also manufactured. Bark-filled composites were characterised by lower TS and WA after both 2 h and 24 h of immersion, as well as lower water contact angles and surface free energy. The bark filler decreased the composites' MORs and MOEs, while greater differences were noticed for variants filled with small particles. The type of filler was the second most important factor contributing to variance in this study, with the filler content being the most important one.

Keywords: wood–polymer composites; filler; bark; sawdust

1. Introduction

Wood–polymer composites (WPCs) provide a cost-effective alternative to both polymer- and wood-based products. While construction is one of the main areas of WPC utilisation, due to their lower thermal stability and material creep, these composites are usually limited to non-load-bearing applications [1]. Other fields in which these composites are widely used include furniture and interior car parts [2]. One of the main advantages of WPCs is the wide spectrum of manufacturing processes that can be used to shape them. Thermoplastic-based composites may be moulded or extruded, as well as machined using equipment dedicated for conventional wood products [3,4].

The physico-mechanical properties of WPCs are usually in the same range as those of a polymer when used as a matrix and wood when used as a filler [5]. Despite this, WPCs should be regarded as neither wood-filled polymers nor polymer-modified wood, but as a completely separate group of materials [6]. WPCs are widely considered to be biocomposites, although most of them are manufactured using petrol-derived polymers, such as polyethylene (PE), polypropylene (PP) and polyvinyl chloride (PVC) [7,8]. Whether WPCs based on non-biodegradable polymers are biocomposites is debatable, but the use of biodegradable polymers (biopolymers), such as polylactic acid (PLA), polyhydroxyalkanoates (PHAs) or thermoplastic starch (TPS), as WPC matrices is also possible [9]. PLA is a plant-derived polymer with properties similar to those of polystyrene (PS) and polyethylene (PE) [10]. It is fully biodegradable, but only under specific conditions, which promotes the long service life of this polymer [11]. PLA is currently regarded as the most extensively researched and used biopolymer [12]. Compared to other commonly used polymers, PLA is characterised by a lower melting point and heat deflection temperature, higher stiffness and lower impact resistance [13].

Wood used in WPCs as a filler is usually a by-product of industrial wood processing in the form of wood shavings, sawdust or wood flour [14]. The choice of wood type and form impacts the physico-mechanical properties of the manufactured composites [15]. Coniferous species are preferable due to their higher slenderness and the simplicity of their anatomical structure [16], while the size of wood particles depends on the WPC's intended use [17]. Overall, there are no limitations to the utilisation of different wood types as fillers in WPCs [18].

According to Gozdecki et al. [19] and Avci et al. [20], bark used as a filler decreases WPCs' mechanical properties, such as their modulus of rupture and modulus of elasticity, compared to wood-flour-filled WPCs. Çetin et al. [21] tested composites based on high-density PE and reported that the addition of Turkish pine bark as a filler reduced the tensile and impact strength, while the bending strength and modulus of elasticity improved. Najafi et al. [22] reported that bark decreases the thickness swelling and water absorption of WPCs, in turn improving their moisture resistance compared to traditional WPCs. These correlations were also confirmed by Farasi [23], who tested a WPC filled with beech bark flour. Obuch et al. [24], as part of the Dendromass4Europe project, showed that it is possible to produce WPC composites with a partial 5% replacement of beech flour with crushed poplar bark. While the impact of the bark filler on the aforementioned properties can be confirmed in the literature, the available research lacks the context of other factors impacting WPCs' performance, such as the filler content or filler particle size. Most of the research also concerns WPCs that are manufactured on the basis of typical petroleum-based thermoplastics (e.g., PE, PP, PVC).

The aim of this study is to evaluate the possibility of manufacturing WPCs based on PLA filled with bark, the impact of the bark filler on selected physico-mechanical properties of WPCs and the significance of that impact compared to the filler content and its particle size.

2. Materials and Methods

Twelve variants of WPC panels based on polylactic acid (PLA Ingeo™ Biopolymer 2003D, NatureWorks LLC, Minnetonka, MN, USA) were manufactured for this study. The manufactured biocomposites varied in terms of type, size and filler content. Coniferous bark and sawdust, obtained from a sawmill, were used as fillers. The fillers were dried to a moisture content of 5% and then ground and sorted into two groups—particles passing through a 2 mm sieve and remaining on a 0.49 mm sieve (big particles) and particles passing through a 0.49 mm sieve (small particles). Photos of selected manufactured composites are shown in Figure 1. The produced variants are shown in Table 1.

Table 1. Composition of individual variants.

Variant	Share of PLA Matrix	Size of Particles	Kind of Particles
I	60%	(0.49–2 mm)	Bark
II			Sawdust
III	50%		Bark
IV			Sawdust
V	40%		Bark
VI			Sawdust
VII	60%	(<0.49 mm)	Bark
VIII			Sawdust
IX	50%		Bark
X			Sawdust
XI	40%		Bark
XII			Sawdust

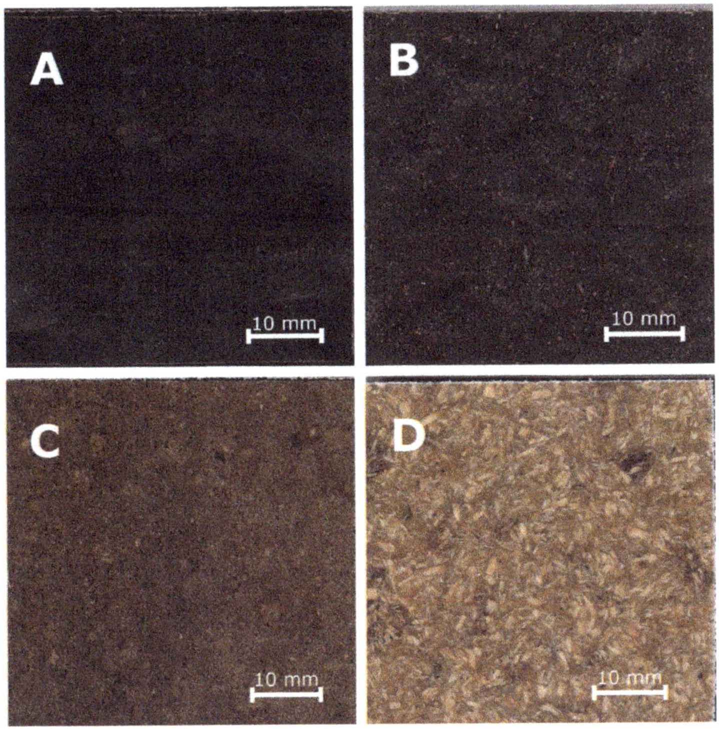

Figure 1. Photos of selected manufactured composites: (**A**) variant XI (60% share of small particles of bark, 40% PLA); (**B**) variant V (60% share of big particles of bark, 40% PLA); (**C**) variant XII (60% share of small particles of sawdust, 40% PLA); (**D**) variant VI (60% share of big particles of sawdust, 40% PLA).

The composites were manufactured in a two-stage process. The first stage consisted of creating a WPC granulate using the Leistritz Extrusionstechnik GmbH (Nürnberg, Germany) extruder and then grinding the composite using a hammer mill. Individual extruder sections were kept at 170–180 °C.

In the second stage, the granulate was used to produce boards sized 300 mm × 300 mm × 2.5 mm by flat pressing in a mould using a single-shelf press. The temperature of the press was set at 200 °C, and the pressure was gradually increased from 0 to a maximum of 1.25 MPa as the materials were plasticizing. The boards were pressed for 6 min, then cooled in the mould for 6 min in a cold press. The manufactured panels were finally conditioned for 7 days at a temperature of 20 ± 2 °C and 65 ± 5% humidity.

For the manufactured panels, the following tests were performed:

- Modulus of rupture (MOR) and modulus of elasticity (MOE) using 3-point flatwise bending tests according to EN 310 [25].
- Density according to EN 323 [26].
- Thickness swelling (TS) and water absorption (WA) after 2 h and 24 h of immersion in water according to EN 317 [27]. Samples were immersed in water with their faces vertical and separated from the edges of the container, as well as each other. The water temperature was maintained at 20 ± 1 °C for the whole duration of the test. After the test time had elapsed, excess water was removed and the samples were weighed and measured.

The tests described above involved ten replicates.

- Wettability (contact angle) was determined by the sessile drop method using a Phoenix 300 (Surface Electro Optics, Suwon City, Korea) contact angle analyser, equipped with microscopic lenses and a digital camera. The contact angles were measured after 20, 40 and 60 s of distilled water droplets touching the surface. Three replicates of these tests were performed.
- Surface free energy was determined by measuring the contact angles of distilled water and diiodomethane after 5 s using the same procedure. The Owens–Wendt [28] method was used to calculate the free surface energy.

Analysis of variance (ANOVA) with a post hoc Tukey test was used to determine the significance of differences between variants, as well as individual factors' contribution to these differences. Both tests were performed in PQStat v.1.8.7 (PQStat Software, Poznań, Poland) with a confidence level of 95%.

3. Results and Discussion

Generally, the average MOR (Figure 2) obtained during the study decreased with an increase in the filler content, while being higher for variants containing sawdust than for variants containing bark. These differences were bigger for variants filled with small particles. The average for variant II was 29% higher than the average for variant I, while the average for variant VIII was 51% higher than the average for variant VII. The difference between variant X and XI was 140%, and between variants XII and XI, it equalled 10%, while the differences between variants filled with 50% and 60% big-particle fillers were not statistically significant.

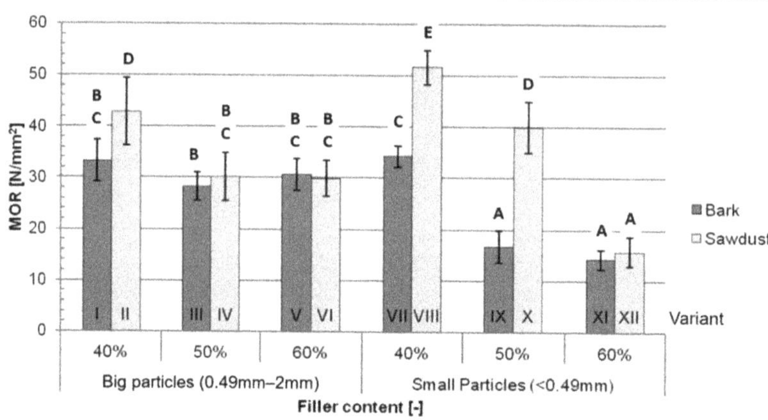

Figure 2. Average MOR for each tested variant: A, B, C, D, E—homogeneous groups, I–XII—composite variant.

The three most significant factors (Table 2) impacting the MOR were filler content (42.57%), type of filler (15.86%) and an interaction between the particle size and filler content (13.27%). These factors contributed to 71.70% of the variance in this test.

Similarly to the averages obtained in the MOR test, bigger differences between the MOE averages (Figure 3) can be seen between variants containing small-particle fillers. The difference between variants with 40% of these fillers is 31%, and between variants filled with 50% and 60% small particles, it is 50% and 76%, respectively. The differences between variants containing big particles are not statistically significant. The average MOE decreases with an increase in small-particle filler content.

Table 2. ANOVA for selected factors affecting MOR, MOE and density of tested composites.

Factor	MOR		MOE		Density	
	p	P	*p*	P	*p*	P
a	0.000	42.57%	0.009	3.72%	0.361	1.32%
b	0.000	15.86%	0.000	23.86%	0.001	7.24%
c	0.000	2.58%	0.002	3.72%	0.204	1.05%
a × b	0.000	6.40%	0.510	0.51%	0.122	2.75%
a × c	0.000	13.27%	0.000	16.71%	0.258	1.77%
b × c	0.000	5.67%	0.000	12.39%	0.000	13.96%
a × b × c	0.000	3.06%	0.340	0.82%	0.361	1.32%
Error		10.58%		38.29%		69.43%

a—filler content, b—type of filler, c—size of particles, *p*—probability of error, P—percentage of contribution.

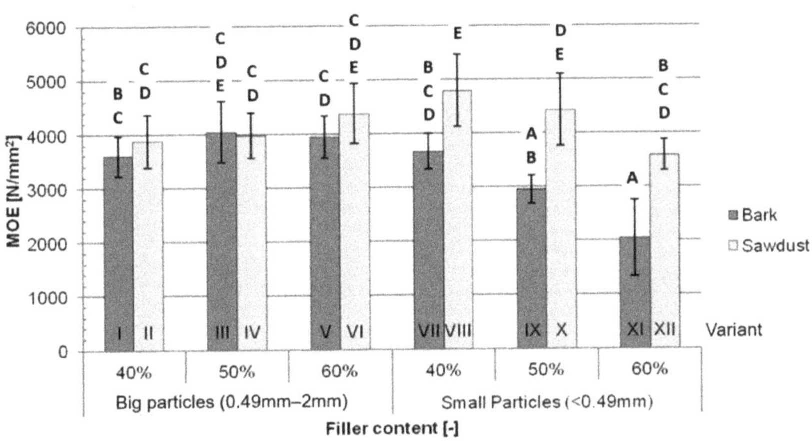

Figure 3. Average MOE for each tested variant: A, B, C, D, E—homogeneous groups, I–XII—composite variant.

The most important factor (Table 2.) impacting the MOE is the type of filler (23.86%), followed by the interactions between the content and the size of filler particles (16.71%) and between the content and type of the filler (12.39%). All three of these factors explain 52.95% of the variance in the MOE test, but each individually impacted the results less than factors that were not considered in this study (Error = 38.29%).

The density measured for the tested variants (Figure 4) ranged from 1061 kg/m^3 for variant VI to 1182 kg/m^3 for variant V, with an average of 1136 kg/m^3. Statistically significant differences in density (Figure 3) were only found for the aforementioned variants.

The type of filler (7.24%) and the interactions between the type of filler and size of particles (13.96%) had minor impacts on the measured density, but the most important factors impacting variance in this test (Table 2) were those that were not considered during this study (Error = 69.43%).

According to Jian et al. [1], both the MOR and MOE of WPCs increase with an increase in filler content up to a maximum, which, depending on composite's formulation, is usually in the range of 40% to 70% filler content. The obtained data suggest that the filler contents used during this study are above the range corresponding to the maximum MOR and MOE for the tested composites. The lower filler content corresponding to the maximum MOR and MOE may be caused by the use of a stiff and brittle matrix, such as PLA, and/or a lack of coupling agent, resulting in a weaker bond between the polymer and fillers. An increase in the tested mechanical properties for the composites filled with small-particle sawdust are in line with the findings of Shah et al. [29] and may be caused by breakage of the fibres being the main failure mechanism. The opposite tendency might be observed for bark-filled composites, which suggests a different failure mechanism. The reason for this difference may be the different polysaccharide composition causing lower adhesion between the filler and matrix. The worse mechanical performance of bark-filled composites might not only be explained by the aforementioned polysaccharide composition, but also by the higher extractives content, more mineral contamination and intrinsically worse mechanical properties of bark itself [30,31]. Compared to traditional high-density fibreboards (HDFs) of similar densities (1050 kg/m^3 and 1150 kg/m^3), all the tested variants are characterised by significantly lower MORs and MOEs than those of HDFs [32]. The ANOVA results show that the type of filler itself does not impact the MOR of the tested composites nearly as significantly as the filler content, which might suggest that, by changing the proportions of matrix and filler, a composite with a high bending strength may be manufactured using bark as a filler. These results could be explained by the aforementioned different failure mechanisms. Lower adhesion between the bark and the polymer, coupled with the worse mechanical properties of bark, would suggest that the failure happens due to breakage in the matrix, and a higher content of polymers (lower filler content) would increase the MOR. Additional research, such as microscopic imaging of the broken samples, would be needed to confirm this explanation.

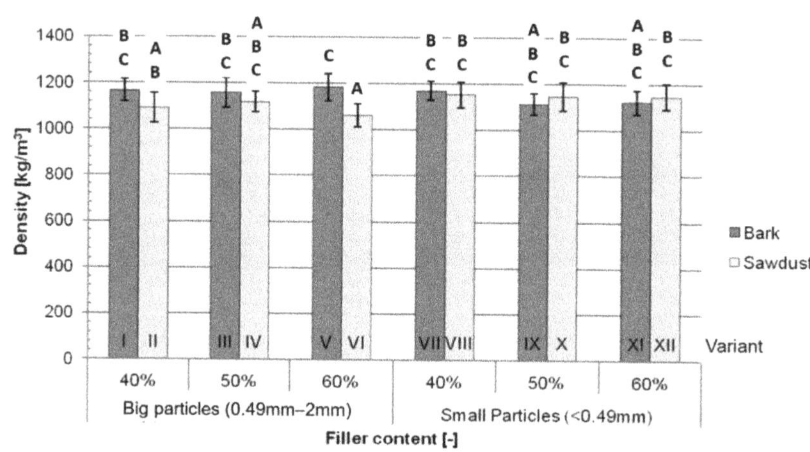

Figure 4. Average density for each tested variant: A, B, C—homogeneous groups, I–XII—composite variant.

Although the decreases in the tested mechanical properties for variants filled with bark compared to variants filled with sawdust are consistent with the literature, this decrease is not statistically significant for variants filled with 50% and 60% big-particle fillers, which shows that the use of bark as a filler might not impact the mechanical properties of certain WPC formulations in a significant way.

The average TS after 2 h of immersion in water (Figure 5) rises along with the increase in the filler content. The differences between the variants containing bark and containing sawdust are not significant for the composites containing 40% and 50% filler. The average TS after 2 h of immersion for the variants filled with 60% sawdust are higher than the average for variants filled with 60% bark, while these differences are bigger between variants XII and XI, which are filled with small particles (408%), than between variants VI and V, filled with big particles (124%).

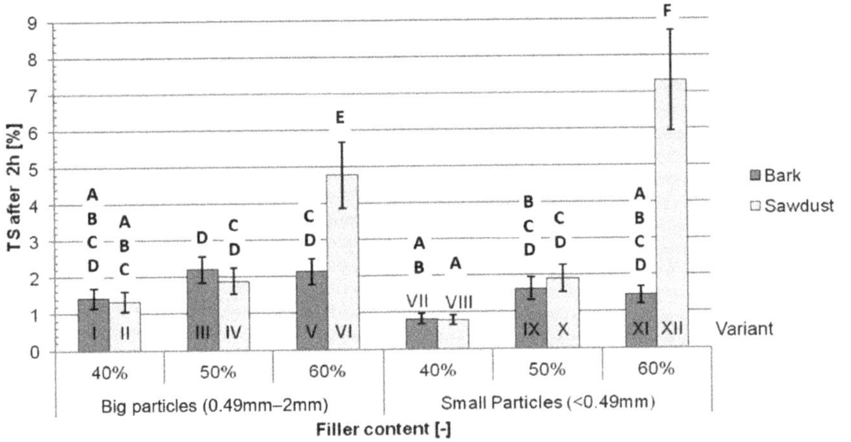

Figure 5. Average thickness swelling after 2 h of immersion in water for each tested variant: A, B, C, D, E, F—homogeneous groups, I–XII—composite variant.

After 24 h of immersion in water, the average TS (Figure 6) for variants IV, VI and XII filled with sawdust is significantly higher than the average TS for variants filled with bark—III, V and XI. The differences between other pairs of variants are not statistically significant. Similarly to the TS after 2 h of immersion, the average TS after 24 h of immersion in water is higher for variants containing more filler.

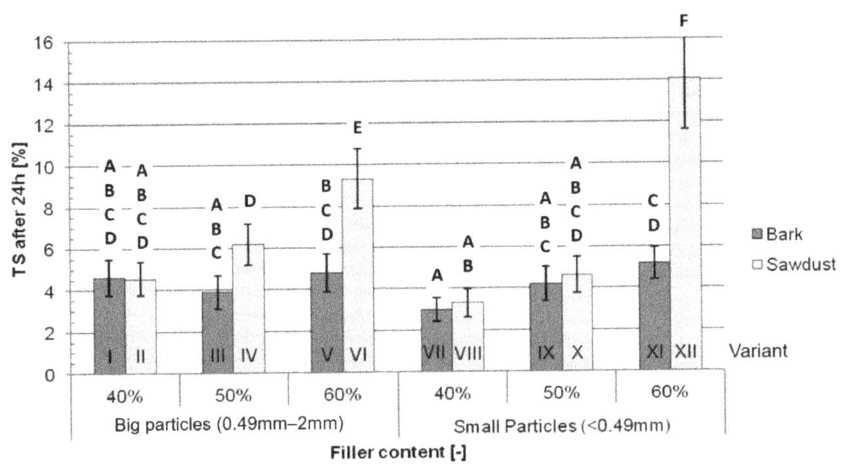

Figure 6. Average thickness swelling after 24 h of immersion in water for each tested variant: A, B, C, D, E, F—homogeneous groups, I–XII—composite variant.

The most important factors impacting the TS (Table 3) both after 2 h and 24 h of immersion are the filler content (40.05% after 2 h and 37.51% after 24 h), type of filler (13.79% after 2 h and 18.63% after 24 h) and interaction between those two factors (29.15% after 2 h and 20.39% after 24 h). Together, the percentage of the impact of these factors is 82.99% after 2 h of immersion and 76.54% after 24 h of immersion. The particle size itself does not significantly impact the TS in either of those tests.

Table 3. ANOVA for selected factors affecting TS and WA after 2 h and 24 h of immersion in water.

Factor	TS after 2 h		TS after 24 h		WA after 2 h		WA after 24 h	
	p	P	*p*	P	*p*	P	*p*	P
a	0.000	40.05%	0.000	37.51%	0.000	46.45%	0.000	54.56%
b	0.000	13.79%	0.000	18.63%	0.000	30.04%	0.000	20.66%
c	0.835	0.00%	0.430	0.06%	0.038	0.28%	0.144	0.10%
a × b	0.000	29.15%	0.000	20.39%	0.000	8.80%	0.000	12.10%
a × c	0.000	3.04%	0.000	7.54%	0.000	5.48%	0.000	7.38%
b × c	0.000	2.93%	0.020	0.58%	0.000	0.94%	0.101	0.12%
a × b × c	0.000	3.43%	0.000	4.13%	0.000	1.15%	0.041	0.29%
Error		7.60%		11.15%		6.87%		4.80%

a—filler content, b—type of filler, c—size of particles, *p*—probability of error, P—percentage of contribution.

The average WA after 2 h of immersion in water (Figure 7) increases with an increase in filler content while being higher for variants filled with sawdust than for those filled with bark. The differences between pairs of variants differentiated by their filler type range from 51% between variants III and IV up to 277% between variants V and VI.

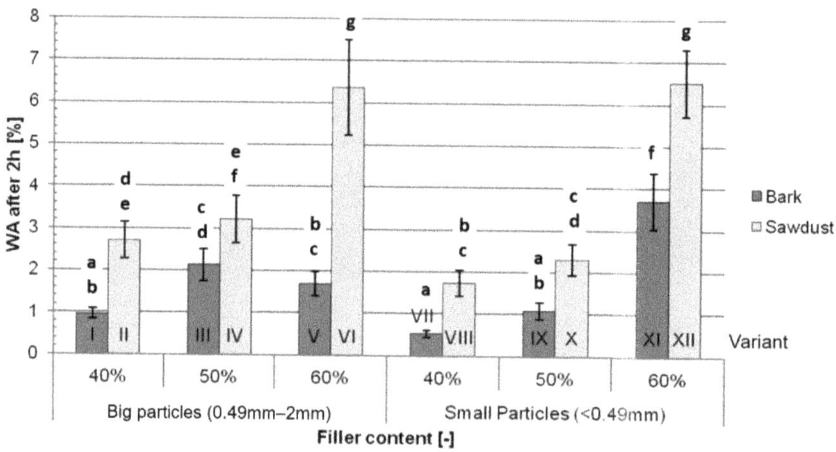

Figure 7. Average water absorption after 2 h of immersion in water for each tested variant: a, b, c, d, e, f, g—homogeneous groups, I–XII—composite variant.

After 24 h of immersion in water, the average WA (Figure 8), similarly to the average WA after 2 h, rises with an increase in filler content. Generally, after 24 h, the WA of the variants filled with sawdust is bigger than the WA after 24 h of the variants filled with bark. The difference between variants I and II is not statistically significant, while other differences range from 40% between variants III and IV to 175% between variants V and VI. Those differences are smaller than the differences between variants filled with sawdust and variants filled with bark that were found in the WA after the 2 h test, which suggests that in this study, sawdust-filled WPCs not only absorb more water than analogical bark-filled composites, they also absorb the water faster.

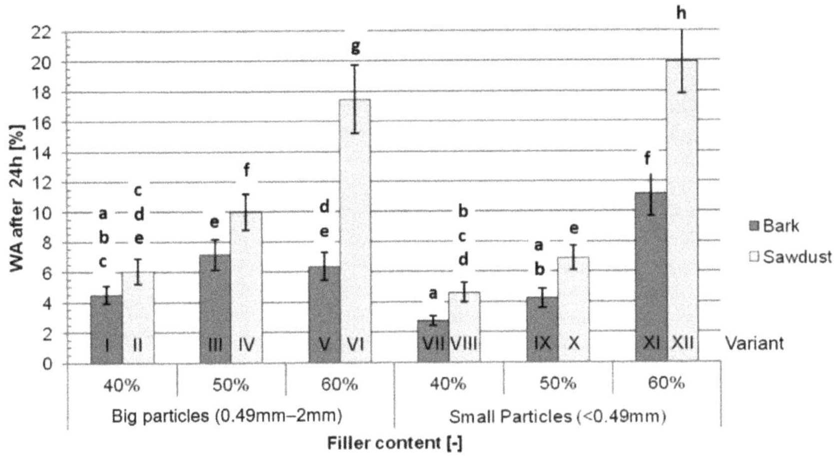

Figure 8. Average water absorption after 24 h of immersion in water for each tested variant: a, b, c, d, e, f, g, h—homogeneous groups, I–XII—composite variant.

The most important factors impacting WA both after 2 h and after 24 h (Table 3) are the filler content (46.45% after 2 h and 54.56% after 24 h), the type of filler (30.04% after 2 h and 20.66% after 24 h) and the interaction between those two factors (8.80% after 2 h and 12.10% after 24 h). All three of those factors contribute to 85.28% of the variance after 2 h and 87.32% of the variance after 24 h in these tests. Similarly to the TS tests, the size of particles itself impacts WA marginally after 2 h (0.28%) and does not impact WA after 24 h ($p > 0.05$).

The fact that TS and WA increase for variants containing higher filler contents is in line with data available in the literature [33] and is caused by higher amounts of exposed fibres, as well as higher amount of fibres overall. According to Bouafif et al. [34], the particle size of the filler has a minimal impact on water absorption and thickness swelling, which may also be noticed in the data obtained during this study. The lower TS and WA values for variants filled with bark are consistent with literature findings. Despite the fact that lower adhesion between the filler and the matrix, which might be inferred from the results of the mechanical tests, promotes water absorption in WPCs, the hydrophilic properties of lignocellulosic fibres themselves have a bigger impact on TS and WA [20,22]. The differences observed during this study may be explained by the different chemical composition of bark, namely a lower content of hydrophilic cellulose and higher content of possibly hydrophobic extractives. Comparing the results obtained during these tests to the TS and WA values of traditional HDF boards of similar densities (1050 kg/m^3 and 1150 kg/m^3), the TS of bark-filled composites, depending on the filler content, range from lower to similar to those of HDFs, while the TS of sawdust-filled composites range from similar to significantly higher than those of HDFs. The WA values of variants with 40% and 50% filler contents and variant 5, filled with 60% big-particle bark, are significantly lower than the WA of HDF. The WA of other variants range from similar to higher than that of HDF [32].

The contact angles of composites filled with big-particle fillers (Figure 9) exhibit not only higher overall values but also bigger differences between variants than the composites filled with small-particle fillers (Figure 10). Apart from variants filled with 60% big-particle fillers, the average contact angles are higher for variants filled with bark than for those filled with sawdust. The differences between the aforementioned variant V and variant VI are not statistically significant. Those findings are true for all tested contact times. The largest differences were measured between variants IX and X (respectively, 19%, 19% and 18% after 20, 40 and 60 s) filled with 50% small-particle fillers. The smallest differences were observed between variants V and VI (respectively, 3%, 3% and 4% after 20, 40 and 40 s), filled with 60% big-particle fillers.

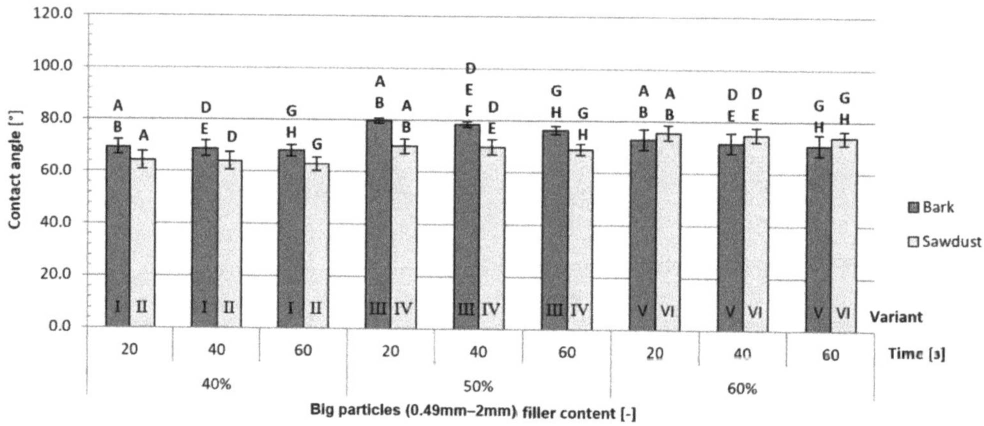

Figure 9. Average contact angles for variants filled with big-particle fillers: A, B, D, E, F, G, H—homogeneous groups, I–XII—composite variant.

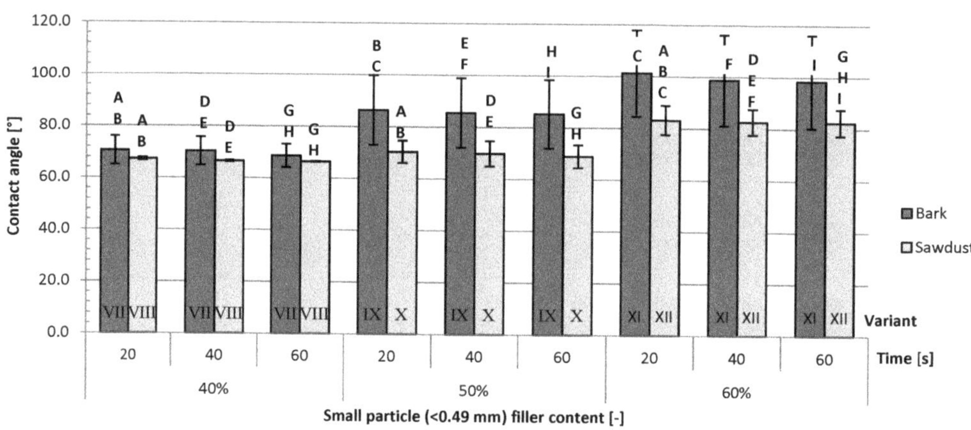

Figure 10. Average contact angles for variants filled with small-particle fillers: A, B, C, D, E, F, G, H, I—homogeneous groups, I–XII—composite variant.

The most important factor impacting the contact angles (Table 4) after either 20, 40 or 60 s is the type of filler. It contributed to 30.56%, 29.35% and 29,66% of the variance, respectively, after 20, 40 and 60 s of contact time. The second most important factor was the filler content for 20 s (12.98%) and 40 s (11.93%) of contact time and the size of the filler particles for 60 s (13.00%) of contact time.

The total surface free energy (Table 5) is higher for the variants filled with sawdust than for the variants filled with bark. The polar component calculated for the variants filled with bark is significantly lower than that of the variants filled with sawdust, with an exception for variants V and VI, filled with 60% large-particle fillers. The disperse component for variants I and VIII filled with bark is higher than that of the analogous variants filled with sawdust, and the inverse is true for all other pairs of variants. The differences in surface free energy are more significant between composites filled with small-particle fillers, which might be explained by the higher specific surface area of smaller particles.

Table 4. ANOVA for selected factors affecting contact angles after 20, 40 and 60 s.

Factor	Contact Angle after 20 s		Contact Angle after 40 s		Contact Angle after 60 s	
	p	P	*p*	P	*p*	P
a	0.002	12.98%	0.004	11.93%	0.005	10.68%
b	0.000	30.56%	0.000	29.35%	0.000	29.66%
c	0.003	11.55%	0.004	11.62%	0.002	13.00%
a × b	0.327	2.45%	0.404	2.20%	0.378	2.28%
a × c	0.086	3.36%	0.093	3.57%	0.093	3.45%
b × c	0.019	9.80%	0.028	9.76%	0.022	10.05%
a × b × c	0.159	4.16%	0.236	3.58%	0.197	3.91%
Error		25.14%		27.99%		26.96%

a—filler content, b—type of filler, c—size of particles, *p*—probability of error, P—percentage of contribution.

Table 5. Surface free energy for each tested variant.

Filler Content	Variant	Surface Free Energy [mJ × m^{-2}] γ^{tot}	γ^{D}	γ^{P}
40% L	I	51.68	45.60	6.07
	II	52.23	43.24	8.99
50% L	III	44.33	41.18	3.15
	IV	52.59	47.45	5.14
60% L	V	47.36	42.17	5.20
	VI	48.75	45.18	3.57
40% S	VII	49.40	44.31	5.10
	VIII	49.50	41.84	7.66
50% S	IX	38.29	36.53	1.77
	X	47.22	40.55	6.67
60% S	XI	37.76	37.75	0.00
	XII	50.24	49.17	1.07

L—large particles (0.49–2 mm), S—small particles (<0.49 mm), dark grey—bark, light grey—sawdust.

Both the contact angle and surface free energy tests show that bark raises the composite's hydrophobicity, which positively impacts its water resistance. This may confirm that lower the TS and WA values of the bark-filled composites are a result of the different chemical composition of bark. While the type of filler is not the most important factor in the TS and WA tests, its impact on those results is major, especially under short-term water exposure. The higher impact of the filler content is caused by low PLA WA (below 1% after 24 h of immersion) and insignificant TS [35]. While also considering the ANOVA results for the contact angle tests, the use of bark as a filler should be considered one of the most significant improvements to PLA-based WPC's water resistance properties.

These results not only show better resistance to water but might also suggest that, for composites used in wet conditions, the impact of a bark filler on the mechanical properties of a tested material might be smaller due to lesser humidity-induced degradation [36].

4. Conclusions

The conducted research allows us to conclude that bark can be an alternative to wood sawdust to use as a filler in WPCs. However, the use of bark as a WPC filler causes a decrease in the MOR and MOE of the produced composite compared to the composite with wood sawdust as a filler. The thickness swelling and water absorption of bark-filled WPCs are lower than those of sawdust-filled WPCs. These differences are more significant for composites with higher filler contents. Similarly, the wettability of bark-filled WPCs is lower than that of sawdust-filled WPCs, both in terms of contact angles and surface free energy.

The use of bark as a filler in WPCs improves the moisture resistance of a composite, especially during short-term water exposure. The lower flexural properties of bark-filled composites compared to sawdust-filled ones may be caused by a different failure mechanism under a flexural load, caused by lower adhesion between the filler and the matrix, which could possibly be mitigated by the use of a coupling agent. It should be noted that the use of bark itself impacts the MOR significantly less than the filler content, and in certain applications, the decrease in its flexural properties is not significant.

Author Contributions: Conceptualization, P.B. and R.A.; methodology, P.B., W.J. and R.A.; validation, P.B., W.J. and S.H.L.; formal analysis, P.B. and R.A.; investigation, P.B. and W.J.; writing—original draft preparation, W.J., P.B., Ł.A. and R.A.; writing—review and editing, S.H.L., P.B., Ł.A. and R.A.; visualisation, R.A.; supervision, P.B. and R.A.; project administration, P.B.; funding acquisition, P.B. All authors have read and agreed to the published version of the manuscript.

Funding: This research was funded by the National Centre for Research and Development, Poland, under the "Strategic research and development program: environment, agriculture, and forestry" (BIOSTRATEG), grant no. BIOSTRATEG3/344303/14/NCBR/2018.

Institutional Review Board Statement: Not applicable.

Data Availability Statement: The original contributions presented in the study are included in the article, further inquiries can be directed to the corresponding author.

Conflicts of Interest: Author Łukasz Adamik was employed by the company Nowy Styl sp. z o.o. The remaining authors declare that the research was conducted in the absence of any commercial or financial relationships that could be construed as a potential conflict of interest.

References

1. Jian, B.; Mohrmann, S.; Li, H.; Li, Y.; Ashraf, M.; Zhou, J.; Zheng, X. A Review on Flexural Properties of Wood-Plastic Composites. *Polymers* **2022**, *14*, 3942. [CrossRef] [PubMed]
2. Chan, C.M.; Vandi, L.J.; Pratt, S.; Halley, P.; Richardson, D.; Werker, A.; Laycock, B. Composites of Wood and Biodegradable Thermoplastics: A Review. *Polym. Rev.* **2018**, *58*, 444–494. [CrossRef]
3. El-Haggar, S.M.; Kamel, M.A. Wood Plastic Composites. *Adv. Compos. Mater. Anal. Nat. Man-Made Mater.* **2011**, 325–344. [CrossRef]
4. Elsheikh, A.H.; Panchal, H.; Shanmugan, S.; Muthuramalingam, T.; El-Kassas, A.M.; Ramesh, B. Recent progresses in wood-plastic composites: Pre-processing treatments, manufacturing techniques, recyclability and eco-friendly assessment. *Clean. Eng. Technol.* **2022**, *8*, 100450. [CrossRef]
5. Falk, R.H.; Vos, D.; Cramer, S.M. The comparative performance of woodfiber-plastic and wood-based panels. In Proceedings of the 5th International Conference on Woodfiber-Plastic Composites, Madison, WI, USA, 26–27 May 1999.
6. *EN 15534-1+A1*; Composites Made from Cellulose-Based Materials and Thermoplastics (Usually Called Wood-Polymer Composites (WPC) or Natural Fibre Composites (NFC))—Part 1: Test Methods for Characterisation of Compounds and Products. BSI: London, UK, 2017.
7. Ratanawilai, T.; Taneerat, K. Alternative polymeric matrices for wood-plastic composites: Effects on mechanical properties and resistance to natural weathering. *Constr. Build. Mater.* **2018**, *172*, 349–357. [CrossRef]
8. Partanen, A.; Carus, M. Biocomposites, find the real alternative to plastic—An examination of biocomposites in the market. *Reinf. Plast.* **2019**, *63*, 317–321. [CrossRef]
9. Quitadamo, A.; Massardier, V.; Valente, M. Eco-Friendly Approach and Potential Biodegradable Polymer Matrix for WPC Composite Materials in Outdoor Application. *Int. J. Polym. Sci.* **2019**, *2019*, 3894370. [CrossRef]
10. Balart, J.F.; García-Sanoguera, D.; Balart, R.; Boronat, T.; Sánchez-Nacher, L. Manufacturing and properties of biobased thermoplastic composites from poly(lactid acid) and hazelnut shell wastes. *Polym. Compos.* **2018**, *39*, 848–857. [CrossRef]
11. Tokiwa, Y.; Calabia, B.P. Biodegradability and biodegradation of poly(lactide). *Appl. Microbiol. Biotechnol.* **2006**, *72*, 244–251. [CrossRef]
12. Farah, S.; Anderson, D.G.; Langer, R. Physical and mechanical properties of PLA, and their functions in widespread applications—A comprehensive review. *Adv. Drug Deliv. Rev.* **2016**, *107*, 367–392. [CrossRef]
13. Auras, R.; Harte, B.; Selke, S. An Overview of Polylactides as Packaging Materials. *Macromol. Biosci.* **2004**, *4*, 835–864. [CrossRef] [PubMed]
14. Das, P.P.; Chaudhary, V. Moving towards the era of bio fibre based polymer composites. *Clean. Eng. Technol.* **2021**, *4*, 100182. [CrossRef]
15. Włodarczyk-Fligier, A.; Polok-Rubiniec, M.; Chmielnicki, B. Kompozyty polimerowe z napełniaczem naturalnym. *Przetwórstwo Tworzyw* **2018**, *24*, 50–57.
16. Mohanty, A.K.; Misra, M.; Drzal, L.T. *Natural Fibers, Biopolymers, and Biocomposites*, 1st ed.; CRC Press: Boca Raton, FL, USA, 2005. [CrossRef]

17. Kuciel, S.; Liber-Kneć, A.; Mikuła, J.S.; Kuźniar, P.; Korniejenko, K.; Łagan, S.D.; Zajchowski, S. *Kompozyty Polimerowe na Osnowie Recyklatów z Włóknami Naturalnymi: Praca Zbiorowa*; Politechnika Krakowska: Kraków, Poland, 2010.
18. Kim, J.-W.; Harper, D.P.; Taylor, A.M. Effect of wood species on the mechanical and thermal properties of wood-plastic composites. *J. Appl. Polym. Sci.* **2009**, *112*, 1378–1385. [CrossRef]
19. Gozdecki, C.; Kociszewski, M.; Mirowski, J.; Wilczynski, A.; Zajchowski, S. Effect of wood bark on wood-plastic composite properties. *Ann. Wars. Univ. Life Sci. SGGW For. Wood Technol.* **2009**, *68*, 278–282.
20. Avci, E.; Acar, M.; Gonultas, O.; Candan, Z. Manufacturing biocomposites using black pine bark and oak bark. *BioResources* **2018**, *13*, 15–26. [CrossRef]
21. Çetin, N.S.; Özmen, N.; Narlıoğlu, N.; Çavuş, V. Effect of bark flour on the mechanical properties of HDPE composites. *Usak Univ. J. Mater. Sci.* **2014**, *3*, 23. [CrossRef]
22. Kazemi Najafi, S.; Kiaefar, A.; Tajvidi, M. Effect of bark flour content on the hygroscopic characteristics of wood-polypropylene composites. *J. Appl. Polym. Sci.* **2008**, *110*, 3116–3120. [CrossRef]
23. Farsi, M. Effect of Nano-SiO_2 and Bark Flour Content on the Physical and Mechanical Properties of Wood–Plastic Composites. *J. Polym. Environ.* **2017**, *25*, 308–314. [CrossRef]
24. Obuch, J.; Hecl, D. *Using SRC Poplar to Produce Bark-Enriched Wood Plastic Composite Profiles*; Dendromass4Europe.eu: Tharandt, Germany, 2022.
25. *EN 310*; Wood-Based Panels—Determination of Modulus of Elasticity in Bending and of Bending Strength. CEN: Brussele, Belgium, 1994.
26. *EN 323*; Wood-Based Panels—Determination of Density. CEN: Brussele, Belgium, 1999.
27. *EN 317*; Particleboards and Fibreboards—Determination of Swelling in Thickness after Immersion in Water. CEN: Brussele, Belgium, 1999.
28. Owens, D.K.; Wendt, R.C. Estimation of the surface free energy of polymers. *J. Appl. Polym. Sci.* **1969**, *13*, 1741–1747. [CrossRef]
29. Shah, B.L.; Selke, S.E.; Walters, M.B.; Heiden, P.A. Effects of wood flour and chitosan on mechanical, chemical, and thermal properties of polylactide. *Polym. Compos.* **2008**, *29*, 655–663. [CrossRef]
30. Gozdecki, C.; Kociszewski, M.; Wilczyński, A.; Mirowski, J. Effect of wood bark content on mechanical properties of wood-polyethylene composite. *Ann. Warsaw Univ. Life Sci. For. Wood Technol.* **2010**, *71*, 203–206.
31. Safdari, V.; Khodadadi, H.; Hosseinihashemi, S.K.; Ganjian, E. The Effects of Poplar Bark and Wood Content on the Mechanical Properties of Wood-Polypropylene Composites. *BioResources* **2011**, *6*, 5180–5192. [CrossRef]
32. Tang, Q.; Fang, L.; Guo, W. UHD fiberboard reinforcement. *BioResources* **2017**, *12*, 6749–6762.
33. Radoor, S.; Karayil, J.; Shivanna, J.M.; Siengchin, S. Water Absorption and Swelling Behaviour of Wood Plastic Composites. *Wood Polym. Compos. Recent Adv. Appl.* **2021**, 195–212. [CrossRef]
34. Bouafif, H.; Koubaa, A.; Perré, P.; Cloutier, A. Effects of fiber characteristics on the physical and mechanical properties of wood plastic composites. *Compos. Part. A Appl. Sci. Manuf.* **2009**, *40*, 1975–1981. [CrossRef]
35. Lay, M.; Thajudin, N.L.N.; Hamid, Z.A.A.; Rusli, A.; Abdullah, M.K.; Shuib, R.K. Comparison of physical and mechanical properties of PLA, ABS and nylon 6 fabricated using fused deposition modeling and injection molding. *Compos. Part B Eng.* **2019**, *176*, 10734. [CrossRef]
36. Tamrakar, S.; Lopez-Anido, R.A. Water absorption of wood polypropylene composite sheet piles and its influence on mechanical properties. *Constr. Build. Mater.* **2011**, *25*, 3977–3988. [CrossRef]

polymers

MDPI

Article

Advanced Electrospun Composites Based on Polycaprolactone Fibers Loaded with Micronized Tungsten Powders for Radiation Shielding

Chiara Giuliani [1,*], Ilaria De Stefano [2], Mariateresa Mancuso [2], Noemi Fiaschini [3], Luis Alexander Hein [3], Daniele Mirabile Gattia [4], Elisa Scatena [5], Eleonora Zenobi [5], Costantino Del Gaudio [6], Federica Galante [7], Giuseppe Felici [7] and Antonio Rinaldi [1,*]

[1] TERIN-DEC-ACEL Laboratory, ENEA—Italian National Agency for New Technologies, Energy and Sustainable Economic Development, Via Anguillarese 301, 00123 Rome, Italy
[2] Division of Biotechnologies (SSPT-BIOTEC), ENEA—Italian National Agency for New Technologies, Energy and Sustainable Economic Development, Via Anguillarese 301, 00123 Rome, Italy; ilaria.destefano@enea.it (I.D.S.); mariateresa.mancuso@enea.it (M.M.)
[3] NANOFABER S.r.l., 00123 Rome, Italy; noemi.fiaschini@nanofaber.com (N.F.)
[4] SSPT-TIMAF-MADD Laboratory, ENEA—Italian National Agency for New Technologies, Energy and Sustainable Economic Development, Via Anguillarese 301, 00123 Rome, Italy; daniele.mirabile@enea.it
[5] E. Amaldi Foundation, Via del Politecnico snc, 00133 Rome, Italy; elisa.scatena@fondazioneamaldi.it (E.S.); eleonora.zenobi@fondazioneamaldi.it (E.Z.)
[6] Italian Space Agency, Via del Politecnico snc, 00133 Rome, Italy; costantino.delgaudio@asi.it
[7] S.I.T. Sordina IORT Technology S.p.A., 04011 Aprilia, Italy; federicagalante95@gmail.com (F.G.); giuseppe.felici@sordina.com (G.F.)
* Correspondence: chiara.giuliani@enea.it (C.G.); antonio.rinaldi@enea.it (A.R.)

Abstract: Exposure to high levels of radiation can cause acute, long-term health effects, such as acute radiation syndrome, cancer, and cardiovascular disease. This is an important occupational hazard in different fields, such as the aerospace and healthcare industry, as well as a crucial burden to overcome to boost space applications and exploration. Protective bulky equipment made of heavy metals is not suitable for many advanced purposes, such as mobile devices, wearable shields, and manned spacecrafts. In the latter case, the in-space manufacturing of protective shields is highly desirable and remains an unmet need. Composites made of polymers and high atomic number fillers are potential means for radiation protection due to their low weight, good flexibility, and good processability. In the present work, we developed electrospun composites based on polycaprolactone (polymer matrix) and tungsten powder for application as shielding materials. Electrospinning is a versatile technology that is easily scalable at an industrial level and allows obtaining very lightweight, flexible sheet materials for wearables. By controlling tungsten powder size, we engineered homogeneous, stable and processable suspensions to fabricate radiation composite shielding sheets. The shielding capability was assessed by an in vivo model on prototype composite sheets containing 80 w% of W filler in a polycaprolactone (PCL) fibrous matrix by means of irradiation tests (X-rays) on mice. The obtained results are promising; as expected, the shielding effectivity of the developed composite material increases with the thickness/number of stacked layers. It is worth noting that a thin barrier consisting of 24 layers of the innovative shielding material reduces the extent of apoptosis by 1.5 times compared to the non-shielded mice.

Keywords: shielding; polycaprolactone (PCL); tungsten powder; polymer composites; electrospinning; radiation protection; aerospace; healthcare industry

1. Introduction

High-energy radiation such as X-ray or gamma ray electromagnetic radiation are often employed or encountered in a wide range of fields, including aerospace and healthcare. Unwanted and/or prolonged exposure to these radiations may be hazardous to health [1,2].

Cosmic radiation, for example, imposes important safety concerns for space exploration missions. Numerous studies have examined the risks associated with exposure to both galactic cosmic rays (GCR) and solar particle events (SPE) [3]. Extended or high-dose radiation exposure can lead to carcinogenesis, cellular mutations, cardiovascular issues, cataracts, and other acute radiation syndromes [1]. Conversely, radiotherapy is frequently employed in medical applications to treat cancer patients. High-energy ionizing radiations (up to tens of MeV) are routinely used to manage tumor growth as part of targeted cancer therapies. The tissues near the treated region are often exposed to penetrative X-rays, which leads to harmful side effects. In addition, X-rays with energies in the keV range are often employed in interventional procedures and in diagnostic radiology such as computed tomography (CT) examination [1,4,5].

Therefore, with the aim to prevent occupational hazards from these kinds of exposures, in the last decade, there has been an increasing demand and research interest in the development of efficient, lightweight, low cost and flexible shielding materials for protection against radiation in the aerospace and medical field [6]. The shielding effectiveness of a given material largely depends on the type of radiation and the range of energies associated with it. Lead (Z = 82) and other high Z materials are known to effectively absorb high-energy radiation and have been commonly used as shielding materials for radiation such as X-rays. [2,4,7,8] Unfortunately, protective equipment made of metals is heavy and bulky, which are often unwanted features for many applications, such as for example for mobile devices, wearable shields, and manned spacecrafts, where there is an unmet need for radioprotective materials that are lightweight and flexible [1].

In this context, polymer composites have become attractive candidates for developing shielding materials for different types of radiation, including high-energy radiation. In particular, polymer composites, prepared by dispersion of metallic or ceramic micro- or nanoparticles (fillers) in the base polymeric matrix, have been found potentially interesting as protective shields [1,2,9–11]. The given combination of the filler-matrix pair determines the shielding effect. The role of the filler(s) is to attenuate effectively the radiation, while the polymer offers the mechanical backbone, ensuring good geometric conformability. Polymer is generally less dense than its metal or ceramic counterpart and can be easily processed. Because of the large surface-area-to-volume ratio, micro and nanoparticles have been reported to show enhanced ability to absorb photons [8]. Lead and its compounds have traditionally been used as attenuation fillers due to their low cost, high density, and effectiveness in reducing ionizing radiation [2]. However, lead is highly toxic, and prolonged exposure can cause serious health problems such as neurological disorders, kidney failure, and a decrease in hemoglobin and red blood cells. As a result, it is crucial to limit the use of lead and explore alternative metals and synthetic materials that are less toxic [12,13]. Eco-friendly shielding materials that can replace lead include tungsten (W), bismuth oxide (Bi_2O_3), barium sulfate ($BaSO_4$), and boron [14–16]. Among these materials, W has an atomic number of 74, which is similar to that of lead (82), offers high density (19.25 g/cm^3), is characterized by low toxicity among the high-Z elements [17], exhibiting a comparable shielding effect [2,18]. For instance, Selyutin et al. studied ultra-high molecular weight polyethylene (UHMWPE) doped with tungsten powder, demonstrating that UHMWPE is resistant to ionizing radiation [19]. The radiation shielding properties of tungsten and epoxy composite were also investigated by Le Chang et al. [20]. Moreover, Soylu et al.'s research showed that the composites' radiation shielding efficiency based on tungsten carbide and ethylene vinyl acetate polymer is similar to lead [21].

In this work, innovative, flexible hybrid composite sheets made of a polycaprolactone (PCL) fibrous matrix containing W powder were prepared via electrospinning. PCL is a biodegradable polymer [22] commonly used for various applications due to its ease of processing [23]. To the best of our knowledge, PCL-based shielding composites are reported in literature only for EMI [24,25] and microwave shielding [26,27]. Electrospinning is a relatively simple, low-cost, and versatile technology that allows obtaining innovative micro- and nano-structured non-woven materials for application in several fields [28]. The developed electrospun composites were structurally and physically characterized, and their ability to shield high-energy radiation (X-rays) was investigated through in vivo tests.

2. Materials and Methods

Commercially available tungsten, W, powder 99+ (fine powder < 20 μm, purity 99%, Sigma Aldrich, St. Louis, MI, USA), polycaprolactone (PCL) polymer (6800D, 80 kDa, Perstorp, Warrington, UK), chloroform (VWR) and dimethylformamide (DMF), (VWR) were used for the preparation of PCL/W based suspensions. Ethanol (Carlo Erba) was used as a penetrating liquid to determine the porosity of the prepared electrospun materials. A commercial polylactid acid (PLA) filament (FILOALFA, Turin, Italy; 1.75 mm diameter) was used for the 3D printing process of the cylindrical samples.

2.1. Optimization of W Powder

W powder was selected as a filler for the preparation of a PCL-based shielding sheet. Commercial W powder was used as the starting material for the filler. Ball milling refined the size of commercial W particles to obtain a stable suspension processable via an electro spinning technique. The ball milling process was carried out in a SPEX 8000 mixer/mill. The W powder was ball-milled in stainless steel vials. The upper cup was allowed to seal the vial by the mean of a viton o-ring. To avoid contamination, the jars were loaded with powder and spheres in a glove box. The powders were ball milled in a slight overpressure of gaseous Argon. After different tests, an optimized powder, in terms of morphological and microstructural properties, was obtained after 26 h of ball milling.

2.2. Electrospinning of PCL/W-Based Shielding Materials

PCL/W sheets were produced via electrospinning using needle-technology electrospinning equipment (Fluidnatek LE100, Bioinicia, Paterna, Spain) outfitted with a flat collector and two-axis emitter motion. The polymer solution was prepared by dissolving the 12% *w*/*w* of PCL in the chloroform/dimethylformamide 65/35 solvent mixture at room temperature. Different amounts of the optimized W powder (10–80 w% with respect to the PCL polymer) were added to the PCL solution to optimize the properties of the resulting microfibrous sheets.

The electrospinning process and the properties of the resulting electrospun materials are highly dependent on process parameters and ambient conditions [29]. The process parameters optimized for the preparation of the PCL-based sheets are summarized in the following table (Table 1).

Table 1. Processing parameters for the electrospinning of PCL PSU/W solutions.

Parameter	Label	Unit	Value
Flow rate	FR	mL/h	6
Voltage at injector	Vi	kV	5
Voltage at collector	Vc	kV	−5
Working distance	d	cm	16
Deposition time	t	min	43
Deposition area	A	cm^2	144

The relative humidity and the temperature in the process chamber were kept constant at average values of 47% (with a mean square error (MSE) of ±2%) and 25 °C (with an MSE of ±0.2 °C), respectively, throughout the entire process.

2.3. Scanning Electron Microscopy (SEM) and Energy Dispersive X-ray Spectroscopy (EDS)

The morphological properties of the PCL/W electrospun material were analyzed using a field emission gun scanning electron microscope (FEG-SEM) model Leo 1530 (ZEISS, Oberkochen, Germany), operating at a low voltage (2 kV) to prevent charging effects and overheating damage to the dielectric polymer. The membrane thickness was assessed by examining the sample's cross-section under the SEM. Additionally, energy-dispersive X-ray spectroscopy (EDS) measurements were performed with an X-MAX detector (AZ-TEC, Oxford, UK) to verify the presence of W in the PCL samples.

2.4. Porosity and Permeability Determination

The porosity of the membranes was assessed using the liquid displacement method [30–32]. Ethanol was chosen as the penetrating liquid due to its ability to permeate the porous membrane and its low density (≈0.790 g/mL). Percent porosity (ε%) was calculated using Equation (1):

$$\varepsilon_{\%} = \frac{(m_3 - m_4 - m_1)}{m_2 - m_4} \tag{1}$$

Weights m_1, m_2, m_3, and m_4 were measured on an ORMA scale (BCA120, Milan, Italy) at 20 °C. For each specimen, m_1 represents the weight of the "dry" sample; m_2 is the weight of a "control" graduated bottle filled with ethanol up to a specified volume; m_3 is the weight of the same bottle after the specimen has been added and immersed in ethanol, with excess ethanol carefully removed to restore the control volume and m_4 is the weight of the bottle after removing the wet sample and leaving residual ethanol.

Air permeability of the electrospun membranes was evaluated using a Gurley densometer (model 4320, Troy, NY, USA). During the test, the time (t) in seconds for a volume (V) of 100 cm^3 of air to pass through a surface area (A) of 1.61 cm^2 was recorded. The percent permeability value (P%) in "cm/s" for each sample was determined using the following Equation (2) [33].

$$P\% = \frac{V}{A \cdot t} \cdot 100 \tag{2}$$

2.5. Mechanical Tests

The tensile properties of the PCL/W-based shielding materials were evaluated by a tensile loading frame (Dynamic Mechanical Analyzer 850, TA Instruments, New Castle, DE, USA) equipped with a 20 N load cell. Rectangular strips were cut and mounted in the loading frame, with a free gauge length of 10–15 mm and a uniform width (w) of 4 mm. The sample thickness (s) was measured by SEM imaging of the cross-section. Load vs. cross-head displacement data was recorded during tensile tests (movable cross-head moving at 0.2 mm/min) at room temperature. The engineering stress (σ, MPa) vs. strain (ε, %) curves were calculated by dividing, respectively, the applied load by the apparent cross-sectional area (A_0 = w × s) and the cross-head displacement by 10, corrected for elongation amount, needed to fully stretch the strip. The Young modulus (E) was estimated from the slope of the linear fit in the elastic region for each stress–strain curve.

2.6. PLA Cylinder Preparation

For the in vivo experimental tests, cylindrical PLA supports were 3D printed by means of the fused deposition modeling (FDM) technique (Raise 3D Inc., Irvine, CA, USA). These samples were designed according to the following geometrical specifications: a cylindrical structure of 18 mm outer diameter and 20 mm height, with an inner cavity of 10 mm diameter and 16 mm height for housing the mice head (Figure 1a,b). The PLA filament was extruded through a 0.4 mm diameter nozzle at 205 °C; the bed temperature was set at 60 °C. All samples were fabricated with the same printing parameters.

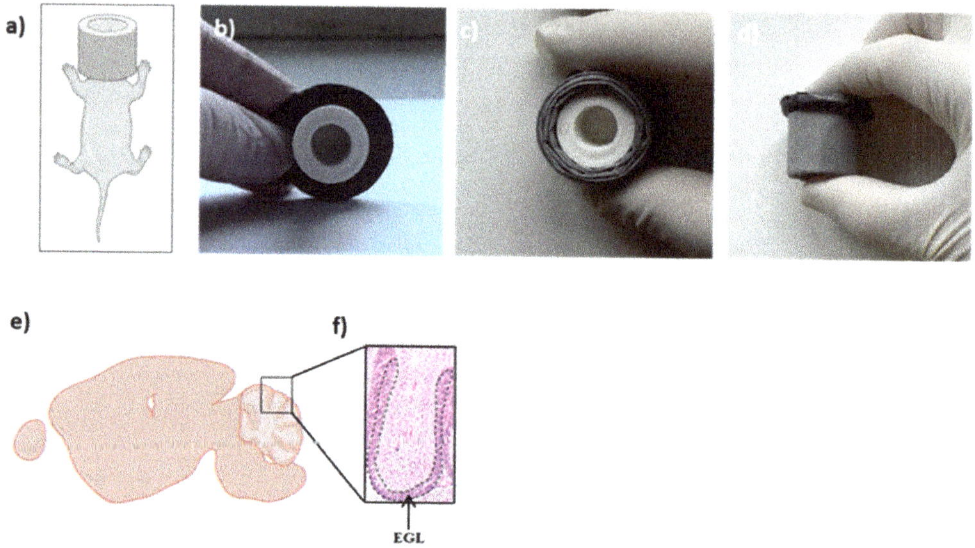

Figure 1. Schematic representation of the experimental setup (**a**), custom-built 3D printed PLA protective device for in vivo test mounted on a 3 mm thick solid lead sheet (**b**); PLA device wrapped up in 12 shielding sheets of PCL/W (**c**); PLA device wrapped up in 24 shielding sheets of PCL/W (**d**); schematic representation of a section of a cerebellum at 2 days of age (**e**); antero-dorsal cardinal lobe of the cerebellum (**f**). The dashed black line externally outlines the EGL. Figure in (**a**,**e**) was obtained by Biorender.

2.7. *Irradiation Shielding Tests*

2.7.1. Mice

Mice lacking one *Ptch1* allele were bred on CD1 background (*Ptch1*$^{+/-}$) and genotyped according to our previous studies [34]. This animal study was performed according to Directive 2010/63/EU of the European Parliament, approved by the local Ethical Committee for Animal Experiments of the ENEA, and authorized by the Italian Ministry of Health (n°740/2021-PR).

2.7.2. Irradiation Shielding Test

Ptch1$^{+/-}$ mice at postnatal day 2 (P2) were shielded and irradiated with 1 Gy of X-ray protecting heads with different kinds of shields (Figure 1a). Specifically, one set of mice was irradiated by shielding the head with PLA cylinder (PLA, n = 6; Figure 1b), another set with PLA cylinder wrapped up in 12 sheets of electrospinning PCL/W (N12, n = 6; Figure 1c), and another set with PLA cylinder wrapped up in 24 sheets of electrospinning PCL/W (N24, n = 6, Figure 1d). Another group of animals was whole-body irradiated (WB, n = 6). An additional group of mice was left unirradiated as a control (CN, n = 6). Irradiation was performed using a Gilardoni CHF 320 G X-ray generator (Gilardoni, Mandello del Lario, Italy), operated at 250 kVp, 15 mA, with filters of 2.0 mm of Al and 0.5 mm of Cu. Protection of shielded parts was verified by dosimetry and Monte Carlo simulation.

2.7.3. Apoptosis Quantification in the Cerebellum

To study the short-term cellular responses to radiation exposure, both control and irradiated mice were euthanized by CO_2 asphyxiation at 4 h post-irradiation. Brains were collected and processed for histological analysis according to standard protocols; 4 µm-thick sections, after Hematoxylin and Eosin staining, were thus observed using light microscopy to quantify the number of apoptotic cells in the cerebellum external granule layer (EGL) (Figure 1e,f) as previously described [35]. The quantitative analysis was performed along the entire antero-dorsal cardinal lobe of the cerebellum and was carried out blindly by two different operators. Samples in which the area of interest could not be analyzed due to poor orientation were excluded from the analysis. The apoptotic rate was calculated as the percentage of pyknotic nuclei relative to the total cell number (Figure 1a–d).

Paraffin brain sections were also processed for the immunohistochemical analysis using an antibody direct against cleaved caspase-3 (Cell Signaling Technology, Danvers, MA, USA; 1:100), an elective marker of apoptosis [36]. Images for quantification were taken using the imaging software NIS-Elements BR 4.00.05 (Nikon Instruments Europe B.V., Amsterdam, The Netherlands).

The colorimetric signal of cleaved caspase 3 was quantified using the HistoQuest software (TissueGnostics, Vienna, VA, Austria) and expressed as the labeled area/total area examined.

2.7.4. Statistics

Analyses were performed using GraphPad Prism 5.0 (GraphPad Software, San Diego, CA, USA). Statistical significance was determined using a two-tailed Student *t*-test. p-values < 0.005 were considered statistically significant.

3. Results and Discussion

3.1. Powder Optimization

In our study, PCL was selected as the polymer matrix for the shielding material due to its good processability, biocompatibility, and thermal stability. The addition of W powder as a filler for the polymer matrix allows us to tune the shielding properties of the material and reduce its weight.

The electrospinning technique offers considerable advantages, such as to easily tune the morphology and the characteristics of the obtained materials and/or to select in wide intervals the polymer matrix, the type and the concentration of the filler. As an electrohydrodynamic process, this technology allows to collect the fibers produced on a counter-electrode without any dispersion of material (unlike wire or powder bed 3D printing technologies, for example). In addition, electrospinning is relatively simple, flexible, and easily scalable at an industrial level and provides very lightweight, flexible and, therefore, wearable TNT materials.

It is worth noting that the particle size of the powder plays a key role in the physical stability of the polymer/W suspension. Therefore, the control of W particle size is essential in achieving a homogeneous, stable, and processable suspension. The selected commercial W powder, with a maximum particle size of 10 µm (Figure 2a,b), was unsuitable for preparing stable suspensions. In fact, W particles remained in suspension with the polymer only with the aid of magnetic stirring, while they settled in a few minutes after stopping the external agitation (Figure 3). In this case, the suspension is not suitable to be processed by electrospinning, requiring continuous stirring in situ to avoid powder settling.

Ball milling is a powerful method to refine powder particle dimensions, particularly in refractory metals like W and oxides. Our goal was thus to optimize the W particle size by ball milling to obtain a stable PCL/W suspension suitable for deposition by electrospinning.

After 26 h of milling, the SEM observation revealed that the diameter of the final particles was more homogeneous and significantly reduced to values equal or less than 1 µm (Figure 2e,f). The obtained powder was mixed with PCL to prepare a stable PCL/W suspension. As shown in Figure 3, the prepared suspension was homogenous and stable for several hours, even without stirring; thus, it was suitable to be processed by the electrospinning technique.

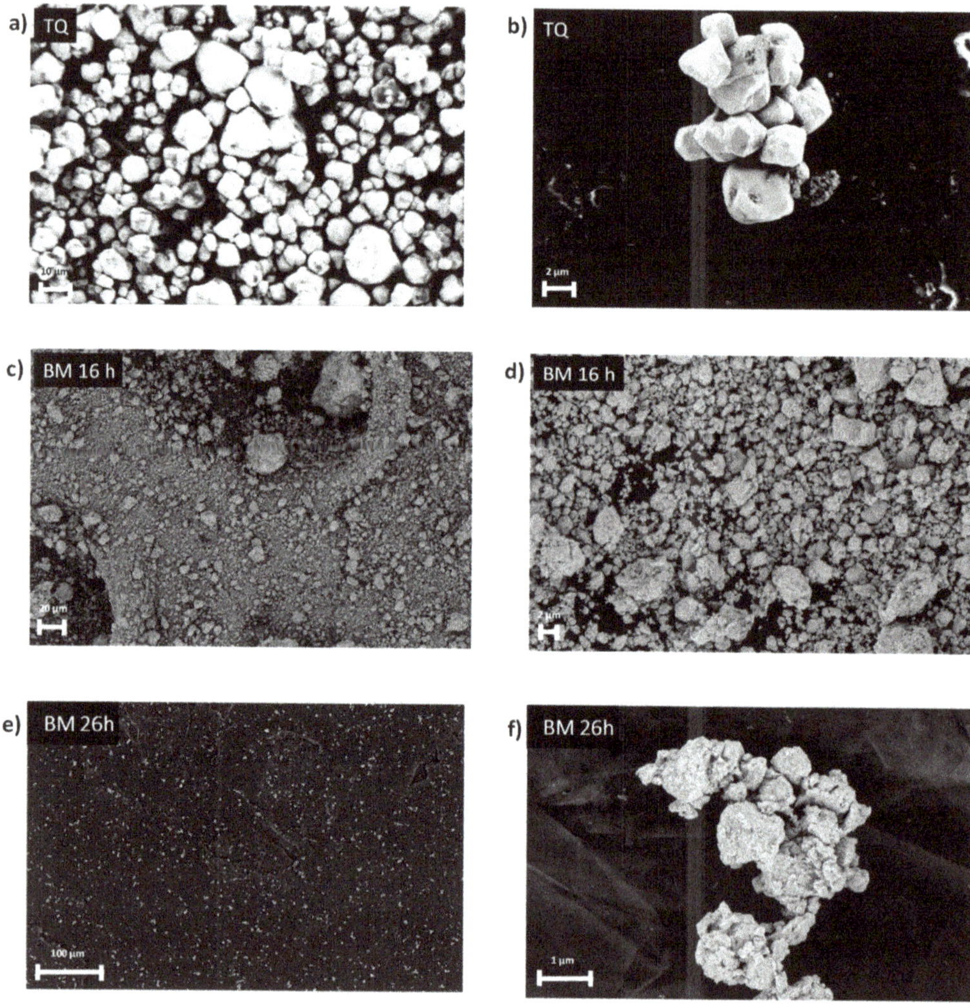

Figure 2. The SEM image of the commercial W powder is as follows: before (**a**,**b**), after 16 h (**c**,**d**) and 26 h (**e**,**f**) of ball milling.

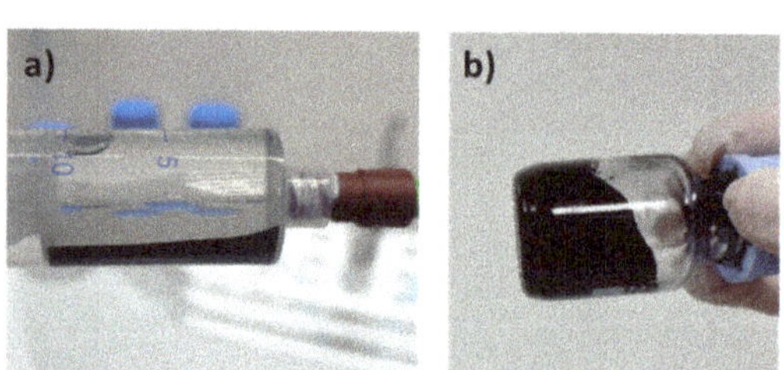

Figure 3. Unstable suspension prepared using PCL and the commercial W powder (**a**), and stable suspension based on PCL and the optimized W powder (**b**).

3.2. Preparation and Microstructural Characterization of PCL/W-Based Shielding Materials

Five batches of PCL/W sheets were prepared via electrospinning and characterized. To tune the electrospun material's final properties, different amounts of optimized W powder (in a 10–80 w% range with respect to the PCL polymer) were successfully added to the PCL solution. For the preparation of the PCL/W sheet here reported, the optimized W powder was ball-milled for 26 h. The prepared suspensions were processed via electrospinning to obtain the micro-fibrous sheets shown in Figure 4.

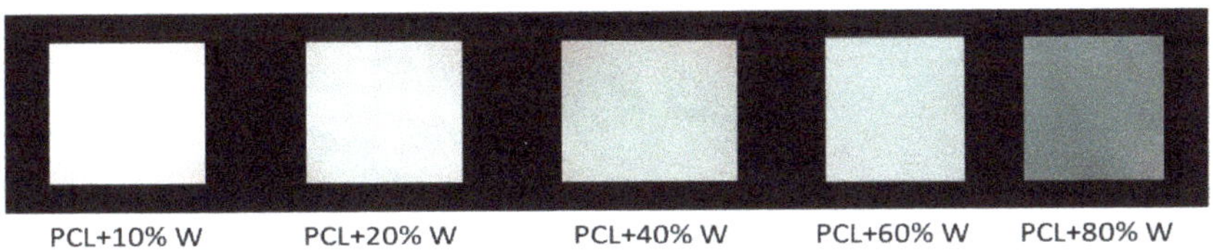

Figure 4. Micro-fibrous PCL sheets characterized by different W content (w% with respect to the PCL polymer).

The increase of the W content is apparent at a glance by visual inspection; the darker the samples, the higher the percentage of the metallic filler.

The morphology of the PCL/W sheets was investigated using SEM. Some representative micrographs of the obtained sheets are reported as an example in Figure 5. SEM images show that the electrospun samples are basically composed of a dense net of randomly oriented fibers. In addition, SEM analysis revealed the presence of the W particles in the PCL micro-fibrous network. It is evident from the images in Figure 5 how the amount of W particles/aggregates increases with increasing the filler percentage, thus confirming the visual appearance of the samples. EDS analysis at high resolution also confirmed the presence of W in the sheets (Figure 6).

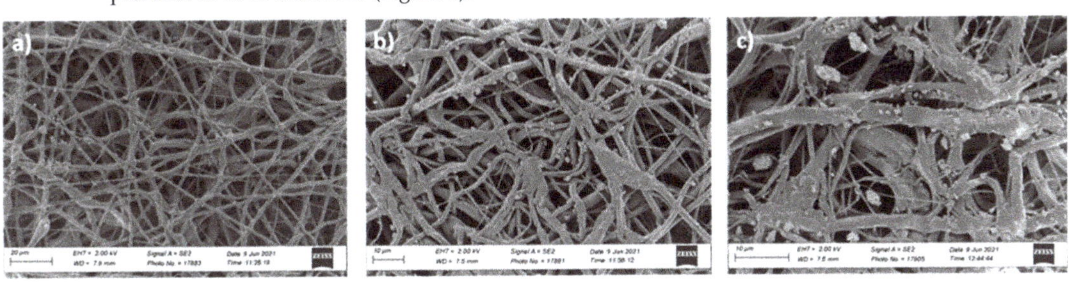

Figure 5. Microfibrous PCL/W sheets characterized as 10 w% (**a**), 40 w% (**b**), and 60 w% (**c**) of W with respect to the PCL polymer.

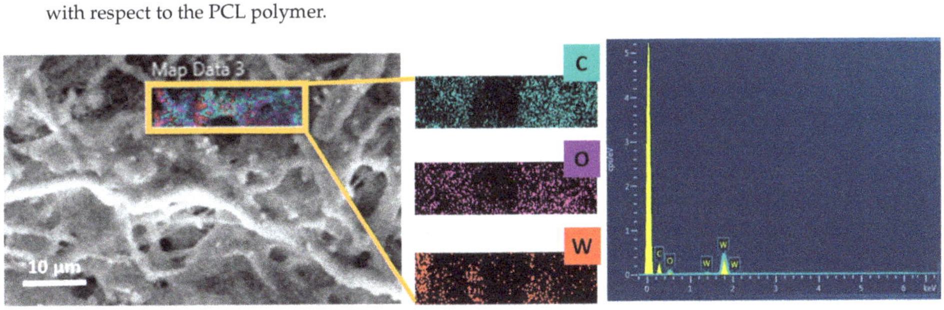

Figure 6. EDS analysis of the PCL/W electrospun sheets.

In addition, the presence and the amount of W filler affect the mean fiber diameter (Table 2). The addition of W powder to PCL increases the diameter of fibers with respect to those of pure polymer. Increasing the filler content by up to 80% means that the mean fiber diameter and standard deviation grow significantly. This leads to an overall structure that is less homogeneous and dense in fibers.

Table 2. Results of the structural characterization of the PCL/W samples containing different amounts of the W filler: the mean fiber diameter (μ), with standard deviation (σ).

Sample	μ (μm)	σ (μm)
PCL	0.60	0.43
PCL/W 10%	0.97	0.37
PCL/W 20%	1.18	0.88
PCL/W 40%	1.17	0.67
PCL/W 60%	1.15	0.89
PCL/W 80%	2.32	1.88

3.3. Porosity and Air Permeability Analysis

A more detailed characterization was carried out on the most promising sample, the one containing the higher percentage of the W filler (PCL/W80); the neat PCL mat was used as a reference. In particular, the porosity and the air permeability of the electrospun PCL/W80 were evaluated, and the average values obtained from three measurements of each membrane are reported in Table 2.

Regarding the analysis of the overall porosity, it was carried out using the volume displacement method [31,32]. The electrospun PCL's porosity was 61%, a value consistent with the values reported in the literature (typically between 60% and 90%) [37]. The porosity describes the volume of voids inside a given volume of a fiber mat. The total volume of voids depends both on the volume and on the number of pores. The porosity of the pure PCL/W80 electrospun sample was found to be lower than that of the neat PCL. This is consistent with the results of the mean fiber diameter analysis. The fiber size is the key parameter that determines the pore size of an electrospun fiber web. An increase in the nanofiber diameter leads to a lower number of pores with larger dimensions [38]. In the specific case of the sole PCL sample, a dense net of thin fibers results in a dense structure of small pores. On the contrary, the PCL/W80 sample is thinner than the PCL and is formed by a less dense net of coarse fibers than the PCL sheets. The result is an overall structure with larger pores than PCL and fewer pores.

Regarding air permeability, the web with low porosity is expected to exhibit poorer transport properties. It must be mentioned that the net morphology significantly influences the permeability of air through it. Pore characteristics, such as dimension, directly influence the air permeability of electrospun webs. The results show that the PCL/W80 sample exhibits higher air permeability than PCL. This can be interpreted as the result of the fact that more resistance in airflow through the smaller pores has occurred.

3.4. Mechanical Characterization

To evaluate the influence of the incorporation of the W filler on mechanical performance of the PCL-based electrospun sheets, tensile tests were performed. Thickness analysis (by SEM cross-sectional analysis) allowed also to evaluate and compare the mechanical properties of the different samples. The obtained results are shown in Table 3.

Young's modulus was lower in the composite sheet than in the pure one, indicating a stiffness decrease due to fillers. Ramier et al. have noticed that hydroxyapatite nanoparticle incorporation within the electrospun poly(3-hydroxybutyrate) fibers (PHB/nHA (blend)) significantly improved the mechanical properties of the fiber mat [38]. Similarly, numerous research articles report an enhancement of mechanical properties, including elastic modulus and tensile strength, after the addition of various carbon fillers [39,40]. On the other hand, in the presence of a high filler content, a reduction in crystallinity may occur. In fact, the formation of filler aggregates would restrain polymer chains' mobility and hinder their ordered arrangement, adversely affecting the crystallization process and thus hindering crystallization [41,42]. Accordingly, we can assume that the lower Young's modulus of the PCL/W80 composite correlates with a reduction in crystallinity due to filler aggregates, as observed in the SEM images (Figure 6).

Table 3. Results of the characterization of the PCL/W80 and pure PCL samples: the average thickness, the Young's modulus (E), porosity, ε, and air permeability.

Sample	Average Thickness (μm) *	E (MPa)	ε (%)	Air Permeability (cm/s)
PCL	156	64 ± 2	61	3.99
PCL/W 80%	120	52 ± 2	51	5.67

* ±2%.

However, it is worth noting that, in the presence of a W content up to 80 w% with respect to the PCL, the electrospun sheet maintains the pure polymer's ductility and plastic behavior.

3.5. Assessment of the Shielding Effect of the Optimized Formulation (PCL/W80): In Vivo Test

The EGL is a region of the cerebellum that undergoes active proliferation and differentiation during early postnatal development. Because of its high proliferative activity, it is used as a model system to study the effects of radiation on developing neural tissue [36,43]. In fact, by assessing the extent of radiation-induced damage, including apoptosis, in the EGL, it is possible to establish a dose-effect relationship, which allows for determining how the severity of damage changes with increasing doses of radiation [35]. Studying the dose-effect relationship in the EGL can thus provide insights into the mechanisms underlying radiation-induced damage to developing neural tissue and potential strategies for mitigating or preventing such damage.

In our experimental setup, we evaluated the apoptotic rate induced by a single dose of 1 Gy of X-rays to determine the effectiveness of PCL/W80 sheets in attenuating DNA damage. As expected, rare apoptotic cells were detected in the cerebella of non-irradiated animals (0.4302% ± 0.17). On the contrary, in WB-irradiated mice, the average percentage of apoptosis was 14.5% ± 0.76, a highly significant value compared to the unirradiated group ($p < 0.0001$; Figure 7a–e). In the EGL of PLA-shielded mice, the percentage of apoptosis did not differ significantly compared to the WB group, suggesting that PLA does not possess shielding capacity (12.70 ± 0.98%, $p = 0.1941$; Figure 7b–e). In both the N12 and N24 configurations, a significant reduction in cell damage is observed compared to the WB group. Notably, the N24 setup reduces the extent of apoptosis to 9.94% ± 1.11 by 1.5 times compared to the irradiated group and by 1.28 times compared to the PLA group. Although this latter value does not reach statistical significance, it suggests that the dose reduction provided by the PCL/W80 shielding is proportional to the thickness of the layer obtained from the sheets.

To validate the data obtained based on morphological criteria, we evaluated apoptosis by adopting an immunohistochemical approach using the cleaved caspase 3 antibody as an apoptotic marker. As shown in Figure 8a,b, apoptosis detected by the antibody is exclusively present in the EGL of the mouse cerebellum. Quantifying the ratio of labeled EGL to the total examined area for each experimental group, results confirmed the effectiveness of PCL/W 80% sheets in attenuating the apoptosis rate compared to that induced in the WB group ($p < 0.0439$ for N12; $p < 0.0145$ for N24) (Figure 8c). Concordantly, the region of interest in the control group showed negative results for caspase 3.

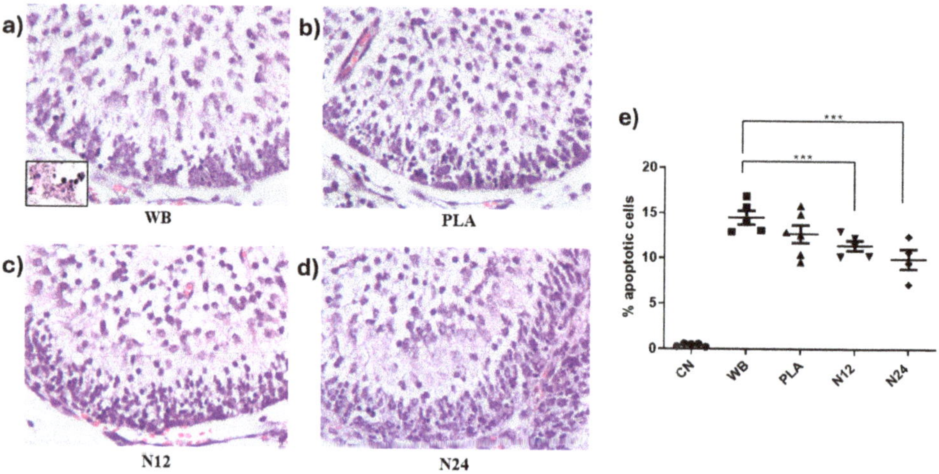

Figure 7. The apoptotic rate in EGL of irradiated mice under different experimental conditions. (**a–d**) Representative images of the EGL in the antero-dorsal cardinal lobe region of the cerebellum at postnatal day 2 (P2); Hematoxylin & Eosin staining; 40× magnification. The inset in (**a**) shows pyknotic nuclei at higher magnification (100×) indicative of apoptosis in WB irradiated mice; (**e**) a graphical representation of the percentage of apoptotic cells in the different experimental groups. *** $p < 0.0001$.

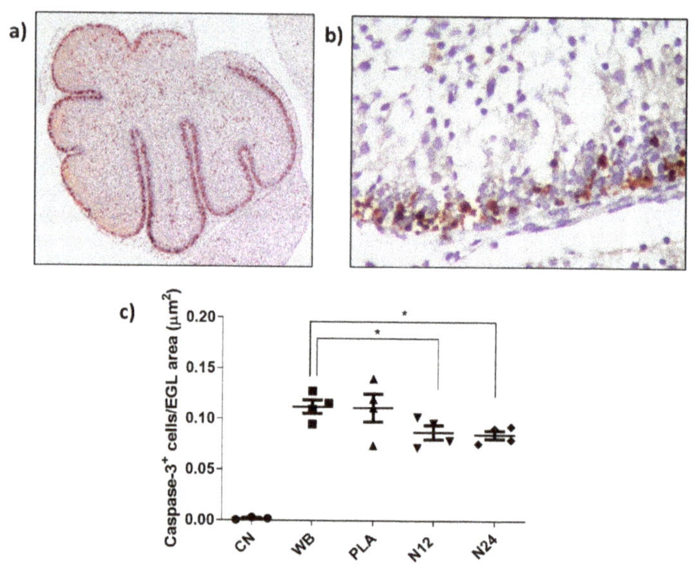

Figure 8. (**a**) Cerebellum at P2 labeled with activated caspase 3; the EGL, where the cells undergoing apoptosis reside, is colored red (2× magnification). (**b**) Detail of the EGL at higher magnification (40×). (**c**) Graphical representation of the quantification of the signal related to the antero-dorsal cardinal lobe region. * $p < 0.05$.

4. Conclusions

In the present study, we proposed the development of shielding materials consisting of an electrospun PCL matrix functionalized with W powder (filler). Particle size plays a key role in the obtainment of homogeneous and stable suspensions processable via electrospinning; the particle dimension of a commercial W powder was refined up to 1 μm by 26 h of ball milling. The obtained powder led to homogenous PCL/W suspensions that were stable for several hours and suitable to be processed by electrospinning technique. Different amounts of the optimized filler were added to the polymer suspension, and the resulting batches were electrospun and characterized. The material containing the higher percentage of the W filler, i.e., 80 w% with respect to the PCL polymer (PCL/W 80%), turned out to be very promising and was selected for subsequent investigation. The shielding properties of the electrospun PCL/W 80% sheets were assessed by means of in vivo tests on mice. Specifically, we evaluated the extent of radiation-induced damage, including apoptosis, in the EGL region of mice cerebellum on both whole body and shielded irradiated mice. Different configurations were considered by increasing the number of PCL/W 80% sheets/thickness of the shielding system. The results showed that a significant reduction in cell damage was observed in the shielded groups compared to the whole-body irradiated mice. In particular, the thin barrier consisting of 24 stacked layers of the developed shielding material was found to reduce the extent of apoptosis by 1.5 times compared to the non-shielded mice. As expected, the dose reduction of the shielding is proportional to the thickness of the layer obtained from the sheets. Thus, these findings, combined with a lightweight, flexible, relatively simple, and easily scalable production methodology at an industrial level, are promising for the application of the developed electrospun material for radiation protection in the aerospace and medical fields. The possibility of implementing this electrospinning manufacturing process also in conditions of micro-gravity or absence of gravity is particularly relevant for in-space manufacturing and space applications.

5. Patents

Italian patent application (102023000027945, PA104151IT01) filed on 22 December 2023.

Author Contributions: Conceptualization, A.R. and M.M.; experimental: electrospinning and SEM, C.G. and N.F.; experimental: mechanical tests, permeability and porosity measurement, C.G. and L.A.H.; experimental: ball milling, D.M.G.; experimental: 3D printing, E.S., E.Z. and C.D.G.; experimental: in vivo validation, M.M. and I.D.S.; resources, A.R. and M.M.; data curation, all authors; writing—original draft preparation, C.G.; writing—review and editing, all authors; supervision, A.R. and M.M.; project administration, A.R. and M.M.; funding acquisition, A.R., C.D.G. and G.F. All authors have read and agreed to the published version of the manuscript.

Funding: This research was funded by Regione Lazio through LazioInnova, Progetto "Schermi flessibili innovativi near-netshape e lavorabili per la schermatura di radiazione cosmica a base di polveri di tungsteno (WSHIELD)", Grant number A0320-2019-28157. N.F. and L.A.H. acknowledge that this research at Nanofaber received funding from the European Commission through the NEWSKIN Project, grant number 862100, and through the I4-GREEN Project, grant number 101084028.

Institutional Review Board Statement: The animal study was performed according to the Directive 2010/63/EU of the Euro-pean Parliament, approved by the local Ethical Committee for Animal Experiments of the ENEA, and authorized by the Italian Ministry of Health (n°740/2021-PR).

Data Availability Statement: The original contributions presented in the study are included in the article, further inquiries can be directed to the corresponding authors.

Conflicts of Interest: Author Giuseppe Felici and author Federica Galante was employed by the company S.I.T. Sordina IORT Technology S.p.A, their research was conducted within the "WSHIELD—Schermi flessibili innovativi near-netshape e lavorabili per la schermatura di radiazione cosmica a base di polveri di tungsteno". Author Noemi Fiaschini and author Luis Alexander Hein were employed by the company Nanofaber S.r.l. Antonio Rinaldi is a co-founder of he company Nanofaber S.r.l. without salary because is a public employee of ENEA. These authors declare that Nanofaber S.r.l. contributed to the methodology and to the characterization of the prototype materials in the scope of the referenced EU projects NEWSKIN and I4GREEN. The authors declare no conflicts of interest. The funders had no role in the design of the study; in the collection, analyses, or interpretation of data; in the writing of the manuscript; or in the decision to publish the results.

References

1. Nambiar, S.; Yeow, J.T. Polymer-composite materials for radiation protection. *ACS Appl. Mater. Interfaces* **2012**, *4*, 5717–5726. [CrossRef] [PubMed]
2. Wu, Y.; Wang, Z. Progress in Ionizing Radiation Shielding Materials. *Adv. Eng. Mater.* **2024**, 2400855. [CrossRef]
3. Thibeault, S.A.; Kang, J.H.; Sauti, G.; Park, C.; Fay, C.C.; King, G.C. Nanomaterials for radiation shielding. *Mrs Bull.* **2015**, *40*, 836–841. [CrossRef]
4. Seon-Chil, K.; Kyung-Rae, D.; Woon-Kwan, C. Medical radiation shielding effect by composition of barium compounds. *Ann. Nucl. Energy* **2012**, *47*, 1–5. [CrossRef]
5. Yin, W.-W.; Zheng, X.-W.; Wang, Z.-Q.; Chen, W.-J.; Tyan, Y.-S.; Chen, T.-R. Ambient and personnel occupational dose assessment in a Hospital's PET/CT center. *Appl. Radiat. Isot.* **2021**, *169*, 109466. [CrossRef]
6. Li, Q.; Wei, Q.; Zheng, W.; Zheng, Y.; Okosi, N.; Wang, Z.; Su, M. Enhanced Radiation Shielding with Conformal Light-Weight Nanoparticle–Polymer Composite. *ACS Appl. Mater. Interfaces* **2018**, *10*, 35510–35515. [CrossRef]
7. Strom, D.J. Ten principles and ten commandments of radiation protection. *Health Phys.* **1996**, *70*, 388–393. [CrossRef]
8. Kim, J.; Seo, D.; Lee, B.C.; Seo, Y.S.; Miller, W.H. Nano-W Dispersed Gamma Radiation Shielding Materials. *Adv. Eng. Mater.* **2014**, *16*, 1083–1089. [CrossRef]
9. Ambika, M.R.; Nagaiah, N.; Suman, S.K. Role of bismuth oxide as a reinforcer on gamma shielding ability of unsaturated polyester based polymer composites. *J. Appl. Polym. Sci.* **2016**, *134*, 44657. [CrossRef]
10. Büyükyıldız, M.; Taşdelen, M.A.; Karabul, Y.; Çağlar, M.; İçelli, O.; Boydaş, E. Measurement of photon interaction parameters of high-performance polymers and their composites. *Radiat. Eff. Defects Solids* **2018**, *173*, 474–488. [CrossRef]
11. Chaitali, V.M.; Zainab, A.; Mohamed, S.B.; Abouzeid, A.T.; Pravina, P.P. Polymeric composite materials for radiation shielding: A review. *Environ. Chem. Lett.* **2021**, *19*, 2057–2090. [CrossRef]
12. Ramirez Cadavid, D.A.; Layman, R.R.; Nishino, T.; Slutzky, J.L.; Li, Z.; Cornish, K. Guayule Natural Rubber Latex and Bi_2O_3 Films for X-ray Attenuating Medical Gloves. *Materials* **2022**, *15*, 1184. [CrossRef] [PubMed]
13. Arash, S.; Payman, R.; Shahram, T.; Masoud, N.; Seyed, M.J.M. Development of Lead-Free Materials for Radiation Shielding in Medical Settings: A Review. *J. Biomed. Phys. Eng.* **2011**, *1*, 17–21. [CrossRef]
14. Maghrabi, H.A.; Vijayan, A.; Deb, P.; Wang, L. Bismuth oxide-coated fabrics for X-ray shielding. *Text. Res. J.* **2016**, *86*, 649–658. [CrossRef]
15. Kim, S.C. Tungsten-Based Hybrid Composite Shield for Medical Radioisotope Defense. *Materials* **2022**, *15*, 1338. [CrossRef]
16. Adlienėa, D.; Gilysa, L.; Griškonisb, E. Development and characterization of new tungsten and tantalum containing composites for radiation shielding in medicineNucl. *Instrum. Methods Phys. Res. Sect. B Beam Interact. Mater. Atoms* **2020**, *467*, 21–26. [CrossRef]
17. Wasel, O.; Freeman, J.L. Comparative Assessment of Tungsten Toxicity in the Absence or Presence of Other Metals. *Toxics* **2018**, *6*, 66. [CrossRef]
18. AbuAlRoos, N.J.; Azman, M.N.; Amin, N.A.B.; Zainon, R. Tungsten-based material as promising new lead-free gamma radiation shielding material in nuclear medicine. *Phys. Med.* **2020**, *78*, 48–57. [CrossRef]
19. Selyutin, G.E.; Gavrilov, Y.Y.; Voskresenskaya, E.N.; Zakharov, V.A.; Nikitin, V.E.; Poluboyarov, V.A. Composite materials based on ultra high molecular polyethylene: Properties, application prospects. *Chem. Sustain. Dev.* **2010**, *18*, 301–314.
20. Chang, L.; Zhang, Y.; Liu, Y.; Fang, J.; Luan, W.; Yang, X.; Zhang, W. Preparation and characterization of tungsten/epoxy composites for γ-rays radiation shielding. *Nucl. Instrum. Methods Phys. Res. Sect. B Beam Interact. Mater. At.* **2015**, *356–357*, 88–93. [CrossRef]
21. Soylu, H.M.; Yurt Lambrecht, F.; Ersöz, O.A. Gamma radiation shielding efficiency of a new lead-free composite material. *J. Radioanal. Nucl. Chem.* **2015**, *305*, 529–534. [CrossRef]
22. Cho, K.; Lee, J.; Xing, P. Enzymatic degradation of blends of poly(ε-caprolactone) and poly(styrene-co-acrylonitrile) by Pseudomonas lipase. *J. Appl. Polym. Sci.* **2002**, *83*, 868–879. [CrossRef]
23. Woodruff, M.A.; Hutmacher, D.W. The return of a forgotten polymer Polycaprolactone in the 21st century. *Prog. Polym. Sci.* **2010**, *35*, 1217–1256. [CrossRef]
24. Pawar, S.P.; Kumar, S.; Misra, A.; Deshmukh, S.; Chatterjee, K.; Bose, S. Enzymatically degradable EMI shielding materials derived from PCL based nanocomposites. *RSC Adv.* **2015**, *5*, 17716. [CrossRef]
25. Xu, P.; Huang, B.; Tang, R. Improved mechanical and EMI shielding properties of PLA/PCL composites by controlling distribution of PIL-modified CNTs. *Adv. Compos. Hybrid Mater.* **2022**, *5*, 991–1002. [CrossRef]
26. Sun, C.; Peng, W.-J.; Huang, M.-L.; Zhao, K.-Y.; Wang, M. Constructing high-efficiency microwave shielding networks in multi-walled carbon nanotube/poly(ε-caprolactone) composites by adding carbon black and graphene nano-plates. *Polym. Int.* **2023**, *72*, 619–628. [CrossRef]
27. Tan, Y.-J.; Li, J.; Tang, X.-H.; Yue, T.-N.; Wang, M. Effect of phase morphology and distribution of multi-walled carbon nanotubes on microwave shielding of poly(L-lactide)/poly(ε-caprolactone) composites. *Compos. Part A Appl. Sci. Manuf.* **2020**, *137*, 106008. [CrossRef]
28. Bhardwaj, N.; Kundu, S.C. Electrospinning: A fascinating fiber fabrication technique. *Biotecnol. Adv.* **2010**, *28*, 325–347. [CrossRef]

29. Seyedmahmoud, R.; Rainer, A.; Mozetic, P.; Giannitelli, S.M.; Trombetta, M.; Traversa, E.; Licoccia, S.; Rinaldi, A. A primer of statistical methods for correlating parameters and properties of electrospun poly (l-lactide) scaffolds for tissue engineering—PART 1: Design of experiments. *J. Biomed. Mater. Res. Part A* **2015**, *103*, 91–102. [CrossRef]
30. Bergamasco, S.; Fiaschini, N.; Hein, L.A.; Brecciaroli, M.; Vitali, R.; Romagnoli, M.; Rinaldi, A. Electrospun PCL Filtration Membranes Enhanced with an Electrosprayed Lignin Coating to Control Wettability and Anti-Bacterial Properties. *Polymers* **2024**, *16*, 674. [CrossRef]
31. Deng, A.; Yang, Y.; Du, S. Tissue Engineering 3D Porous Scaffolds Prepared from Electrospun Recombinant Human Collagen (RHC) Polypeptides/Chitosan Nanofibers. *Appl. Sci.* **2021**, *11*, 5096. [CrossRef]
32. Di Carli, M.; Aurora, A.; Rinaldi, A.; Fiaschini, N.; Prosini, P.P. Preparation of Electrospun Membranes and Their Use as Separators in Lithium Batteries. *Batteries* **2023**, *9*, 201. [CrossRef]
33. Pozio, A.; Cemmi, A.; Carewska, M.; Paoletti, C.; Zaza, F. Characterization of gas diffusion electrodes for polymer electrolyte fuel cells. *J. Fuel Cell Sci. Technol.* **2010**, *7*, 041003. [CrossRef]
34. Pazzaglia, S.; Tanori, M.; Mancuso, M.; Rebessi, S.; Leonardi, S.; Di Majo, V.; Covelli, V.; Atkinson, M.J.; Hahn, H.; Saran, A. Linking DNA damage to medulloblastoma tumorigenesis in patched heterozygous knockout mice. *Oncogene* **2006**, *25*, 1165–1173. [CrossRef]
35. Mancuso, M.; Giardullo, P.; Leonardi, S.; Pasquali, E.; Casciati, A.; De Stefano, I.; Tanori, M.; Pazzaglia, S.; Saran, A. Dose and spatial effects in long-distance radiation signaling in vivo: Implications for abscopal tumorigenesis. *Int. J. Radiat. Oncol. Biol. Phys.* **2013**, *85*, 813–819. [CrossRef]
36. Mancuso, M.; Pasquali, E.; Leonardi, S.; Tanori, M.; Rebessi, S.; Di Majo, V.; Pazzaglia, S.; Toni, M.P.; Pimpinella, M.; Covelli, V.; et al. Oncogenic bystander radiation effects inPatchedheterozygous mouse cerebellum. *Proc. Natl. Acad. Sci. USA* **2008**, *105*, 12445–12450. [CrossRef]
37. Liu, W.; Walker, G.; Price, S.; Yang, X.; Li, J.; Bunt, C. Electrospun membranes as a porous barrier for molecular transport: Membrane characterization and release assessment. *Pharmaceutics* **2021**, *13*, 916. [CrossRef]
38. Ramier, J.; Bouderlique, T.; Stoilova, O.; Manolova, N.; Rashkov, I.; Langlois, V.; Renard, E.; Albanese, P.; Grande, D. Biocomposite scaffolds based on electrospun poly(3-hydroxybutyrate) nanofibers and electrosprayed hydroxyapatite nanoparticles for bone tissue engineering applications. *Mater. Sci. Eng. C* **2014**, *38*, 161–169. [CrossRef]
39. Movahedi, M.; Karbasi, S. A core-shell electrospun scaffold including extracellular matrix and chitosan to promote articular cartilage tissue regeneration. *Polym. Plast. Technol. Mater.* **2020**, *59*, 417–429. [CrossRef]
40. Mattioli-Belmonte, M.; Vozzi, G.; Whulanza, Y.; Seggiani, M.; Fantauzzi, V.; Orsini, G.; Ahluwalia, A. Tuning polycaprolactone–carbon nanotube composites for bone tissue engineering scaffolds. *Mater. Sci. Eng. C* **2012**, *32*, 152–159. [CrossRef]
41. Ho, M.H.; Li, S.Y.; Ciou, C.Y.; Wu, T. The morphology and degradation behavior of electrospun poly(3-hydroxybutyrate)/Magnetite and poly(3-hydroxybutyrate-co-3-hydroxyvalerate)/Magnetite composites. *J. Appl. Polym. Sci.* **2014**, *131*, 41070. [CrossRef]
42. Tang, C.Y.; Chen, D.Z.; Tsui, C.P.; Uskokovic, P.S.; Yu, P.H.; Leung, M.C. Nonisothermal melt-crystallization kinetics of hydroxyapatite-filled poly (3-hydroxybutyrate) composites. *J. Appl. Polym. Sci.* **2006**, *102*, 5388–5395. [CrossRef]
43. Nowak, E.; Etienne, O.; Millet, P.; Lages, C.S.; Mathieu, C.; Mouthon, M.A.; Boussin, F.D. Radiation-induced H2AX phosphorylation and neural precursor apoptosis in the developing brain of mice. *Radiat Res.* **2006**, *165*, 155–164. [CrossRef]

polymers

MDPI

Article

Effect of Vibration Pretreatment–Microwave Curing Process Parameters on the Mechanical Performance of Resin-Based Composites

Dechao Zhang [1], Lihua Zhan [1,2,*], Bolin Ma [1,*], Jinzhan Guo [1], Wentao Jin [1], Xin Hu [1], Shunming Yao [3] and Guangming Dai [4]

1 Light Alloys Research Institute, Central South University, Changsha 410083, China; zhangdechao@csu.edu.cn (D.Z.); guruci@163.com (J.G.); jwt2242444825@163.com (W.J.); ladihu4@gmail.com (X.H.)
2 State Key Laboratory of Precision Manufacturing for Extreme Service Performance, Central South University, Changsha 410083, China
3 College of Mechanical and Electrical Engineering, Central South University, Changsha 410083, China; yaoshunming@csu.edu.cn
4 College of Mechanical and Energy Engineering, Zhejiang University of Science and Technology, Hangzhou 310023, China; daigm@zust.edu.cn
* Correspondence: yjs-cast@csu.edu.cn (L.Z.); mabolin_csu@163.com (B.M.)

Abstract: The vibration pretreatment–microwave curing process can achieve high-quality molding under low-pressure conditions and is widely used in the curing of resin-based composites. This study investigated the effects of the vibration pretreatment process parameters on the void content and the fiber weight fraction of T700/TRE231; specifically, their influence on the interlaminar shear strength and impact strength of the composite. Initially, an orthogonal experimental design was employed with interlaminar shear strength as the optimization target, where vibration acceleration was determined as the primary factor and dwell time as the secondary factor. Concurrently, thermogravimetric analysis (TGA) was performed based on process parameters that corresponded to the extremum of interlaminar shear strength, revealing a 2.17% difference in fiber weight fraction among specimens with varying parameters, indicating a minimal effect of fiber weight fraction on mechanical properties. Optical digital microscope (ODM) analysis identified interlaminar large-size voids in specimens treated with vibration energy of 5 g and 15 g, while specimens subjected to a vibration energy of 10 g exhibited numerous small-sized voids within layers, suggesting that vibration acceleration influences void escape pathways. Finally, impact testing revealed the effect of the vibration pretreatment process parameters on the impact strength, implying a positive correlation between interlaminar shear strength and impact strength.

Keywords: vibration pretreatment; microwave curing; thermogravimetric analysis; void content; microscopic morphology; impact strength

1. Introduction

Due to the advantages of high specific strength, high specific stiffness [1], fatigue resistance, and designable material mechanical properties, carbon-fiber-reinforced resin matrix composites are widely used in aerospace [2], marine [3,4], rail transportation [5,6], military [7,8], and medical fields [9,10]. However, composite components are extremely sensitive to interlaminar delamination performance and impact resistance in the actual service process [11]. The mechanical properties of components are closely related to the molding process. Therefore, it is necessary to identify the influence of the curing process parameters on their mechanical properties to guide the design of composite components, reduce costs, and ensure product quality. At present, the autoclave curing process is still the most important molding method for composites, but its high energy consumption, long curing cycle [12], uneven temperature field distribution inside the components, and hysteresis in temperature control limit the development of autoclave curing technology. Scholars have gradually shifted their focus to the research of out-of-autoclave curing and molding technologies, such as UV curing [13,14], electron beam curing [15,16], laser curing [17,18], and microwave curing [19,20].

However, the microwave curing process has attracted extensive attention from researchers and scholars due to its unique "in-volume" and "selective heating" characteristics [21]. In the existing research on the microwave curing technology of composites, the main focus is on comparing the mechanical properties of the specimen formed using microwave curing and conventional thermal curing processes (the study refers exclusively to air circulation heating). For example, concerning the interfacial properties of composites, Hang et al. [22] found that the microwave curing process improved the fiber/resin interfacial bonding properties, increasing shear strength and flexural strength by 11.5% and 6.6%, respectively. Similar conclusions were also drawn by Haider et al. [23]. In terms of impact strength, De Vergara et al. [24] used a drop hammer to test microwave-cured basalt fiber composites with an increase in maximum load and penetration threshold by 17.8% and 14.5%, respectively. Dasari et al. [25] also found that microwave-cured laminates exhibited a 6% increase in peak impact force and an 18% increase in rebound energy. Chen et al. [26] compared the Charpy impact strength between microwave-cured and conventional thermal-cured specimens, concluding that microwave curing offered productivity advantages. However, both processes still had room for improvement in terms of the mechanical properties, affected mainly by void content. Nevertheless, microwave curing generally demonstrated superior mechanical properties compared to those under the conventional thermal curing process.

Therefore, in order to reduce the void content during the curing process of composites, research scholars have introduced the vibration energy into the curing process of composites. Muric-Nesic et al. [27] maintained a vibration frequency of 30 Hz for 30 min during the UV curing process of composites, and the void content was reduced to 0.3%. In order to explore the effects of vibrational forms, some studies have been conducted by introducing random vibrations during the curing process. Guan et al. [28] introduced random vibration into the microwave curing process, and the data showed that the experimental void content and mechanical properties were comparable to those achieved by the standard autoclave process. Yang et al. [29] also employed random vibration in the thermal curing process, and the average interfacial shear strength of the samples increased by 48.26% compared to the thermal curing process. However, the enhancement of mechanical properties was from the perspective of void content and the influence of fiber weight content due to changes in the process parameters not considered. For instance, the results of Xiang et al. [30] showed that the mechanical properties first increased and then decreased with the increase in fiber volume fraction. The same conclusion was obtained by Yang et al. [31], who conducted three-point bending tests on FeCrAl(f)/HA composite. Guo et al. [32] conducted a study focusing on the impact of short carbon fiber volume fraction on the impact resistance of composite, and the results showed that the impact resistance gradually increased as the fiber volume fraction increased from 0% to 11%. It is evident that fiber weight fraction has a significant effect on interlaminar shear and impact strengths. In conclusion, the recent studies have not fully explored the effect of vibration pretreatment–microwave process parameters on the fiber weight fraction of composites, nor have they adequately considered the combined effect of fiber weight fraction and void content on mechanical properties.

Therefore, this study employed orthogonal experimental design to investigate the effects of pretreatment temperature, dwell time, and vibration acceleration on the interlaminar shear strength of T700/TRE231 composite materials. TGA tests were conducted based on parameters corresponding to peak interlaminar shear strength, elucidating the influence of various fiber weight fractions on interlaminar shear strength. Concurrently, ODM was used to characterize the effects on the microstructure of voids. Finally, impact performance was evaluated using a drop hammer impact testing platform to assess the effects of the vibration pretreatment process parameters.

2. Materials and Methods

2.1. Materials and Equipment

The composite prepreg utilized in this study was a T700/TRE231, containing 58% fiber and 42% resin by volume, manufactured by Changzhou Tianqi Xinxin Technology Co., Ltd. in Changzhou, China. The prepreg has an areal density of 130 g/m^2 and a single-layer thickness of 0.125 mm. The laminate size was 200 mm × 200 mm, with a plying orientation of $[0/90/0/90/0/90/0/90]_S$.

The eighth layer of the composite laminate was embedded with a temperature measuring fiber to facilitate the subsequent monitoring and regulation of microwave curing temperature. The vibration pretreatment and microwave equipment were consistent with the literature [12], as shown in Figure 1. The laminate was placed into an octagonal microwave heating furnace for curing after vibration pretreatment.

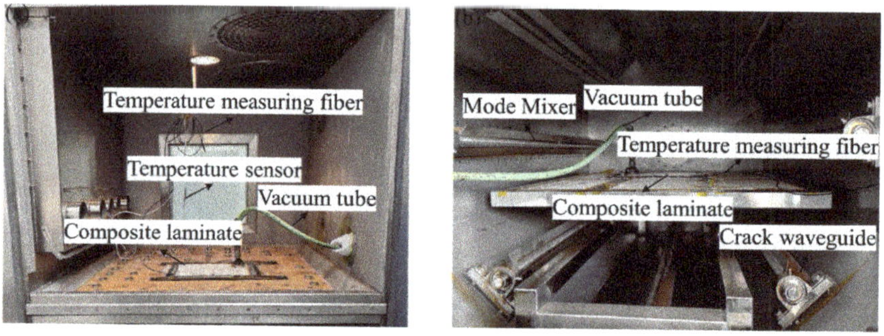

Figure 1. Vibration pretreatment–microwave curing equipment; (**a**) composite laminates in the vibration pretreatment equipment; (**b**) composite laminates in the microwave furnace.

2.2. Selection of Process Parameters and Curing Process Curve

Investigating the effects of the vibration pretreatment process parameters (vibration pretreatment temperature, dwell time, and vibration acceleration) on the interlaminar shear strength and impact performance of T700/TRE231, three levels were selected for each parameter. Vibration exhibits optimal void suppression when resin viscosity is at its lowest [12]. Therefore, for this study, pretreatment temperature parameters were chosen as 80 °C, 90 °C, and 100 °C, respectively. Based on prior research [28,29], the study range for the vibration acceleration parameters was between 5 and 15 g. This choice was influenced by the observation that when the vibration energy is too low, its void reduction effect is not significant, and when it is too high, it may lead to destructive outcomes. Hence, 5 g, 10 g, and 15 g acceleration parameters were selected. The process of void reduction took time, and a dwell time of 10, 30, and 50 min were selected, as depicted in Table 1.

Table 1. Orthogonal experimental design factors and levels (1 g = 9.8 m/s^2).

Factor / Level	Pretreatment Temperature/°C	Dwell Time/min	Vibration Acceleration/g
I	80	10	5
II	90	30	10
III	100	50	15

Figure 2 shows the vibration pretreatment–microwave curing process curve for the T700/TRE231 composite. The curve includes two parts: vibration and microwave heating. For the vibration pretreatment section, the laminate is heated up to the pretreatment temperature at a rate of 2 °C/min and maintained for a certain dwell time. The design parameters are shown in Table 1. The microwave heating section continues to heat at the same rate until reaching 130 °C and dwell time for 120 min. The curing process is carried out under the vacuum bagging pressure, followed by natural cooling in the furnace.

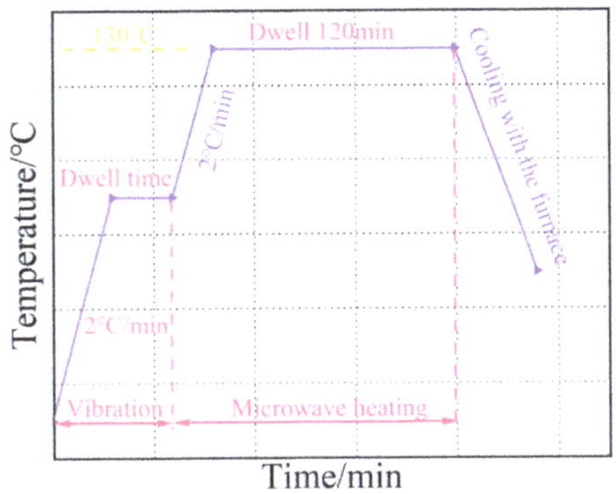

Figure 2. Vibration pretreatment–microwave curing process curve for T700/TRE231.

2.3. Interlaminar Shear Strength Test

To explore the impact of different process parameters on the interfacial strength of composite, the interlaminar shear strength was evaluated using the short beam three-point bending method. The testing process strictly followed the standards of the People's Republic of China JC/T 773-2010 [33]. The testing process schematic and specimen dimensions were shown in Figure 3. The equipment used was the CMT5105 model produced by Sansi Taijie Electrical Equipment Co., China, in Zhuhai, Guangdong. When the indenter first contacted the surface of the specimen, a preload force of 0–30 N was maintained. Then, the indenter was lowered at a rate of 1 mm/min until the delamination failure of the sample occurred. The specimens were supported across a span of (10 ± 0.3) mm, with a bottom support radius of 2 mm and a punch radius of 5 mm. Four specimens were tested to determine the interlaminar shear strength of the process. The interlaminar shear strength was calculated using Equation (1):

$$\tau_{In} = \frac{3F_{max}}{4bh} \tag{1}$$

where τ_{In} represents interlaminar shear strength, and F_{max} means the maximum load. b and h are the width and thickness of the specimen, respectively.

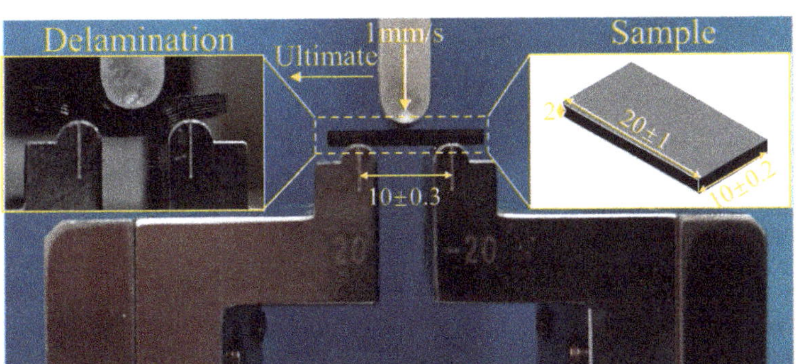

Figure 3. The testing process schematic and specimen dimensions of three-point bending.

2.4. Characterization of Void Morphology

ODM was employed to investigate the void microstructure of samples subjected to vibration pretreatment–microwave curing processes. For each set of processing parameters, samples were collected from three distinct locations and averaged to ensure statistical accuracy. The samples, sized 10 × 10 mm, were shot in a 0° direction. Void analysis was conducted using Image Pro Plus 6.0 software, and void content was quantified according to Equation (2). The sampling area is illustrated in Figure 4.

$$\gamma = \frac{S_v}{S_a} \tag{2}$$

where γ represents the void content of the specimen, and S_v means the area of sample cross-sectional voids. S_a is the area of the specimen cross-section.

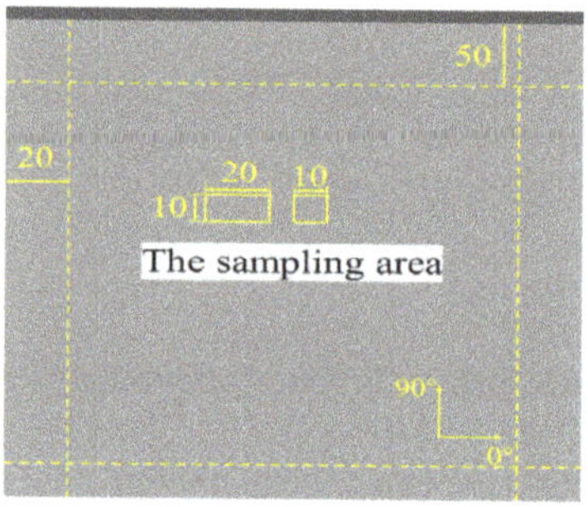

Figure 4. Sampling area for three-point bending and ODM specimens.

2.5. Evaluation of Fiber Weight Fraction in Composite Laminates

To investigate the impact of vibration pretreatment parameters on the fiber weight fraction in laminates. The Platinum Elmer Corporate Management (Shanghai, China) Co., TGA/STA8000-FTIR-GCMS-ATD thermogravimetric-IR-gas chromatography analyzer was utilized. In order to make the experimental data more accurate, the samples were precisely cut, weighing approximately (20 ± 1) mg. The test conditions were as follows: the samples underwent controlled heating at a rate of 10 °C/min until reaching (785 ± 5) °C [34]; a continuous flow of nitrogen gas at a rate of 50 mL/min was maintained [35]. Figure 5 illustrates the variation in the weight of TRE231 resin with temperature. Resin decomposes rapidly at around (300–400) °C. At the end of the test, the residual resin percentage was 18.926%.

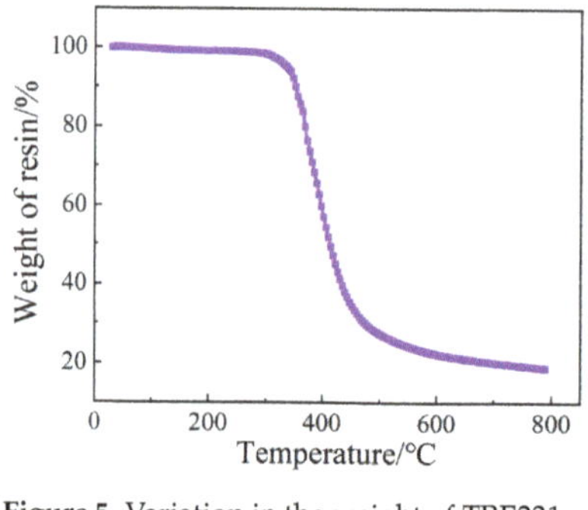

Figure 5. Variation in the weight of TRE231 resin with temperature.

The fiber weight fraction of laminates processed could be accurately determined using Equation (3) [29,36]:

$$C_f = (\frac{Res_{coms} - Res_{resin}}{1 - Res_{resin}}) \times 100\% \quad (3)$$

where C_f represents the fiber weight fraction, and Res_{coms} indicates composite residual weight fraction. Res_{resin} is the resin residual weight fraction.

2.6. Testing Impact Properties of Composite Laminates

In various engineering fields such as aerospace, composite components need to withstand external loads at certain speeds, including collisions and impacts. Therefore, the impact performance of composite components was crucial for material design and selection. To investigate the effects of vibration pretreatment–microwave curing process parameters on the dynamic mechanical properties of composite laminates, a simply supported beam impact test method was employed. During the experiment, the relationship between impact force and time was recorded by sensors. The tests were conducted according to the standards of the People's Republic of China GB/T 1043.1-2008 [37]. The HIT2000F device, manufactured by Zwick/Roell in Ulm, Germany, with a 9.294 kg hammer and a 2 mm diameter impactor was used, as shown in Figure 6a. The experiment was conducted at 25 °C with an impact energy of 20 J. The specimens were characterized for strength using notch-free through-the-thickness impact testing [38,39]. The specimen dimensions and span were (60 ± 2) mm × (10 ± 0.2) mm × (2 ± 0.2) mm (length × width × thickness) and (40 ± 0.5) mm, respectively, as shown in Figure 6b. Figure 6c illustrates the impact test process, and the calculation of simply supported beam impact strength for specimens was performed using Equation (4):

$$I_s = \frac{E_e}{hb} \times 10^3 \quad (4)$$

where I_s represents impact strength (kJ/m^2), and E_e means energy absorbed in the destruction of the corrected specimen (J). b and h are the width and thickness of the specimen, respectively (mm).

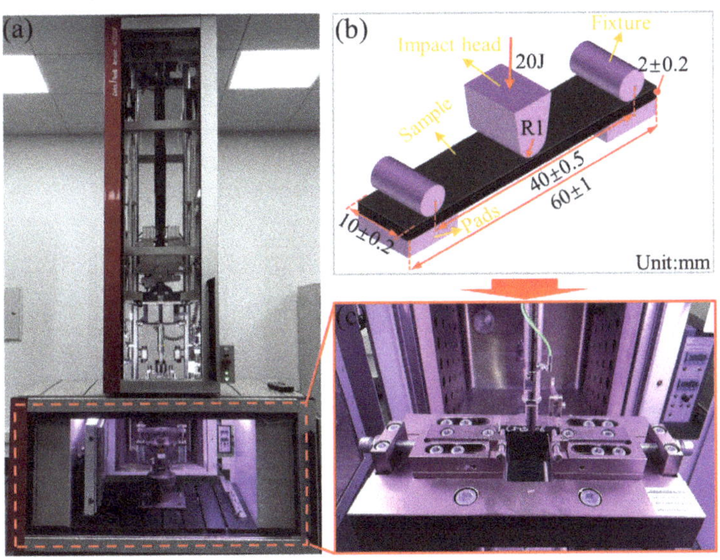

Figure 6. Drop hammer impact test equipment and impact schematic; (**a**) drop hammer impact equipment; (**b**) specimen dimensions and impact clamping; (**c**) simply supported beam impact process.

3. Results and Discussion

3.1. Analysis of the Optimized Vibration Pretreatment Process Parameters

As shown in Figure 7, the curve illustrates the variation in the viscosity of TRE231 resin with temperature. The test conditions for the curve were as follows: the temperature was increased from 50 °C to 130 °C at a heating rate of 2 °C/min, and the shear rate was 10 s^{-1}. From the trend of the curve, the viscosity of the resin at 50 °C is 294.8 Pa·s, indicating a relatively high viscosity. As the temperature increases from 50 °C to 60 °C, the rate of viscosity change peaks and gradually decreases, reaching 98 Pa·s at 60 °C. The viscosity of the resin changes by 200.8% with just a 10 °C temperature difference, highlighting the high sensitivity of resin viscosity to temperature.

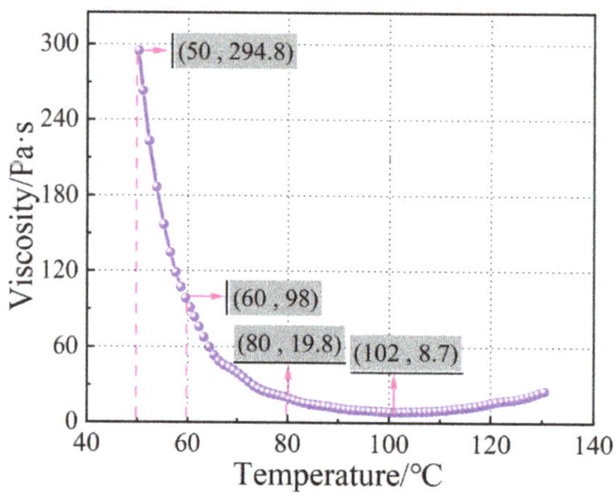

Figure 7. The temperature–viscosity curve of TRE231 resin.

When the temperature rises to 80 °C, the resin viscosity decreases to 19.8 Pa·s. This can be attributed to the increased thermal motion of resin molecules with rising temperatures, leading to an increase in the average kinetic energy of the molecules and a weakening of intermolecular interactions. This enhances molecular mobility and facilitates easier movement through barriers between molecules, resulting in a reduction in resin viscosity. The viscosity of the resin changes slowly between 90 and 110 °C, reaching a minimum viscosity of 8.7 Pa·s at 102 °C. As the temperature continues to rise, molecular cross-linking reactions begin in the resin, gradually forming a three-dimensional network structure. The increase in molecular weight and complexity of the resin system limits the free radicals in the resin, leading to a subsequent increase in viscosity [40].

Table 2 shows the orthogonal experimental results of the interlaminar shear strength and void content for different vibration pretreatment process parameters. It can be seen that there is a negative correlation between void content and interlaminar shear strength [41], meaning that the higher the void content, the lower the interlaminar shear strength. The reason for the above phenomenon is that high void content reduces the actual contact area between materials. Additionally, voids induce stress concentration under applied forces. These factors collectively result in a significant reduction in interlaminar shear strength. It is evident that when the vibration acceleration is set at 10 g, the interlaminar shear strength consistently reaches its maximum value and has the lowest void content. Therefore, the vibration acceleration factor is initially identified as the primary factor influencing both interlaminar shear strength and void content, which is consistent with the previous research conclusion [12]. The order of dwell time and pretreatment temperature is opposite, and this discrepancy may be attributed to differences in resin systems. Based on the parameter analysis results in this study, the order of influence of vibration pretreatment process parameters on interlaminar shear strength is as follows: vibration acceleration > dwell time > pretreatment temperature. The optimized vibration pretreatment process parameters derived from the orthogonal experimental results indicate a microwave curing process with a pretreatment temperature of 90 °C, a dwell time of 30 min, and a vibration acceleration of 10 g (referred to as the 90 °C–30 min–10 g microwave curing process for brevity).

Table 2. Orthogonal experimental results of $L_9(3^4)$.

Test No.	Pretreatment Temperature/°C	Dwell Time/min	Vibration Acceleration/g	Interlaminar Strength/MPa	Void Content/%
1	80	10	5	48.52	0.98
2	80	30	10	58.18	0.57
3	80	50	15	53.89	0.71
4	90	10	10	59.44	0.44
5	90	30	15	55.82	0.66
6	90	50	5	48.60	0.80
7	100	10	15	52.23	0.77
8	100	30	5	51.65	0.78
9	100	50	10	55.58	0.67
Level I	53.53	53.40	49.59		
Level II	54.62	55.22	57.73		
Level III	53.15	52.69	53.98		
Extreme variance	1.47	2.53	8.14		

3.2. *Effect of Vibration Pretreatment Process Parameters on Fiber Weight Fraction*

To elucidate the influence of vibration pretreatment process parameters on the fiber weight fraction in composite laminates, TGA was conducted on the pretreatment processes corresponding to the maximum and minimum interlaminar shear strengths at vibration pretreatment temperatures. Figure 8 illustrates the weight retention rate of composite specimens with temperature changes under different vibration pretreatment process parameters. At lower temperatures, the weight retention rate slightly decreases, mainly due to the conversion of moisture in the composite into water vapor [29] and expulsion as the temperature rises. When the temperature reaches 300–400 °C, the rate of change in weight retention rate gradually increases to a maximum and then decreases. At this stage, the resin begins to reach its optimal decomposition temperature, and the rate of resin decomposition gradually accelerates until it reaches a maximum. As the temperature continues to rise and with time, the resin content in the specimen rapidly decreases. It leads to a rapid overall decrease in weight retention rate. The remaining resin in the specimen completely decomposes, and the variation in weight retention rate eventually stabilizes. The values enclosed within the box in the figure represent the residual weight fraction of the sample following testing. This includes both the residual weight fraction of the resin (i.e., the carbon remaining after combustion) and the weight fraction of the fibers.

The fiber weight fraction in the prepreg is 67%. According to Figure 9, the minimum fiber weight fraction under different vibration pretreatment process parameters is approximately 67.45%. This indicates that the composite laminates indeed undergo a compaction process during pretreatment. The TRE231 resin has a low viscosity in the range of 8.7 to 19.8 Pa·s at 80–100 °C. The introduction of a vibrational energy field results in a good impregnation effect, and excess resin has been squeezed out of the laminate. It can also be observed from the figure that within the range of pretreatment parameters study in this research, the fiber weight fraction has changed by 2.17%, suggesting that the variation in fiber weight fraction has a minimal impact on interlaminar shear strength.

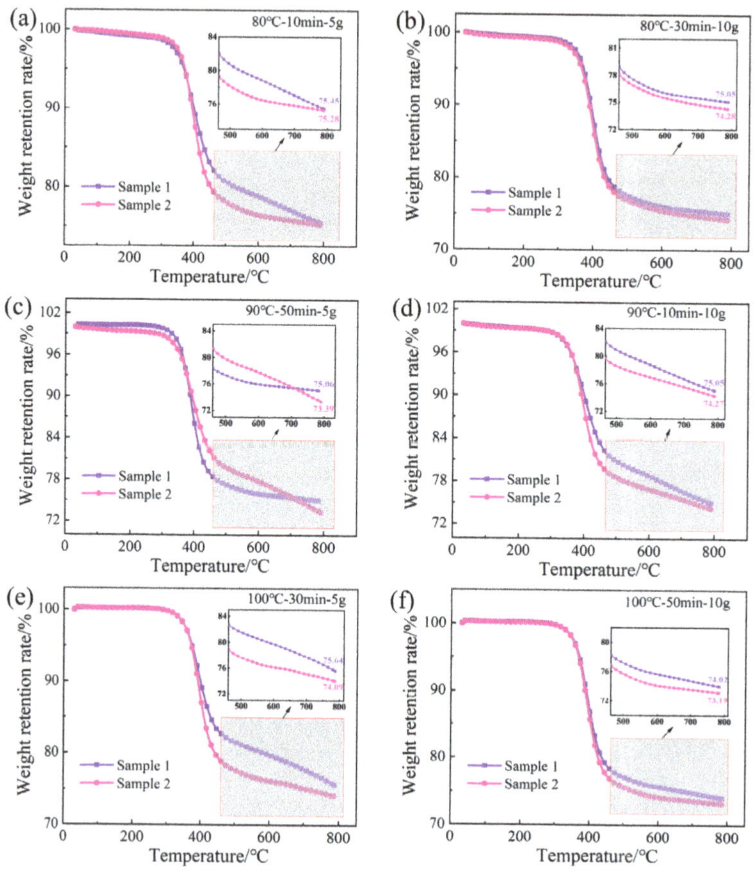

Figure 8. Weight retention rate of samples under vibration pretreatment parameters.

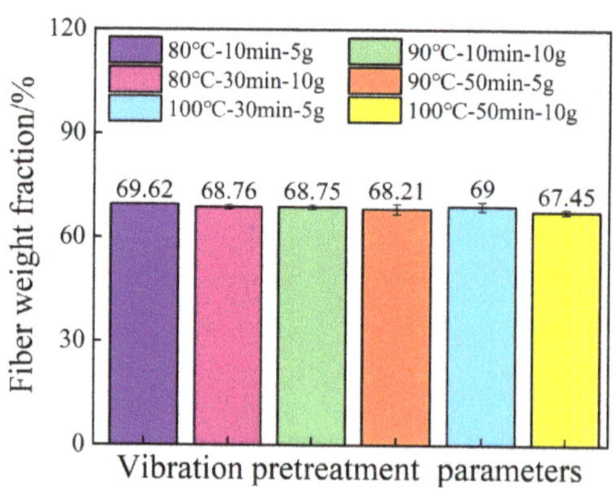

Figure 9. The fiber weight fraction of samples under vibration pretreatment parameters.

3.3. *Effect of Vibration Pretreatment Process Parameters on Void Micromorphology*

Figure 10 depicts the microstructure of the specimen under different vibration pretreatment process parameters. According to Section 3.1, pretreatment temperature and dwell time have minimal impact on interlaminar shear strength. Therefore, only the microscopic morphology under variable acceleration and optimized pris examined. It is noticeable that when the vibration pretreatment parameters are set at 80 °C–10 min–5 g, there are numerous interlaminar elliptical and circular voids and a significant number of small-sized intralaminar voids. This situation is markedly different from the previous one, where voids were primarily distributed between the layers [12]. As the vibration energy increases to 10 g, interlaminar voids disappear, replaced by elliptical, circular, and small-sized intralaminar voids. When vibration energy is absent, the internal void pressure, hydrostatic pressure and vacuum bag pressure inside the laminate are in equilibrium, and the voids are distributed in the resin with a certain diameter. The introduction of vibration energy disrupts this equilibrium. According to Equation (5) [27], the alternating positive and negative forces induced by vibration cause large interlaminar voids to fracture into smaller voids under the influence of negative vibration forces, ultimately shifting from interlaminar to intralaminar spaces.

$$P_v - P_r - P_e - P_{vibr} = \frac{K\sigma_v}{R_v} \quad (5)$$

where P_v represents internal pressure of the void, and P_r is hydrostatic pressure. P_e is vacuum bag pressure, and P_{vibr} represents pressure generated by vibration. K is the surface tension coefficient, and σ_v, R_v are void surface tension and radius.

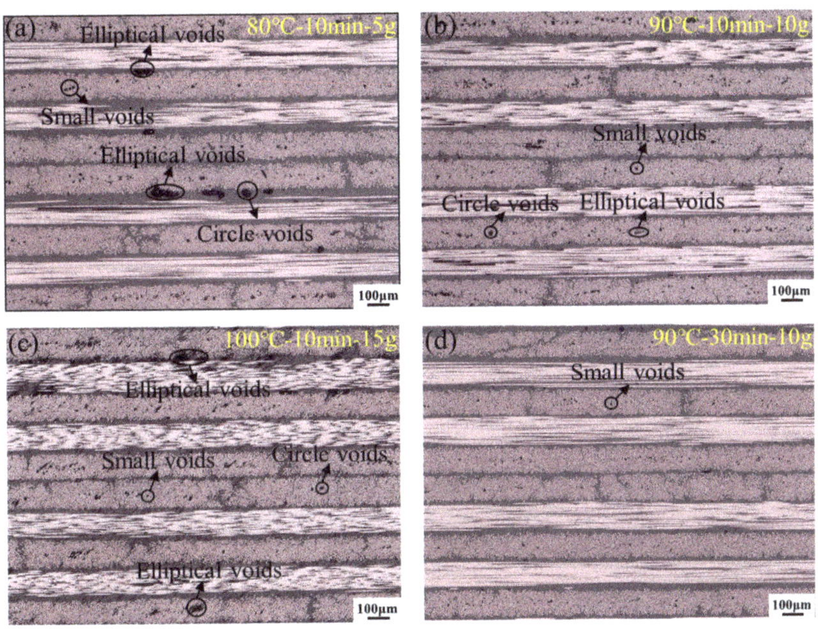

Figure 10. Microscopic morphology of samples under vibration pretreatment parameters.

The original small voids partially dissolve into the resin, as shown in Figure 10c. When the vibration energy increases to 15 g, interlaminar voids reappear, indicating that the magnitude of vibration energy can affect the escape path of voids. Figure 10d describes the void distribution and morphological features under the optimized vibration pretreatment process parameters. Compared to Figure 10a, large elliptical and circular voids have completely disappeared, and more small-sized intralaminar voids are observed.

3.4. Effect of Vibration Pretreatment Process Parameters on Impact Strength

Figure 11 illustrates the impact load–time curves and impact strengths of simply supported beams for specimens with different pretreatment process parameters. In Figure 11a–c, the variation of load with time exhibits oscillatory linear growth until reaching the maximum load [42]. This oscillation is mainly caused by the force generated when the impact head contacts the specimen. Additionally, since composites are linear elastic materials, the impact force gradually increases with time until reaching the maximum load. At this stage, the composite laminate remains undamaged and stays the elastic region. When the impact force exceeds the elastic limit of the composite material, fiber fracture, delamination, and matrix cracking start to occur [1]. The decrease in stiffness and the weakened load-bearing capacity of the composite laminate leads to a sudden drop in load. Subsequent oscillations may be caused by behaviors such as the energy absorption and crack propagation of the composite laminate under impact loading.

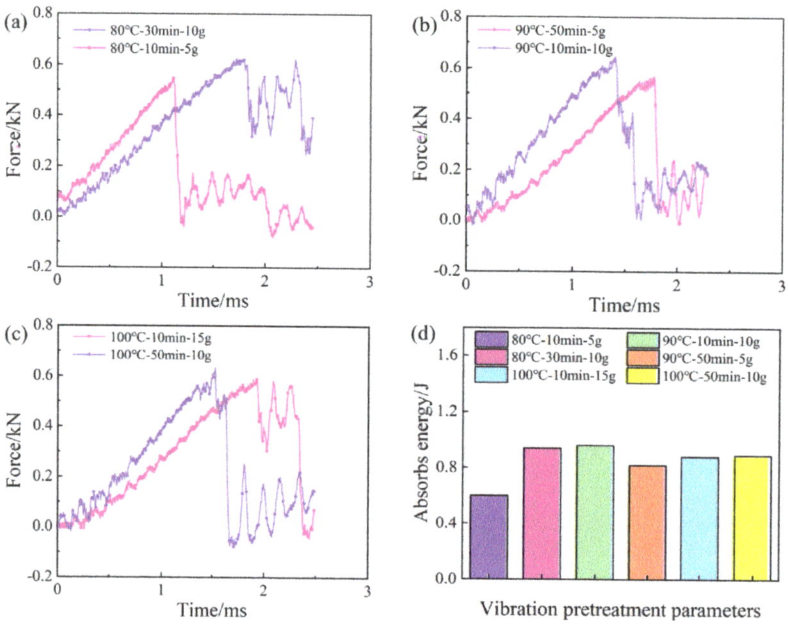

Figure 11. The impact time–force curves and impact strength of samples under pretreatment parameters.

Figure 11d shows the impact energy absorption of simply supported beam specimens under different vibration pretreatment–microwave curing process parameters. It can be observed that the specimen absorbs the highest impact energy when the vibration acceleration is 10 g at the same pretreatment temperature. This is primarily influenced by the internal void content of the laminate. In terms of void content, when the vibration acceleration is 10 g, there is a lower void content in the laminate with almost no large voids visible compared to conditions at 5 g and 15 g. This indicates that the internal structure of the laminate is more compact, increasing the number of connection points and contact surfaces to effectively transmit and disperse impact energy, thereby enhancing the load-bearing capacity. Additionally, the low void content helps reduce the initiation points of cracks, shortening the path for energy absorption and dissipation, thereby slowing down the rate of crack propagation.

Table 3 presents the impact strength results calculated according to Equation (4), which indicate a positive correlation between impact strength and interlaminar shear strength. This correlation primarily arises because interlaminar shear strength denotes the bonding strength between the layers of composite laminates and their ability to resist external shear forces. When the interlaminar shear strength of composite laminates is high, it effectively prevents crack expansion and enhances the overall strength and toughness.

Impact strength, on the other hand, refers to the material's load-bearing capacity under external impact forces. Composite laminates with higher interlaminar shear strength are better at absorbing and dispersing impact energy, which slows down the damage process and increases their ability to withstand peak impact loads. Therefore, the higher the interlaminar shear strength in composite laminates, the more effectively they prevent crack propagation and enhance their impact resistance.

Table 3. Impact strength of different vibration pretreatment–microwave process parameters.

Pretreatment Temperature/°C	Dwell Time/min	Vibration Acceleration/g	Interlaminar Strength/MPa	Absorbs Energy/J	Impact Strength/(kJ/m^2)
80	10	5	48.52	0.60	29.94
80	30	10	58.18	0.94	47.14
90	10	10	59.44	0.96	47.95
90	50	5	48.60	0.82	40.96
100	10	15	52.23	0.88	44.00
100	50	10	55.58	0.89	44.19

4. Conclusions

In this study, the influence of vibration pretreatment–microwave curing process parameters on the interlaminar shear strength of T700/TRE231 composite materials was firstly explored through an orthogonal experimental design. Based on the extremum of interlaminar shear strength, thermal analysis was conducted on samples corresponding to process parameters, clarifying the effect of fiber weight fraction on interlaminar shear strength. ODM was used to examine the void morphology of samples treated with different vibration pretreatment parameters. Finally, the impact performance of samples was evaluated through a simply supported beam drop hammer impact test. The main findings of this study are as follows:

(1) The order of pretreatment process parameters affecting the interlaminar shear strength of T700/TRE231 is as follows: vibration acceleration > dwell time > pretreatment temperature. The optimized parameters are a pretreatment temperature of 90 °C, a dwell time of 30 min, and a vibration acceleration of 10 g.

(2) Within the range of vibration pretreatment process parameters studied in this paper, the minimum fiber weight fraction increased by 0.45% over the initial prepreg, indicating that the pretreatment process produced a compaction effect. However, the maximum range of variation in fiber weight fraction is only 2.17%. Therefore, the influence of pretreatment process parameters on the fiber weight fraction is relatively slight.

(3) When the vibration energy is 5 g, large circular and elliptical voids are present in the samples. Increasing the vibration energy to 10 g caused large voids between layers to fracture into smaller voids and moved into within the layers. Further increasing the energy to 15 g reintroduces interlaminar voids, indicating that vibration acceleration can alter the escape path of voids.

(4) The simply supported beam impact tests of samples treated with different process parameters revealed that samples subjected to a vibration energy of 10 g exhibited higher impact strength, mainly attributed to lower void content. Additionally, the impact strength showed a positive correlation with interlaminar shear strength.

Author Contributions: Conceptualization, D.Z. and L.Z.; methodology, D.Z.; software, D.Z.; validation, J.G., W.J., X.H. and S.Y.; formal analysis, D.Z.; investigation, D.Z., W.J., G.D. and S.Y.; resources, D.Z.; writing—original draft preparation, D.Z.; writing—review and editing, L.Z. and B.M. All authors have read and agreed to the published version of the manuscript.

Funding: This work is supported by the National Natural Science Foundation of China (grant numbers 52175373 and U22A20190), the National Key Research and Development Program of China (grant number 2018YFA0702800), and the Project of the State Key Laboratory of Precision Manufacturing for Extreme Service Performance, Central South University (grant number ZZYJKT2021-03).

Institutional Review Board Statement: Not applicable.

Data Availability Statement: The data that support the findings of this study are available from the corresponding author upon reasonable request.

Acknowledgments: The authors also gratefully acknowledge the composite research team members of Central South University for their support and useful discussions in this research.

Conflicts of Interest: The authors declare no conflicts of interest.

References

1. Tian, K.; Tay, T.E.; Tan, V.B.C.; Haris, A.; Chew, E.Q.; Pham, V.N.H.; Huang, J.Z.; Raju, K.; Sugahara, T.; Fujihara, K.; et al. Improving the impact performance and residual strength of carbon fibre reinforced polymer composite through intralaminar hybridization. *Compos. Part A Appl. Sci. Manuf.* **2023**, *171*, 107590. [CrossRef]
2. Kalusuraman, G.; Kumaran, S.T.; Balamurugan, K.; Sivashanmugam, N.; Sivaprakasam, P.; Kurniawan, R.; Ezhilmaran, V. Vibration studies on fiber reinforced composites—A review. *J. Nat. Fibers* **2023**, *20*, 2157361. [CrossRef]
3. Ojha, S.; Bisaria, H.; Mohanty, S.; Kanny, K. Fabrication and characterization of light weight PVC foam based E-Glass reinforced polyester sandwich composites. *Proc. Inst. Mech. Eng. Part C-J. Mech. Eng. Sci.* **2023**, *237*, 5042–5051. [CrossRef]
4. Nan, J.J.; Zhi, C.; Meng, J.G.; Miao, M.H.; Yu, L.J. Seawater aging effect on fiber-reinforced polymer composites: Mechanical properties, aging mechanism, and life prediction. *Text. Res. J.* **2023**, *93*, 3393–3413. [CrossRef]
5. Wang, A.N.; Liu, X.G.; Yue, Q.R.; Xian, G.J. Tensile properties hybrid effect of unidirectional flax/carbon fiber hybrid reinforced polymer composites. *J. Mater. Res. Technol.* **2023**, *24*, 1373–1389. [CrossRef]
6. Huang, X.H.; Su, S.Y.; Xu, Z.D.; Miao, Q.S.; Li, W.F.; Wang, L.X. Advanced composite materials for structure strengthening and resilience improvement. *Buildings* **2023**, *13*, 2406. [CrossRef]
7. Jiao, J.K.; Cheng, X.Y.; Wang, J.L.; Sheng, L.Y.; Zhang, Y.M.; Xu, J.H.; Jing, C.H.; Sun, S.Y.; Xia, H.B.; Ru, H.L. A review of research progress on machining carbon fiber-reinforced composites with lasers. *Micromachines* **2023**, *14*, 24. [CrossRef]
8. Zhu, T.L.; Wang, Z. Research and application prospect of short carbon fiber reinforced ceramic composites. *J. Eur. Ceram. Soc.* **2023**, *43*, 6699–6717. [CrossRef]
9. Devi, G.; Nagabhooshanam, N.; Chokkalingam, M.; Sahu, S.K. EMI shielding of cobalt, red onion husk biochar and carbon short fiber-PVA composite on X and Ku band frequencies. *Polym. Compos.* **2022**, *43*, 5996–6003. [CrossRef]
10. Gandhi, Y.; Minak, G. A review on topology optimization strategies for additively manufactured continuous fiber-reinforced composite structures. *Appl. Sci.* **2022**, *12*, 11211. [CrossRef]
11. Liu, J.W.; Cai, M.; Guo, Y.W.; Ma, Q.H.; Sun, B.Z. The static and dynamic mechanical properties analysis of ramie/carbon fiber-reinforced composites. *Polym. Compos.* **2024**, *45*, 6575–6587. [CrossRef]
12. Zhang, D.C.; Zhan, L.H.; Guan, C.L.; Guo, J.Z.; Ma, B.L.; Dai, G.M.; Yao, S.M. Optimization of vibration pretreatment microwave curing in composite laminate molding process. *Polymers* **2023**, *15*, 296. [CrossRef] [PubMed]
13. Meng, Q.S.; Yang, Y.C.; Han, S.S.; Meng, F.Z.; Liu, T.Q. Preparation of high-performance bismuthene thermoelectric composites doped with graphene using UV-curing 3D printing technology. *Polym. Compos.* **2024**, *45*, 8176–8186. [CrossRef]
14. Kim, C.S.; Jang, J.; Im, H.G.; Yoon, S.; Kang, D.J. Preparation and performance of alumina/epoxy-siloxane composites: A comparative study on thermal- and photo-curing process. *Heliyon* **2024**, *10*, 27580. [CrossRef] [PubMed]
15. Kumar, M.; Saini, J.S.; Singh, K.; Bhunia, H. Comparison of different failure criteria in numerical modeling of electron beam cured polymer nanocomposites. *J. Compos. Mater.* **2024**, *58*, 533–545. [CrossRef]
16. Kim, H.B.; Oh, S.H.; Jeong, Y.G.; Kim, H.S.; Park, B.G.; Park, J.S. Effects of heat treatment on the microstructure and flexural mechanical properties of carbon fiber reinforced composite cured by electron beam. *Mater. Today Commun.* **2023**, *37*, 107181. [CrossRef]
17. Mandic, V.N.; Par, M.; Marovic, D.; Rakic, M.; Tarle, Z.; Sever, E.K. Blue laser for polymerization of bulk-fill composites: Influence on polymerization kinetics. *Nanomaterials* **2023**, *13*, 303. [CrossRef]
18. Jiang, Y.; Zhao, W.W.; Yu, W.; Yu, Z.Q.; Xiao, X.Y.; Zhou, W.H.; Liu, X.Q. Free-standing laser-induced graphene heaters for efficient curing and repairing of composites. *J. Mater. Sci.* **2023**, *58*, 2604–2618. [CrossRef]
19. Li, Y.G.; Cheng, L.B.; Zhou, J. Curing multidirectional carbon fiber reinforced polymer composites with indirect microwave heating. *Int. J. Adv. Manuf. Technol.* **2018**, *97*, 1137–1147. [CrossRef]
20. Rao, S.; Chiranjeevi, M.C.; Prakash, M.R. Vacuum-assisted microwave processing of glass-epoxy composite laminates using novel microwave absorbing molds. *Polym. Compos.* **2018**, *39*, 1152–1160. [CrossRef]
21. Guan, C.L.; Zhan, L.H.; Zhang, D.C.; Yao, S.M.; Zhong, S.C.; Wang, B. Microwave uniformity regulation and its influence on temperature field distribution of composite materials. *J. Cent. South Univ.* **2023**, *30*, 3374–3394. [CrossRef]

22. Hang, X.; Li, Y.G.; Hao, X.Z.; Li, N.Y.; Wen, Y.Y. Effects of temperature profiles of microwave curing processes on mechanical properties of carbon fibre-reinforced composites. *Proc. Inst. Mech. Eng. Part B J. Eng. Manuf.* **2017**, *231*, 1332–1340. [CrossRef]
23. Haider, I.; Gul, I.H.; Umer, M.A.; Baig, M.M. Silica-fiber-reinforced composites for microelectronic applications: Effects of curing routes. *Materials* **2023**, *16*, 1790. [CrossRef] [PubMed]
24. de Vergara, U.L.; Sarrionandia, M.; Gondra, K.; Aurrekoetxea, J. Impact behaviour of basalt fibre reinforced furan composites cured under microwave and thermal conditions. *Compos. Part B Eng.* **2014**, *66*, 156–161. [CrossRef]
25. Dasari, S.K.; Rangapuram, M.; Fashanu, O.; Chandrashekhara, K.; Iyyer, N.; Phan, N. Manufacturing and experimental evaluation of microwave cured carbon/epoxy composites. *Appl. Compos. Mater.* **2021**, *28*, 2087–2103. [CrossRef]
26. Chen, Y.L.; Li, Y.; You, Y.Y.; Xiao, J.; Song, Q.H. Research on mechanical properties of epoxy/glass fiber composites cured by microwave radiation. *J. Reinf. Plast. Compos.* **2014**, *13*, 1441–1451. [CrossRef]
27. Muric-Nesic, J.; Compston, P.; Noble, N.; Stachurski, Z.H. Effect of low frequency vibrations on void content in composite materials. *Compos. Part A-Appl. Sci. Manuf.* **2009**, *40*, 548–551.
28. Guan, C.L.; Zhan, L.H.; Dai, G.M.; Wu, X.T.; Xiao, Y. A unique method for curing composite materials by introducing vibration treatment into the hybrid heating process. *J. Cent. South Univ.* **2021**, *28*, 2961–2972. [CrossRef]
29. Yang, X.B.; Zhan, L.H.; Peng, Y.F.; Liu, C.; Xiong, R. Interface controlled micro- and macro-mechanical properties of vibration processed carbon fiber/epoxy composites. *Polymers* **2021**, *13*, 2764. [CrossRef]
30. Xiang, Y.X.; Shen, K.; Wu, H.; He, Z.C.; Li, X.K. Preparation of fixed length carbon fiber reinforced plastic composite sheets with isotropic mechanical properties. *New Carbon Mater.* **2022**, *36*, 1188–1194. [CrossRef]
31. Yang, H.L.; Li, J.; Fang, H.C.; Zhang, X.; Zhou, Z.C.; Tong, X.Y.; Ruan, J.M. Influence of FeCrAl fiber on microstructure and mechanical properties of FeCrAl(f)/HA composites. *Trans. Nonferrous Met. Soc. China* **2013**, *23*, 711–717. [CrossRef]
32. Guo, L.J.; Li, H.J.; Xue, H.; Li, K.Z.; Fu, Y.W. Mechanical properties of short carbon fiber reinforced pitch-based carbon/carbon composites. *New Carbon Mater.* **2006**, *21*, 36–42.
33. *JC/T 773-2010*; Fibre-Reinforced Plastics Composites. Determination of Apparent Interlaminar Shear Strength by Short-Beam Method. The Standardization Administration of the People's Republic of China: Beijing, China, 2010.
34. Zuo, P.Y.; Tcharkhtchi, A.; Fitoussi, J.; Shirinbayan, M.; Bakir, F. Multiscale thermal study of virgin and aged polyphenylene sulfide reinforced by glass fiber. *J. Appl. Polym. Sci.* **2020**, *137*, e49031.
35. Moon, C.R.; Bang, B.R.; Choi, W.J.; Kang, G.H.; Park, S.Y. A technique for determining fiber content in FRP by thermogravimetric analyzer. *Polym. Test.* **2005**, *24*, 376–380. [CrossRef]
36. Meier, R.; Kahraman, I.; Seyhan, A.T.; Zaremba, S.; Drechsler, K. Evaluating vibration assisted vacuum infusion processing of hexagonal bo-ron nitride sheet modified carbon fabric/epox composites in terms of interlaminar shear strength and void content. *Compos. Sci. Technol.* **2016**, *128*, 94–103. [CrossRef]
37. *GB/T 1043.1-2008;* Plastics—Determination of Charpy Impact Properties—Part 1: Non–Instrumented Impact Test. AQSIQ; SAC: Beijing, China, 2008.
38. Tesfay, A.G.; Kahsay, M.B.; Kumar, P.S.S. Effect of carbon and glass fillers on tensile and impact strength, water absorption, and degradation properties of sisal/polyester composites. *J. Nat. Fibers* **2023**, *20*, 2202886. [CrossRef]
39. Yang, K.; Guan, J.; Numata, K.; Wu, C.G.; Wu, S.J.; Shao, Z.Z.; Ritchie, R.O. Integrating tough antheraea pernyi silk and strong carbon fibres for impact-critical structural composites. *Nat. Commun.* **2019**, *10*, 3786. [CrossRef] [PubMed]
40. Yan, Y.H.; Shi, X.M.; Liu, J.G.; Zhao, T.; Yu, Y.Z. Thermosetting resin system based on novolak and bismaleimide for resin-transfer molding. *J. Appl. Polym. Sci.* **2002**, *83*, 1654–1657. [CrossRef]
41. Zhu, H.Y.; Li, D.H.; Zhang, D.X.; Wu, B.C.; Chen, Y.Y. Influence of voids on interlaminar shear strength of carbon/epoxy fabric laminates. *Trans. Nonferrous Met. Soc. China* **2009**, *19*, S470–S475. [CrossRef]
42. Lei, Z.; Ma, J.W.; Sun, W.; Yin, B.; Liew, K.M. Low-velocity impact and compression-after-impact behaviors of twill woven carbon fiber/glass fiber hybrid composite laminates with flame retardant epoxy resin. *Compos. Struct.* **2023**, *321*, 117253. [CrossRef]

polymers

MDPI

Article

Zn-Al Ferrite/Polypyrrole Nanocomposites: Structure and Dielectric and Magnetic Properties for Microwave Applications

Huda F. Khalil [1,*], Sherif G. Elsharkawy [2], Nouf F. AL-Harby [3,*] and Mervette El-Batouti [4]

1 Electronic Materials Department, Advanced Technology and New Material Institute (ATNMI), City of Scientific Research and Technological Applications (SRTA-City), Alexandria 21934, Egypt
2 Basic and Applied Sciences, College of Engineering and Technology, AASTMT, Alexandria 21934, Egypt; ssharkawy@aast.edu
3 Department of Chemistry, College of Science, Qassim University, Buraydah 51452, Saudi Arabia
4 Chemistry Department, Faculty of Science, Alexandria University, Alexandria 21934, Egypt; mervette_b@yahoo.com
* Correspondence: hudafarid85@gmail.com (H.F.K.); hrbien@qu.edu.sa (N.F.A.-H.)

Abstract: In this study, Zn-Al ferrite/polypyrrole (PPy) nanocomposites were synthesized and thoroughly characterized to explore their potential for microwave applications. X-ray diffraction analysis confirmed the presence of ZnO, $AlFeO_3$, and Fe_2O_3 phases, with the crystal size decreasing from 31 nm to 19.6 nm as aluminum content increased. High-resolution transmission electron microscopy (HR-TEM) revealed a distinctive core–shell morphology, where the polypyrrole encapsulates the $ZnAl_xFe_{2-x}O_4$ particles. Magnetic measurements showed that decreasing aluminum concentration led to a reduction in both saturation magnetization (Ms) from 75 emu/g to 36 emu/g and remanent magnetization (Mr) from 2.26 emu/g to 2.00 emu/g. Dielectric analysis indicated that both the real (ε') and imaginary (ε'') components of dielectric permittivity decreased with increasing frequency, particularly between 10 and 14 GHz. Furthermore, electrical modulus analysis highlighted the significant impact of aluminum doping on relaxation time (τIP), indicating the presence of interface polarization. Impedance spectroscopy results underscored the dominance of interface polarization at lower frequencies and the presence of strong conduction paths at higher frequencies. These combined magnetic and dielectric loss mechanisms suggest that the Zn-Al ferrite/polypyrrole nanocomposite is a promising candidate for advanced microwave absorption applications.

Keywords: Zn-Al ferrite; polypyrrole; nanocomposites; magnetic properties; dielectric properties; microwave absorption; impedance spectroscopy; core–shell structure

1. Introduction

Microwave technology has become increasingly important in various applications, including telecommunications, radar systems, and energy transmission. However, the efficiency of microwave energy utilization remains a critical challenge. Many researchers have made efforts to develop advanced microwave-absorbing materials that can be used in the field of microwave applications [1]. Nanocomposites have been raised among these materials, combining different elements for more refined and improved performance. Then, further investigations have been made to include polymers in addition to these nanocomposites, for example, Mn-Zn ferrites [2], Cu-Zn ferrites [3], $ZnFe_2O_4$ [4], and $Zn_{0.5}Ni_{0.4}Cr_{0.1}Fe_2O_4$ [5], all these nanocomposites are fabricated by combining the ferrite with conductive polymers like polyaniline (PANI) and polypyrrole (PPy). These nanocomposites have shown excellent microwave absorption performance, which is attributed to the effects of magnetic and dielectric losses.

Throughout this work, spinel ferrites, such as zinc–aluminum (Zn-Al) ferrite, have been given significant attention due to their excellent dielectric and magnetic properties, making them suitable for microwave absorption applications [6,7]. Polypyrrole, PPy [8], has been widely used with various ferrites to increase the magnetization saturation (Ms), and their ability to absorb electromagnetic waves makes them suitable for shielding electronic devices from unwanted interference. Accurately mixed ferrite powder with PPy produces nanocomposites characterized by highly dielectric and magnetic properties. Understanding the interplay between the spinel ferrite and conductive polymer components will contribute to the development of advanced microwave-absorbing materials, ultimately leading to improved efficiency and performance in various microwave-based technologies. The improved structure of the Zn-Al ferrite/PPy nanocomposite leads to better electromagnetic interference (EMI) shielding performance [9]. Moreover, this structure constitutes various applications, including radar stealth technology, wireless communication systems, and electromagnetic compatibility. In this work, the microwave absorption properties of the Zn-Al ferrite/PPy nanocomposite will be explored, and the potential for effective microwave absorption and shielding in diverse microwave-based applications will be discussed. This will be achieved by investigating the structural, dielectric, and magnetic properties of Zn-Al ferrite/PPy nanocomposites, with the goal of advancing the efficiency of microwave applications. The research will focus on the synthesis, characterization, and evaluation of the nanocomposites' performance, providing insights into the underlying mechanisms responsible for their enhanced microwave absorption capabilities.

2. Experimental Techniques and Procedures

$ZnAl_xFe_{2-x}O_4$/polypyrrole nanocomposite with concentrations of x (wt.%) = 0.0, 0.2, 0.4, 0.6, and 0.8 were synthesized using solid-state reaction technique [8]. The zinc oxide (ZnO), aluminum oxide (Al_2O_3), iron oxide (Fe_2O_3), and polypyrrole (PPy) powders were weighed according to the desired composition ratio (wt.%) for each sample. ZnO, Al_2O_3, and Fe_2O_3 powders were obtained from Sigma-Aldrich; also, PPy was synthesized using a chemical polymerization method. A calculated amount of PPy powder, approximately 30% of the total weight of the nanocomposite, was added to the weighed ZnO, Al_2O_3, and Fe_2O_3 powders. Then, the mixture was further ground to evenly distribute the PPy within the $ZnAl_xFe_{2-x}O_4$ matrix with ball milling. Finally, the mixture was calcinated at 850° for 7 h. After the calcination process, the resultant nanocomposite powder was removed from the furnace and allowed to cool down to room temperature. The synthesized $ZnAl_xFe_{2-x}O_4$/PPy nanocomposite was characterized using X-ray diffraction (XRD) using X'Pert Philips (Eindhoven, The Netherlands) powder diffractometer with CuKα radiation (λ = 1.54056 Å) in the range $10° \leq 2\theta \leq 80°$. Additionally, the morphology of the specimens was investigated using High-Resolution Transmission Electron Microscopy (TEM), JEOL-JEM-2100 (Tokyo, Japan) equipment was operated at 200 kV, and included a SAED (Selected Area Electron Diffraction) mode. The magnetic hysteresis loops were measured at room temperature using a vibrating sample magnetometer (VSM) operating system 7400 Series-Lake Shore Cryotronics (Westerville, OH, USA). The frequency-dependent electromagnetic properties and microwave shielding measurements were carried out on a vector network analyzer (model 8510–45 MHz to 26.56 GHz, USA) by waveguide transmission line technique. A vector network analyzer (VNA) for the microwave measurements was used. The measurements were carried out using the free space method, which is ideal for minimizing interference from external sources and ensuring accurate results. The configuration involved positioning the sample between two horn antennas in a free-space environment to measure the S-parameters (S11 and S21). The S-parameters were then used to extract the complex permittivity (ε) and permeability (μ) of the material. The extraction was performed using the Nicolson–Ross–Weir (NRW) method, a well-established technique for determining electromagnetic properties from S-parameter measurements. All instruments and methodologies used in this study were carefully selected to ensure high precision and reliability of the results.

3. Results and Discussion

3.1. XRD Analysis

The XRD peaks of $ZnAl_xFe_{2-x}O_4$/polypyrrole nanocomposites with concentrations of x (wt.%) = 0.0, 0.2, 0.4, 0.6, and 0.8 are shown in Figure 1a. The XRD of the nanocomposite with x = 0.0 (wt.%) showed strong diffraction at 2θ = 31.82°, 34.21°, 36.33°, 47.49°, 56.52°, 62.90°, 67.95°, 69.01°, and 78.32°, corresponding to 100, 002, 200, 102, 125, 103, 112, 201, and 202 planes, respectively. The nanocomposite structure of Zn-Al ferrite/PPy contains three phases: the ZnO hexagonal phase (JCPDS No: 01-079-2205), $AlFeO_3$ orthorhombic phase (JCPDS No: 01-084-2153), and Fe_2O_3 rhombohedral phase (JCPDS No. 01-084-0308). Figure 1b illustrates the crystallite size (D) nm, and Figure 1c shows the lattice microstrain (ε) and dislocation density (δ)nm^{-3} for $ZnAl_xFe_{2-x}O_4$/polypyrrole nanocomposites with concentrations of x (wt.%) = 0.0, 0.2, 0.4, 0.6, and 0.8. The corresponding results are tabulated below in Table 1 and were obtained using the following formulae [10,11]:

$$D = \frac{k\lambda}{\beta cos\theta} \quad (1)$$

$$\varepsilon = \frac{\beta \cos\theta}{4} \quad (2)$$

$$\delta = \frac{1}{D^2} \quad (3)$$

where (k) is the Scherer's constant, typically taken as 0.89; (λ) is the wavelength of the X-rays; (β) is the full width at half-maximum (FWHM) of the diffraction peak; and θ is the Bragg's angle. From Figure 1b, the crystallite size, D, was observed to decrease from 30.01 to 19.89 nm as the Al^{3+} substitution ratios x (wt.%) were increased from 0.0 to 0.8, showing a 66.27% loss in the total crystallite size. This reduction in size could be referenced to the disruption of the crystal growth process due to the differing atomic size, bonding, and diffusion characteristics of the substituted Al^{3+} ions compared to the original Fe^{3+} ions. The inhibition of the diffusion of atoms necessary for continued crystal growth is caused by the altered atomic interactions and mobility introduced by the Al^{3+} substitution. Moreover, the crystal structure destabilizes and hinders long-range crystal order due to the incorporation of smaller Al^{3+} ions. The increased lattice strain and defects introduced by the size mismatch between the substituted Al^{3+} and original Fe^{3+} ions interrupts the normal crystal nucleation and growth mechanisms. The presence of the Al^{3+} ions disrupts the regular crystal formation processes [12,13]. Doping can alter the nucleation and growth kinetics of crystallites, leading to variations in size. Conversely, the lattice microstrain and dislocation density exhibited an opposing trend, as illustrated in Figure 1c. Certainly, the opposing trend observed suggests an intriguing aspect of the material's behavior. This can be attributed to the introduction of Al^{3+} as a dopant; the incorporation of foreign atoms into the crystal lattice may cause lattice distortion or mismatches, contributing to an increase in lattice strain [14,15].

Table 1. Crystalline size, D (nm); microstrain, ε; and dislocation, δ (nm^{-3}).

x (wt.%)	D (nm)	ε	δ (nm^{-3})
0.0	30.01	0.00072	0.00111
0.2	27.13	0.00107	0.00136
0.4	23.90	0.00135	0.00175
0.6	21.02	0.00169	0.00226
0.8	19.89	0.00175	0.00253

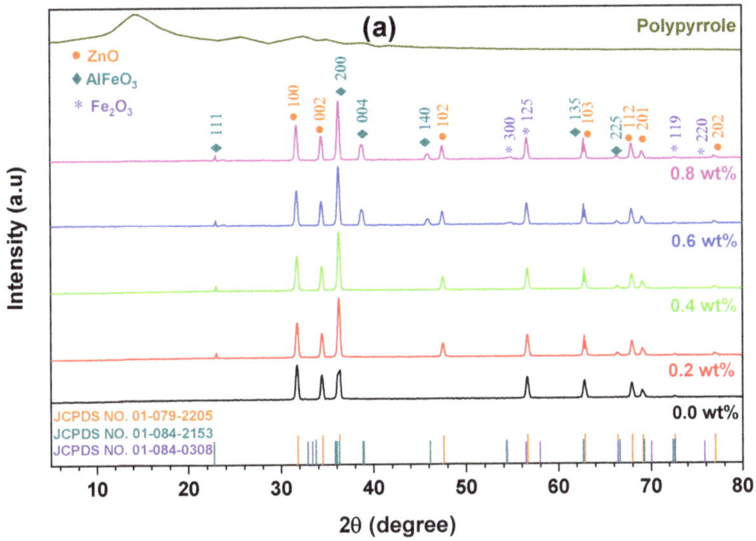

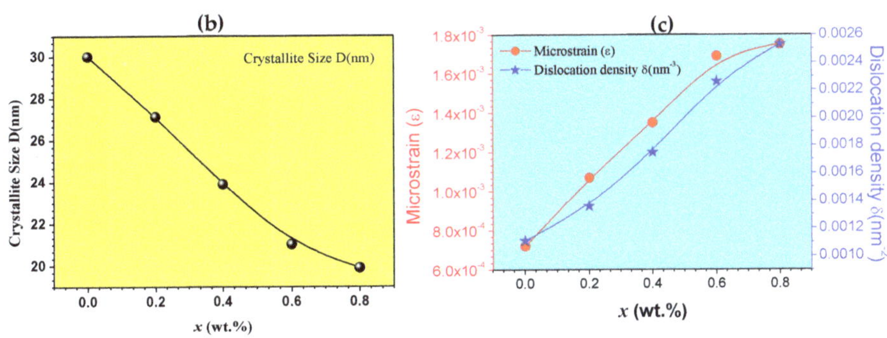

Figure 1. (**a**) XRD of $ZnAl_xFe_{2-x}O_4$/polypyrrole nanocomposite with concentrations of x (wt.%) = (0.0, 0.2, 0.4, 0.6, and 0.8). (**b**) crystallite size D (nm), (**c**) microstrain (ε), and dislocation density δ (nm^{-3}) of the nanocomposite.

3.2. Morphology

The morphology of the $ZnAl_xFe_{2-x}O_4$/polypyrrole composite powders was characterized using transmission electron microscopy (TEM), as shown in Figure 2a–e. For each compound, a TEM image (100 nm scale), an HR-TEM image (10 nm scale) (see Figure 2a′–e′), another HR-TEM image (1 nm scale) (see Figure 2a″–e″), and a SAED diagram (see Figure 2A–E) are presented. TEM patterns show a significant difference in particle shape upon the substitution of Zn for Al in the spinel framework. Figure 2a,d show that most of the $ZnAl_xFe_{2-x}O_4$ particles at concentrations of x = 0.0 and 0.6 are faceted with a hexagonal shape [16]. These nanoparticles adopted a well-defined polyhedral morphology with an alignment of 100 nm. This implies that the high calcination temperature (850 °C) not only affects the shape but also the size of the particles. However, in Figure 2e, primary particles have irregular shapes and are agglomerated, a morphology inside the composite. This indicates that the coexistence of the $ZnAl_xFe_{2-x}O_4$ spinel and the secondary Fe_2O_3 phase results in a polycrystalline sample with a large proportion of irregularly shaped particles, where HR-TEM Figure 2b″,d″ enlarged from the selected area show lattice fringes with interplanar distances of 0.307, 0.339, and 0.348 nm for $ZnAl_{0.2}Fe_{1.8}O_4$, $ZnAl_{0.4}Fe_{1.6}O_4$, and $ZnAl_{0.6}Fe_{1.4}O_4$ samples, respectively. These match well with the (111) plane of the Fd3m spinel phase. A careful examination of the HR-TEM of Zn-rich $ZnAl_{0.8}Fe_{1.2}O_4$ in Figure 2e exposes the presence of a matrix corresponding to the Fe_2O_3 and Al_2O_3 phases. The SAED patterns in Figure 2A–E provide important insights into the crystal structures; where $x = 0.0$ wt.%, the diffraction spots are indicative of a well-defined crystalline structure. The rings in the pattern correspond to specific planes of the spinel structure. In Figure 2B, for the $ZnAl_{0.2}Fe_{1.8}O_4$/polypyrrole composite, the diffraction spots are slightly less sharp than in Figure 2A, indicating a minor structural disorder. For Figure 2C–E, the SAED pattern shows more pronounced rings, suggesting an increase in structural disorder. The presence of rings alongside spots indicates that the samples contain regions of crystallinity as well as areas with less order, possibly due to the presence of secondary phases.

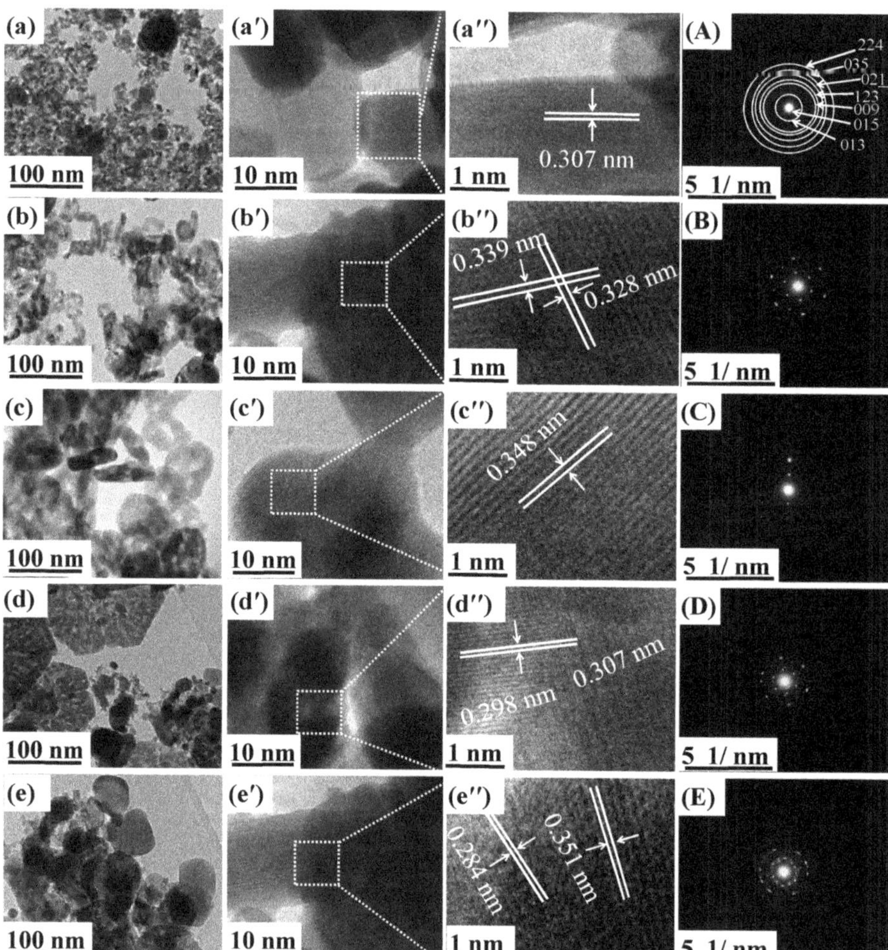

Figure 2. HR-TEM images of $ZnAl_xFe_{2-x}O_4$/polypyrrole nanocomposite with concentrations of x (wt.%) = 0.0, 0.2, 0.4, 0.6, and 0.8, (**a–e**) alignment 100 nm, (**a′–e′**) alignment 10 nm, (**a″–e″**) alignment 1 nm, and (**A–E**) SEAD.

3.3. Magnetic Properties

Figure 3 shows the magnetic hysteresis loop for the $ZnAl_xFe_{2-x}O_4$/polypyrrole nanocomposite, with key magnetic properties at ambient temperature, such as loop area, saturation magnetization (Ms), remanence magnetization (Mr), the ratio of Mr to Ms (squareness), and coercivity (Hc), presented in Table 2. Given the polycrystalline nature of the structure with a size under 10 nm, it is likely to possess a mono-domain configuration, facilitating alignment with the external magnetic field with relative ease [17]. A notably low coercivity of 124.53 KOe suggests minimal resistance to domain alignment, with the unsaturated magnetization further implying a singular domain structure. The minimum value of Mr/Ms composites superparamagnetic-like properties. Polypyrrole, within this context, acts as a barrier to domain alignment under an external magnetic field but facilitates domain rotation when the field's strength increases. The composite film's squareness value, standing at a mere 0.106, reflects its soft magnetic ferrite nature [18]. The incorporation of polypyrrole, particularly at an x value of 0.6, degrades the magnetic attributes of the composites, leading to decreased saturation and remanence magnetizations. Specifically, the Ms and Mr readings for $x = 0.6$ are markedly reduced, registering at 0.4575 emu/g and 48.75×10^{-3} emu/g, respectively, an effect attributed to the diminished A-B exchange interaction caused by polypyrrole.

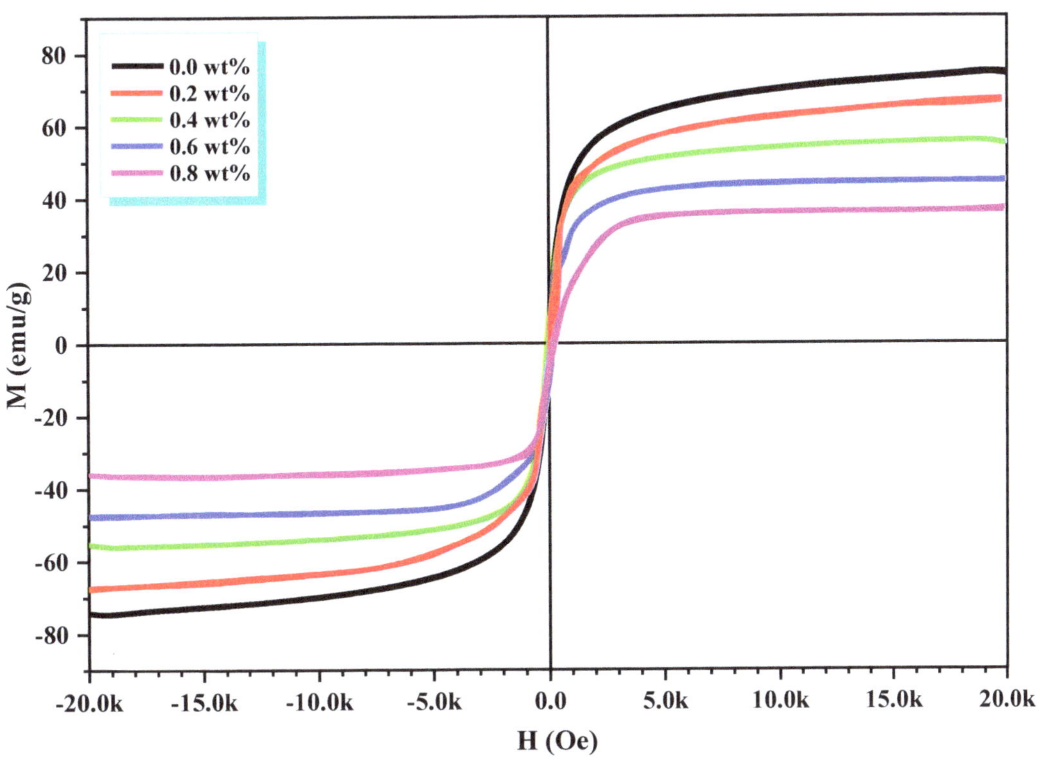

Figure 3. The hysteresis loops for $ZnAl_xFe_{2-x}O_4$/polypyrrole nanocomposite at x = 0.0, 0.2, 0.4, 0.6, and 0.8 wt.%.

Table 2. Saturation magnetization (Ms), remanent magnetization (Mr), coercivity (Hc), (Mr/Ms), and magnetic moment (μB) of $ZnAl_xFe_{2-x}O_4$/polypyrrole nanocomposite with concentrations of x (wt.%) = 0.0, 0.2, 0.4, 0.6, and 0.8.

x (wt.%)	Ms (emu/g)	Mr (emu/g)	Hc (Oe)	Mr/Ms	μB
0.0	75	2.26	38.91	0.030	10.79
0.2	67	3.87	38.52	0.057	9.17
0.4	55	2.19	38.86	0.039	7.15
0.6	45	1.48	39.17	0.032	5.54
0.8	36	2.00	39.64	0.055	4.18

4. Dielectric and Microwave Properties

The dielectric properties of the composite depend on the frequency subjected, which is closely linked to the polarizability and orientation of the present dipoles. As the frequency of the applied field increases, the dipoles of the system struggle to match the changing field faster, giving decline to the real and imaginary parts of the dielectric constant (ε') and (ε''), respectively. The value of ε' represents the strength of the dielectric polarization, which indicates the ability of the material to store electricity, while ε'' indicates the energy transfer to the dielectric material [19]. These values vary with the dielectric loss tangent ($\tan\delta = \varepsilon''/\varepsilon'$) at different frequencies. The frequency-dependent dielectric response of polypyrrole/Zn-Al ferrite nanocomposites with different Al content, as shown in Figure 4, was measured at ambient temperature in the frequency range of 10–17 GHz. The observation of a dielectric diffusion pattern shows a significant decrease in the values of ε' and ε'', especially in the range of 10–15 GHz [20]. From this point on, ε' decreases linearly up to 17 GHz, while ε'' and $\tan\delta$ remain constant, unaffected by frequency changes. The initial phase decreases ε' and ε'' are mainly due to the frequency elevation affecting the interface polarization, which is caused by charge accumulation at the interfaces between the conducting and dielectric blocks under an alternating electric field. Moreover, charge hopping plays an important role in Zn-Al ferrite. Nanostructures, with a large surface area compared to their size, exhibit numerous defects, dipoles, and unpaired bonds, enhancing the dipolar electron polarization effect and thus enhancing interface polarization. Generally, Zn-Al ferrite nanoparticles exhibit higher ε', ε'', and $\tan\delta$ values compared to polypyrrole/Zn-Al ferrite nanocomposites. $ZnAl_xFe_{2-x}O_4$/polypyrrole composites, characterized by a dense granular structure coated with polypyrrole chains on ferrite particles, exhibit variable electrical properties due to interactions between ferrite and polypyrrole to initiate polymerization, which can be degraded polymer chains and block electrical conductivity [21]. The encapsulation of ferrite nanoparticles in polypyrrole lowers the electron density, resulting in decreased conductivity. The addition of ZnAl ferrite to the polypyrrole matrix creates an additional space charge at the interface. Increasing the amount of Al increases the number of free valence electrons, which affects the conductivity properties of the composite. Figure 4c,d presents the frequency-dependent behavior of the electric modulus components, μ' and μ'', along with their representation on a Cole–Cole plot (M′ versus M″). The M′ component showcases a progressive increase up to a frequency of f = 17 GHz, with enhanced values for the sample enriched with a higher Al concentration. Notably, at lower frequencies, the M′ values are non-zero, suggesting that the diminished ε' values in these samples are not influenced by electrode polarization (EP) effects. There's a clear correlation observed between the rising M′ values and the decreasing ε' values. Space charge accumulation within electrical insulation emerges as a significant issue, as it amplifies the internal electric fields, potentially leading to partial electric discharges or dielectric breakdown, a process intricately linked to charge migration and storage [22]. The extraction of permittivity (ε) and permeability (μ) from S-parameters is indeed complex, as it depends heavily on the extraction model used. We have now explicitly described the NRW (Nicolson–Ross–Weir) method utilized for the extraction of ε and μ values from the measured S-parameters. Additionally, we have provided a brief comparison with other methods, such as the NRW (Nicolson–Ross–Weir) method [23] and the Boughriet method [24,25], to highlight the robustness and limitations of our chosen model. The sensitivity of the extracted parameters to the model selection has been discussed, acknowledging that different models could yield varying results. However, the NRW method is widely recognized for its reliability in the microwave frequency range, which justifies its use in this study.

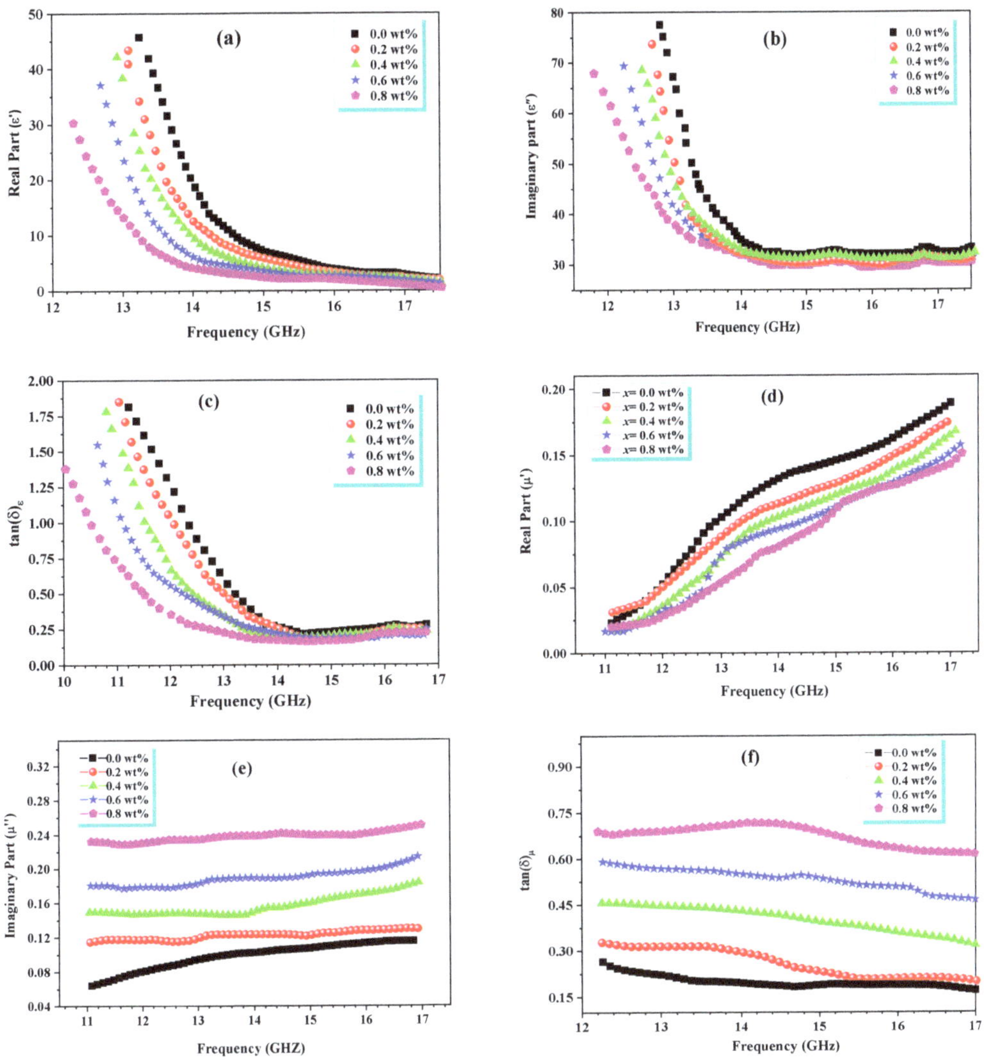

Figure 4. Frequency dependence of (**a**) real part of complex permittivity; (**b**) imaginary part of complex permittivity; (**c**) tan $(\delta)_\varepsilon$ for $ZnAl_xFe_{2-x}O_4$/polypyrrole nanocomposite; (**d**) real part of complex permeability; (**e**) imaginary part of complex permeability; (**f**) tan $(\delta)_\mu$ for $ZnAl_xFe_{2-x}O_4$/polypyrrole nanocomposite at x = (0.0, 0.2, 0.4, 0.6, 0.8) wt.%.

The M'' component delineates two distinct relaxation peaks across different frequency bands. The initial peak, found in the lower frequency domain, underscores the dominant role of interfacial polarization (IP) relaxation phenomena attributed to charge build-up at the interfaces between the polypyrrole framework and the Zn-Al ferrite particles within the core–shell configurations. The subsequent peak, observed at higher frequencies, is attributed to α-relaxation, indicative of the local dynamics of polymer chain segments in polypyrrole, particularly dipolar polarization. The intensity of IP peaks surpasses that of the α-relaxation peaks significantly.

In addition to the NRW method used in this study, other methods, such as the Boughriet method, are also commonly employed for extracting permittivity and permeability from S-parameters. The Boughriet method, while effective in certain contexts, has limitations in terms of its accuracy at higher frequencies and for materials with complex magnetic properties. In contrast, the NRW method is widely recognized for its robustness and reliability, particularly in the microwave frequency range, making it a suitable choice for our study. However, it is important to note that the choice of extraction method can significantly impact the results, and care must be taken to select a method that aligns with the specific characteristics of the material under investigation.

The choice between these methods can significantly influence the results obtained in material characterization. The NRW method's robustness makes it a preferred option for studies involving materials in the microwave frequency range, while the Boughriet method may be more applicable in situations where its specific advantages can be leveraged.

The Cole–Cole plot exhibits a semi-circular trajectory, deviating from the Debye model, followed by a linear increase. This semicircle's radius inversely relates to the composite's electrical conductivity; a larger radius suggests reduced conductivity [26]. The relaxation times (τIP) for samples with x = 0.0 wt.% and 0.2 wt.% are measured at 12,160.33 ms and 17,054.75 ms, respectively, illustrating the significant influence of Al^{3+} content on τIP values. The τIP values for Zn-Al ferrite samples are 816 ms and 2289 ms for grain boundary relaxation and 6.19 ms and 41 ms for grain relaxation, respectively. However, the τIP values for x = 0.4, 0.6, and 0.8 wt.% polypyrrole/Zn-Al ferrite nanocomposites are substantially higher, showing an increase of 7 to 15 times that of the grain boundary relaxation time, highlighting the profound effect of incorporating Al^{3+} into the matrix.

The impedance spectroscopy has been utilized to explore the electrical characteristics of Zn-Al ferrite/polypyrrole nanocomposites, focusing on the analysis of changes in both the real (Z') and imaginary (Z'') components of the impedance across different frequencies. These variations are graphically represented in a Cole–Cole plot, as illustrated in Figure 5, with measurements conducted at room temperature. To comprehend the dynamics between Z' and Z'' over the entire frequency spectrum, it is essential to consider the behaviors of ε' and ε''. Integrating ferrite nanoparticles within the polypyrrole matrix markedly reduces both Z' and Z'' in the lower frequency domain (11–13 GHz), affecting the polymer matrix's polymerization state and thus disrupting the continuity of its chains. This disruption leads to the emergence of interfacial polarization [24]. Nevertheless, as frequency increases, the significance of interfacial polarization wanes, giving way to the prominence of the charge-hopping mechanism within the Zn-Al ferrite. Between frequencies of 14 and 17 GHz, both Z' and Z'' demonstrate stability, suggesting that the conductivity of polypyrrole predominates without interference from Zn ferrite, thus establishing a stable conduction pathway. The Cole–Cole plot reveals a semicircle that characterizes the Debye relaxation phenomena, essential for understanding the attenuation of electromagnetic waves. Z' is indicative of the dissipation component, which accounts for the energy conversion from the input field to thermal energy, subsequently resulting in loss. On the other hand, Z'' represents the impedance's imaginary component, reflecting the non-dissipative segment where the system temporarily stores the energy derived from the field, mimicking an ideal capacitor scenario. The semicircle displayed on the Cole–Cole plot is indicative of various Debye relaxation processes arising due to the interfacial polarization between the Zn-Al ferrite nanoparticles and the conductive polypyrrole matrix, leading to a textbook example of Debye relaxation. The impedance, or Nyquist diagram, delineates the relationship between Z' and Z'' across varying frequencies (Figure 5), revealing a steady linear escalation across the full frequency range. This trend suggests a direct relationship between the increase in the dissipative component's ratio and the impedance, which denotes the non-dissipative component. With an increase in frequency, there is a reduction in the energy loss, implying a decrease in the energy transition from the input field to the thermal reservoir, thereby allowing more energy to be conserved within the system.

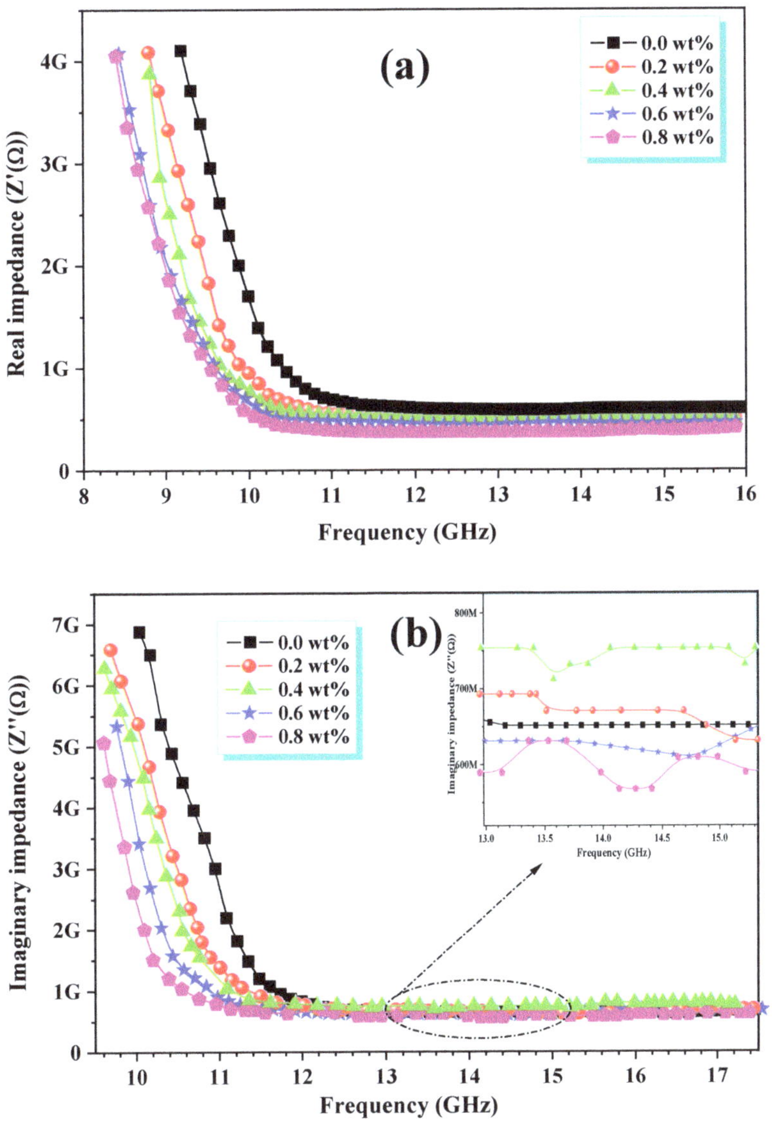

Figure 5. Frequency dependence of (**a**) the real impedance spectra and (**b**) the imaginary impedance spectra for $ZnAl_xFe_{2-x}O_4$/polypyrrole nanocomposite at x = 0.0, 0.2, 0.4, 0.6, and 0.8 wt.%.

The potential applications of Zn-Al/polypyrrole nanocomposites in advancing microwave application efficiency are promising due to their excellent microwave absorption performance. The synergistic effects of magnetic losses and dielectric losses contribute to their efficient electromagnetic wave absorption properties [27]. For instance, $ZnFe_2O_4$/polypyrrole nanocomposites exhibit broadband electromagnetic wave absorption at 8–18 GHz, with magnetic losses being the main microwave absorption mechanism [28]. Similarly, $ZnFe_2O_4/SiO_2$/polypyrrole nanocomposites show efficient electromagnetic wave absorption properties in the K and Ka band regions, with magnetic losses and dielectric losses being the primary microwave absorption mechanisms [29]. In addition, the nanocomposites of polypyrrole with Zn-Al-ferrite exhibit a core–shell structure, and the possible bonding effect between metal cations and PPy results in a decrease in conductivity [30]. Microwave measurements are essential for evaluating the electromagnetic interference (EMI) shielding effectiveness (SE) of materials. The SE refers to the ability of a shielding material to attenuate or reduce the propagation of electromagnetic waves. It quantifies the level of attenuation provided by the shielding material and is expressed using Equations (1) and (2):

$$SE(dB) = -10\log\left(\frac{P_t}{P_i}\right) \tag{4}$$

$$SE_T = SE_R + SE_A + SE_M \tag{5}$$

where P_i represents the power of the incident electromagnetic wave, and P_t represents the power of the transmitted electromagnetic wave [31–33]. The total effectiveness of the electromagnetic shielding is being measured, by comparing the power of the electromagnetic waves before they encounter the shielding (the incident power, P_i) versus the power of the waves that are able to transmit through the shielding (the transmitted power, P_t). For a shielding material, the total shielding effectiveness is the sum of the contribution due to reflection (SE_R), absorption (SE_A), and multiple reflections (SE_M). From the scattering parameters S_{11} and S_{21} of a vector network analyzer (measured by waveguide transmission line technique), the reflection coefficient (R) and transmission coefficient (T) can be obtained as $R = |S_{11}|^2$ and $T = |S_{21}|^2$. Using the parameters R and T, the absorption coefficient (A) can be evaluated as $A + R + T = 1$ [34,35]. With negligible multiple reflections between both interfaces of the material, the relative intensity of the effectively incident EM wave is based on the factor $(1 - R)$, and the effective absorption can be expressed as $A_{eff} = (1 - R - T)/(1 - R)$. The reflection and effective absorption can be expressed in decibels as SE_R and SE_A [36–41]:

$$SE_R = 10\log(1 - R) \tag{6}$$

$$SE_A = 10\log\left(1 - A_{eff}\right) = 10\log\left(\frac{T}{1 - R}\right) \tag{7}$$

When negligible multiple reflections occur between the interfaces of the material, the relative intensity of the effectively incident electromagnetic wave can be estimated using the factor $(1 - R)$. This factor determines the effective absorption of the material (Figure 6).

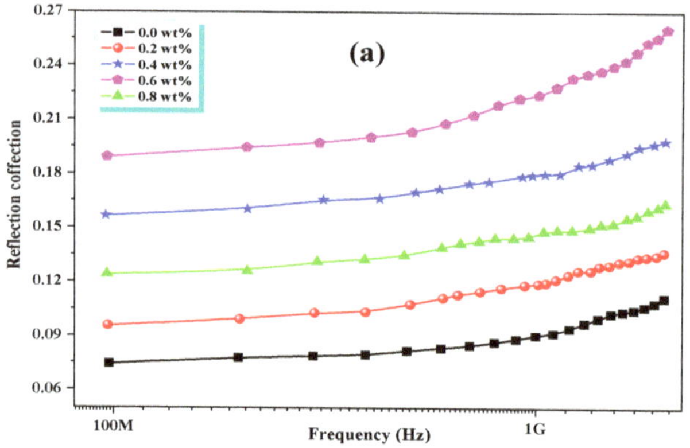

Figure 6. *Cont.*

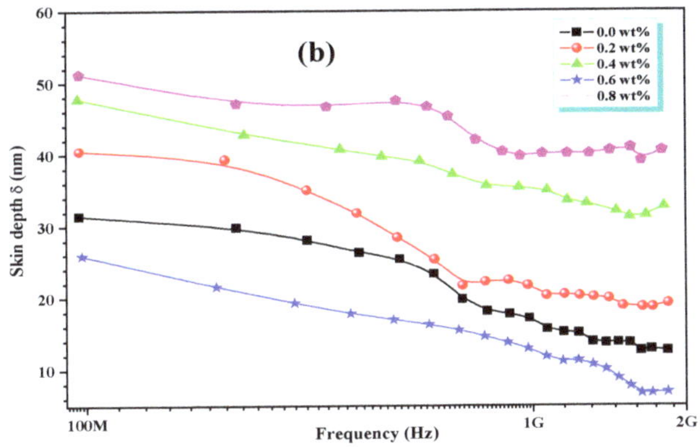

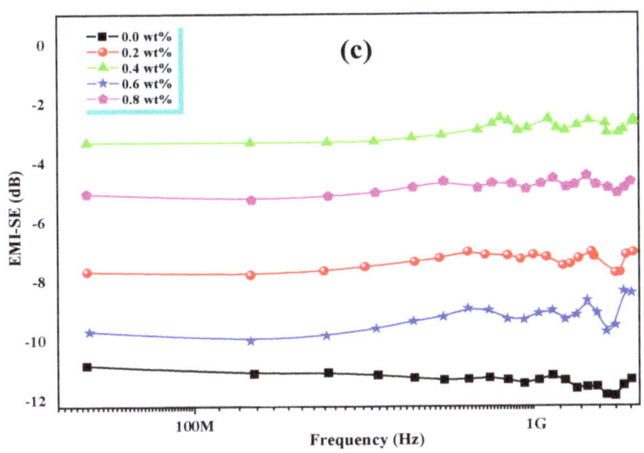

Figure 6. (**a**) Reflection coefficient, (**b**) skin depth, and (**c**) EMI-SE as a function of frequency for $ZnAl_xFe_{2-x}O_4$/polypyrrole nanocomposite at x = 0.0, 0.2, 0.4, 0.6, and 0.8 wt.%.

5. Conclusions

This study presents the successful synthesis and comprehensive characterization of $ZnAl_xFe_{2-x}O_4$//polypyrrole nanocomposites with varying Al content (x = 0.0, 0.2, 0.4, 0.6, 0.8 wt.%). The addition of Al led to significant changes in both structural and magnetic properties, notably reducing the crystal size from 31 nm to 19.6 nm and decreasing the saturation magnetization (Ms) from 75 emu/g to 36 emu/g. The core–shell morphology observed in TEM analysis significantly influenced the dielectric behavior, with both real (ε') and imaginary (ε'') dielectric constants decreasing as the frequency increased.

The dielectric properties, coupled with the results from impedance spectroscopy, underscore the role of interface polarization, with the relaxation time being notably affected by the Al content. The stability of both Z' and Z'' within the 14–17 GHz frequency range suggests that the conductivity of polypyrrole predominates without interference from the Zn ferrite, making these nanocomposites particularly promising for microwave absorption applications. Furthermore, the electromagnetic parameters, ε and μ, extracted using the Nicolson–Ross–Weir method, confirmed the material's potential for advanced microwave technologies. The combination of enhanced magnetic–dielectric characteristics points to Zn-Al ferrite/polypyrrole nanocomposites as viable candidates for applications in telecommunications, electromagnetic interference shielding, and other microwave-related technologies.

The Zn-Al ferrite/polypyrrole nanocomposite presents a promising material with enhanced magnetic and dielectric properties compared to previously studied materials.

Its unique combination of properties makes it suitable for a wide range of applications, including magnetic data storage, microwave devices, and energy storage devices.

Author Contributions: Conceptualization, methodology, and validation, H.F.K.; resources, M.E.-B.; writing—original draft preparation, S.G.E. and H.F.K.; writing—review and editing, H.F.K. and S.G.E.; visualization, S.G.E.; supervision, M.E.-B.; investigation, resources, funding acquisition, N.F.A.-H. All authors have read and agreed to the published version of the manuscript.

Funding: This research received no external funding.

Institutional Review Board Statement: We choose to exclude this statement because the study did not require ethical approval.

Data Availability Statement: The authors confirm that the data supporting the findings of this study are available within the article.

Acknowledgments: The researchers would like to thank the Deanship of Graduate Studies and Scientific Research at Qassim University for financial support (QU-APC-2024-9/1).

Conflicts of Interest: The authors declare no conflict of interest.

References

1. Elmahaishi, M.F.; Azis, R.S.; Ismail, I.; Muhammad, F.D. A review on electromagnetic microwave absorption properties: Their materials and performance. *J. Mater. Res. Technol.* **2022**, *20*, 2188–2220. [CrossRef]
2. Klimov, A.S.; Bakeev, I.Y.; Dolgova, A.V.; Kazakov, A.V.; Korablev, N.S.; Zenin, A.A. Features of Electron Beam Processing of Mn-Zn Ferrites in the Fore-Vacuum Pressure Range in Continuous and Pulse Modes. *Coatings* **2023**, *13*, 1766. [CrossRef]
3. Vissurkhanova, Y.A.; Ivanova, N.M.; Soboleva, Y.A.; Muldakhmetov, Z.M.; Abulyaissova, L.K.; Minaev, B.F. Fe-Cu composites preparation using Cu-Zn ferrite and their electrocatalytic application. *Mater. Lett.* **2023**, *333*, 133521. [CrossRef]
4. Joulaei, M.; Hedayati, K.; Ghanbari, D. Investigation of magnetic, mechanical and flame retardant properties of polymeric nanocomposites: Green synthesis of $MgFe_2O_4$ by lime and orange extracts. *Compos. Part B Eng.* **2019**, *176*, 107345. [CrossRef]
5. Hajlaoui, M.E.; Dhahri, R.; Hnainia, N.; Benchaabane, A.; Dhahri, E.; Khirouni, K. Conductivity and giant permittivity study of $Zn0.5Ni0.5Fe_2O_4$ spinel ferrite as a function of frequency and temperature. *RSC Adv.* **2019**, *9*, 32395–32402. [CrossRef]
6. Ahmad, M.; Amin, K.; Rehman, A.U.; Wahab, A.; Wabaidur, S.M. Detailed investigation of Mn-substituted Zn ferrites for microwave applications up to 6 GHz. *Mater. Sci. Technol.* **2024**, *40*, 479–492. [CrossRef]
7. Ajeel, K.I.; Kareem, Q.S. Synthesis and Characteristics of Polyaniline (PANI) Filled by Graphene (PANI/GR) nano-Films. *J. Phys. Conf. Ser.* **2019**, *1234*, 012020. [CrossRef]
8. Rybicki, T.; Stempien, Z.; Karbownik, I. EMI Shielding and Absorption of Electroconductive Textiles with PANI and PPy Conductive Polymers and Numerical Model Approach. *Energies* **2021**, *14*, 7746. [CrossRef]
9. Mamatha, G.M.; Pradipkumar Dixit RHari Krishna Girish Kumar, S. Polymer based composites for electromagnetic interference (EMI) shielding: The role of magnetic fillers in effective attenuation of microwaves, a review. *Hybrid Adv.* **2024**, *6*, 100200.
10. Khalil, H.F.; Malidarreh, R.B.; Alabsy, M.T.; Hassan, A.M.; El-Khatib, A.M.; Issa, S.A.; Zakaly, H.M. Structural, Morphological, and γ-ray Attenuation properties of m-type hexaferrite $BaFe_{12}O_{19}$ doped with V_2O_5, Ce_2O_3 and Bi_2O_3 for Radiation Shielding applications. *Ceram. Int.* **2024**, *50*, 33771–33780. [CrossRef]
11. Devi, R.; Patra, J.; Tapadia, K.; Chang, J.-K.; Maharana, T. Arrangement of $ZnFe_2O_4$@PPy nanoparticles on carbon cloth for highly efficient symmetric supercapacitor. *J. Taiwan Inst. Chem. Eng.* **2022**, *138*, 104474. [CrossRef]
12. Elahi, A.; Shakoor, A.; Irfan, M.; Mahmood, K.; Niaz, N.A.; Awan, M.S.; Bashir, T. Polypyrrole and its nanocomposites with $Zn0.5Ni0.4Cr0.1Fe_2O_4$ ferrite: Preparation and electromagnetic properties. *J. Mater. Sci. Mater. Electron.* **2016**, *27*, 6964–6973. [CrossRef]
13. Khan, M.Z.; Gul, I.H.; Baig, M.M.; Khan, A.N. Comprehensive study on structural, electrical, magnetic and photocatalytic degradation properties of Al^{3+} ions substituted nickel ferrites nanoparticles. *J. Alloys Compd.* **2020**, *848*, 155795. [CrossRef]
14. Carrara, S. Towards New Efficient Nanostructured Hybrid Materials for ECL Applications. Ph.D. Thesis, Université de Strasbourg, Strasbourg, France, 2017.
15. Khafagy, R.M. Synthesis, characterization, magnetic and electrical properties of the novel conductive and magnetic Polyaniline/$MgFe_2O_4$ nanocomposite having the core–shell structure. *J. Alloys Compd.* **2021**, *509*, 9849–9857. [CrossRef]
16. Ben Farhat, L.; Ben Ahmed, S.; Ezzine, S.; Amami, M. Particle size dependent structural, magnetic and electrical properties of Cr-doped lead-free multiferroic $AlFeO_3$ prepared by co-precipitation and solid state method. *Mater. Chem. Phys.* **2020**, *255*, 123631. [CrossRef]
17. Srinivasulu, T.; Saritha, K.; Reddy, K.R. Synthesis and characterization of Fe-doped ZnO thin films deposited by chemical spray pyrolysis. *Mod. Electron. Mater.* **2017**, *3*, 76–85. [CrossRef]
18. Nigam, A.; Pawar, S. Structural, magnetic, and antimicrobial properties of zinc doped magnesium ferrite for drug delivery applications. *Ceram. Int.* **2020**, *46*, 4058–4064. [CrossRef]

19. Soin, N. *Magnetic Nanoparticles—Piezoelectric Polymer Nanocomposites for Energy Harvesting, Magnetic Nanostructured Materials*; Elsevier: Amsterdam, The Netherlands, 2018; pp. 295–322.
20. Wan, C.; Bowen, C.R. Multiscale-structuring of polyvinylidene fluoride for energy harvesting: The impact of molecular-, micro- and macro-structure. *J. Mater. Chem. A* **2017**, *5*, 3091–3128. [CrossRef]
21. Wu, Y.; Hsu, S.L.; Honeker, C.; Bravet, D.J.; Williams, D.S. The Role of Surface Charge of Nucleation Agents on the Crystallization Behavior of Poly(vinylidene fluoride). *J. Phys. Chem. B* **2012**, *116*, 7379–7388. [CrossRef]
22. Ma, Y.; Tong, W.; Wang, W.; An, Q.; Zhang, Y. Montmorillonite/PVDF-HFP-based energy conversion and storage films with enhanced piezoelectric and dielectric properties. *Compos. Sci. Technol.* **2018**, *168*, 397–403. [CrossRef]
23. Zhang, S.; Tong, W.; Wang, J.; Wang, W.; Wang, Z.; Zhang, Y. Modified sepiolite/PVDF-HFP composite film with enhanced piezoelectric and dielectric properties. *J. Appl. Polym. Sci.* **2019**, *137*, 48412. [CrossRef]
24. Kusuma, D.Y.; Nguyen, C.A.; Lee, P.S. Enhanced ferroelectric switching characteristics of P(VDF-TrFE) for organic memory devices. *J. Phys. Chem. B* **2010**, *114*, 13289–13293. [CrossRef] [PubMed]
25. Lei, D.; Hu, N.; Wu, L.; Alamusi; Ning, H.; Wang, Y.; Jin, Z.; Liu, Y. Improvement of the piezoe-lectricity of PVDF-HFP by $CoFe_2O_4$ nanoparticles. *Nano Mater. Sci.* **2023**, *6*, 201–210. [CrossRef]
26. Durgaprasad, P.; Hemalatha, J. Magnetoelectric investigations on poly (vinylidene fluoride)/$CoFe_2O_4$ flexible electrospun membranes. *J. Magn. Magn. Mater.* **2018**, *448*, 94–99. [CrossRef]
27. Rothwell, E.J.; Frasch, J.L.; Ellison, S.M.; Chahal, P.; Ouedraogo, R.O. Analysis of the Nicol-son-Ross-Weir Method for Characterizing the Electromagnetic Properties of Engineered Materials. *Prog. Electromagn. Res.* **2016**, *157*, 31–47. [CrossRef]
28. Costa, F.; Borgese, M.; Degiorgi, M.; Monorchio, A. Electromagnetic Characterisation of Materials by Using Transmission/Reflection (T/R) Devices. *Electronics* **2017**, *6*, 95. [CrossRef]
29. Barroso, J.J.; de Paula, A.L. Retrieval of Permittivity and Permeability of Homogeneous Materials from Scattering Parameters. *J. Electromagn. Waves Appl.* **2010**, *24*, 1563–1574. [CrossRef]
30. Indrakanti, R.; Brahmaji Rao, V.; Udaya Kiran, C. Optical parameters of gallium nitride doped ferrite–polypyrrole nanocomposites. *J. Mater. Sci. Mater. Electron.* **2020**, *31*, 3238–3244. [CrossRef]
31. Martins, P.; Costa, C.M.; Lanceros-Mendez, S. Nucleation of electroactive β-phase poly(vinilidene fluoride) with $CoFe_2O_4$ and $NiFe_2O_4$ nanofillers: A new method for the preparation of multiferroic nanocomposites. *Appl. Phys. A* **2010**, *103*, 233–237. [CrossRef]
32. Manohar, A.; Geleta, D.D.; Krishnamoorthi, C.; Lee, J. Synthesis, characterization and magnetic hyperthermia properties of nearly monodisperse $CoFe_2O_4$ nanoparticles. *Ceram. Int.* **2020**, *46*, 28035–28041. [CrossRef]
33. Kolhar, P.; Sannakki, B. Investigation of the electrical properties of synthesised polyaniline-magnesium ferrite compo-sites. *Mater. Today Proc.* **2023**, *422*, 2214–7853. [CrossRef]
34. Ge, Y.; Li, C.; Waterhouse, G.I.N.; Zhang, Z.; Yu, L. $ZnFe_2O_4$@SiO_2@Polypyrrole nanocompo-sites with efficient electromagnetic wave absorption properties in the K and Ka band regions. *Ceram. Int.* **2021**, *47*, 1728–1739. [CrossRef]
35. Li, F.; Zhuang, L.; Zhan, W.; Zhou, M.; Sui, G.; Zho, A.; Bai, G.; Xia, W.; Yang, X. Desirable micro-wave absorption performance of $ZnFe_2O_4$@ZnO@rGO nanocomposites based on controllable permittivity and permeability. *Ceram. Int.* **2020**, *46*, 21744–21751. [CrossRef]
36. Dong, S.; Zhang, X.; Li, X.; Chen, J.; Hu, P.; Han, J. SiC whiskers-reduced graphene oxide composites decorated with MnO nanoparticles for tunable microwave absorption. *Chem. Eng. J.* **2020**, *392*, 123817. [CrossRef]
37. Liu, P.-J.; Yao, Z.-J.; Ng, V.M.H.; Zhou, J.-T.; Yang, Z.-H.; Kong, L.-B. Enhanced Microwave Absorption Properties of Double-Layer Absorbers Based on Spherical NiO and Co0.2Ni0.4Zn0.4Fe_2O_4 Ferrite Composites. *Acta Met. Sin.* **2018**, *31*, 171–179. [CrossRef]
38. Choi, M.; Lee, S.; Kim, J. Clustering effect on the frequency-dependent magnetic properties of Fe–Co micro hollow fiber composites. *IEEE Trans. Magn.* **2017**, *53*, 1–5. [CrossRef]
39. Wang, J.-H.; Wu, R.-Z.; Feng, J.; Zhang, J.-H.; Hou, L.-G.; Liu, M.-D. Recent advances of electromagnetic interference shielding Mg matrix materials and their processings: A review. *Trans. Nonferrous Met. Soc. China* **2022**, *32*, 1385–1404. [CrossRef]
40. Wang, J.; Li, Y.; Wu, R.; Xu, L.; Zhang, Z.; Feng, J.; Zhang, J.; Hou, L.; Jiao, Y. X-band shielding properties of Mg−9Li matrix composite containing Ni0.4Zn0.4Co0.2Fe_2O_4 fabricated by multi-layer composite rolling. *J. Alloys Compd.* **2020**, *843*, 156053. [CrossRef]
41. Kruželák, J.; Kvasničáková, A.; Hložeková, K.; Hudec, I. Progress in polymers and polymer composites used as efficient materials for EMI shielding. *Nanoscale Adv.* **2021**, *3*, 123–172. [CrossRef]

polymers

MDPI

Communication

Properties of EPDM Nanocomposites Reinforced with Modified Montmorillonite

Zhanxu Li [1], Zilong Chen [1], Weichong Sun [1], Yangling Liu [1], Xiong Wang [2,*], Jun Lin [1], Jian Wang [1] and Shaojian He [1,*]

[1] State Key Laboratory of Alternate Electrical Power System with Renewable Energy Sources, North China Electric Power University, Beijing 102206, China; zhanxuli0614@163.com (Z.L.); zilongchen86@163.com (Z.C.); 13811157813@163.com (W.S.); liuyangling010223@163.com (Y.L.); jun.lin@ncepu.edu.cn (J.L.); wangjian31791@ncepu.edu.cn (J.W.)

[2] South China Institute of Environmental Sciences, Ministry of Ecology and Environment, Guangzhou 510655, China

* Correspondence: wangxiong@scies.org (X.W.); heshaojian@ncepu.edu.cn (S.H.)

Abstract: Considering the dilemma of obtaining economic and high-performance composites based on non-polar and main-chain-saturated ethylene propylene diene monomer rubber (EPDM), we proposed an effective and universal filler modification and nanocomposite preparation method. Specifically, the montmorillonite (MMT) surface was coated with polydopamine (PDA) to obtain DMMT, which was confirmed by XRD, XPS, FTIR, and TGA. After compounding DMMT gel with the solid EPDM via the gel compounding method, a silane coupling agent, vinyltrimethoxysilane, was introduced to construct covalent interactions between rubber and filler. Compared with the unmodified MMT filler EPDM, the EPDM/DMMT nanocomposite showed much fewer filler aggregates in the matrix. The highest tensile strength of the composites reached 6.5 MPa with 10 phr DMMT, almost 200% higher than that of pure EPDM.

Keywords: ethylene propylene diene monomer; montmorillonite; polydopamine; vinyltrimethoxysilane; gel compounding method

1. Introduction

Ethylene propylene diene monomer rubber (EPDM) is widely used in automobiles, seals, and other applications due to its excellent heat and weather resistance [1–3]. Since vulcanized EPDM made without reinforcement agents has a low mechanical strength, it cannot be applied in the engineering field unless effective reinforcing treatment is carried out [4]. Nonetheless, it is difficult to construct a strong interface interaction in non-polar EPDM composites by directly incorporating reinforcing agents. The lack of polar functional groups makes it difficult to uniformly disperse reinforcing agents in the EPDM matrix, resulting in an unsatisfactory reinforcement effect. Consequently, the organic modification of filler is a common strategy to overcome these deficiencies. For example, coupling agents are often used to modify silica [5], and grafting strategies are often employed to modify carbon black [6]. Nevertheless, those fillers are relatively expensive, and at the same time, the modification processes are often complex, which limits their application in engineering.

As a result of its low price, high aspect ratio, and good mechanical properties, montmorillonite (MMT) is expected to replace traditional reinforcing agents, such as carbon black and silica, in elastomers [7–10]. To obtain high-performance polymer/MMT nanocomposites, good dispersion of the MMT layer and suitable polymer–MMT interfacial interactions are essential [11–14]. MMT, however, is easily aggregated in polymer matrixes due to the poor compatibility between hydrophilic montmorillonite and hydrophobic matrixes [15–17]. In particular, for EPDM/MMT composites, it is more difficult to achieve the uniform dispersion of MMT in EPDM due to the lack of polar groups [18,19]. Because the EPDM main chain is highly saturated, it is difficult to design the interface between EPDM and montmorillonite, especially the covalent interface. In the above cases, the method of modifying the filler with a compatibilizer is generally used to change the surface adhesion. For example, multifunctional modifiers [20] and grafted polymer brushes [21] can be used.

Polydopamine (PDA), a substance that forms from dopamine in aqueous solutions by self-polymerization, adheres closely to various substrates and is widely used to modify surfaces [22–25]. Furthermore, hydroxyl groups used in PDA coating offer the potential to introduce multiple functional groups through reactions, allowing further surface modification [26,27]. PDA-coated MMT (DMMT) layers are easily exfoliated in water, and so directly blending rubber latex with a DMMT suspension to prepare composites is a feasible way to reduce DMMT aggregation and functional group embedding [28,29]. It is of practical significance to find a method with which to prepare EPDM/DMMT composites with good dispersion of DMMT; since EPDM lacks a latex form, this will also provide a new path for the preparation of solid rubber/lamellate filler composites.

In previous studies [30–32], we developed an effective method, called the gel compounding method, that can obtain good filler dispersion and high-performance nanocomposites. Herein, we proposed an effective and universal method with which to prepare high-performance EPDM/MMT nanocomposites using DMMT. An EPDM/DMMT nanocomposite was obtained by the gel compounding method. Specifically, the general experimental process is as follows: EPDM was compounded with vinyltrimethoxysilane (VTMS) and DMMT gel successively using a two-roll mill. The mixture was vacuum-dried and then we followed the conventional rubber processing procedure to prepare EPDM/DMMT nanocomposite. Finally, an EPDM/DMMT nanocomposite with superior DMMT dispersion and high interface strength was obtained. This provides a feasible path for the preparation of high-performance non-polar rubber/sheet filler composites.

2. Experimental Section

2.1. Materials

Sodium montmorillonite (MMT) with a cation exchange capacity of 93 mequiv/100 g was supplied by Liufangzi Clay Factory, Jilin, China. EPDM 4045 (ENB type) was supplied by Jilin Chemical Industry Co., Ltd., Jilin, China. Dopamine hydrochloride (Dopa, 98%) was purchased from Innochem and tris(hydroxymethyl)aminomethane (Tris, 99.8%) was purchased from Alfar Aser. Other compounding ingredients obtained, including vinyltrimethoxysilane (VTMS), zinc oxide (ZnO), magnesium oxide (MgO), N-isopropyl-N′-phenyl-p-phenylenediamine (4010NA), triallyl isocyanurate (TAIC), and dicumyl peroxide (DCP), were commercial-grade products. All materials were used as received.

2.2. Preparation of DMMT Gel

Purified MMT was obtained according to the procedure outlined in our previous work [33]. Then, a given amount of purified MMT was stirred in deionized water under vigorous stirring conditions for 2 h to obtain a suspension with a concentration of 2 wt%. Dopamine hydrochloride (1.5 g/L) and Tris (1.2 g/L, with the pH adjusted to 8.5) were added into the suspension, followed by continuous agitation for 4 h at ambient temperature. The product was centrifuged at 10,000 rpm for 15 min and subsequently washed with deionized water at least five times. After centrifugation, a product named DMMT gel with a concentration of about 10 wt% was obtained, which was directly blended with EPDM. Purified MMT was swelled in deionized water at a concentration of 10 wt% by vigorous stirring for 2 h to obtain DMMT gel.

2.3. Preparation of EPDM/DMMT Nanocomposites

Then, a given amount of DMMT gel was mixed with EPDM in a two-roll mill to obtain an EPDM/DMMT gel composite. To remove the water, the gel compound was rolled into a thin sheet and subsequently vacuum-dried at 80 °C for 48 h to obtain an EPDM/DMMT nanocomposite. Then, an EPDM/DMMT nanocomposite was mixed with the curing agents. Finally, an EPDM/DMMT nanocomposite was vulcanized at 160 °C with optimal vulcanization time to obtain an EPDM/DMMT nanocomposite. For comparison, EPDM/MMT nanocomposite was prepared by replacing DMMT gel with 10 wt% MMT gel according to the above process. The recipes (parts by weight) for EPDM composites are shown in Table 1. And the interfacial co-crosslinking mechanism of EPDM/DMMT-10 and mechanism for the polymerization of dopamine are illustrated in Scheme 1.

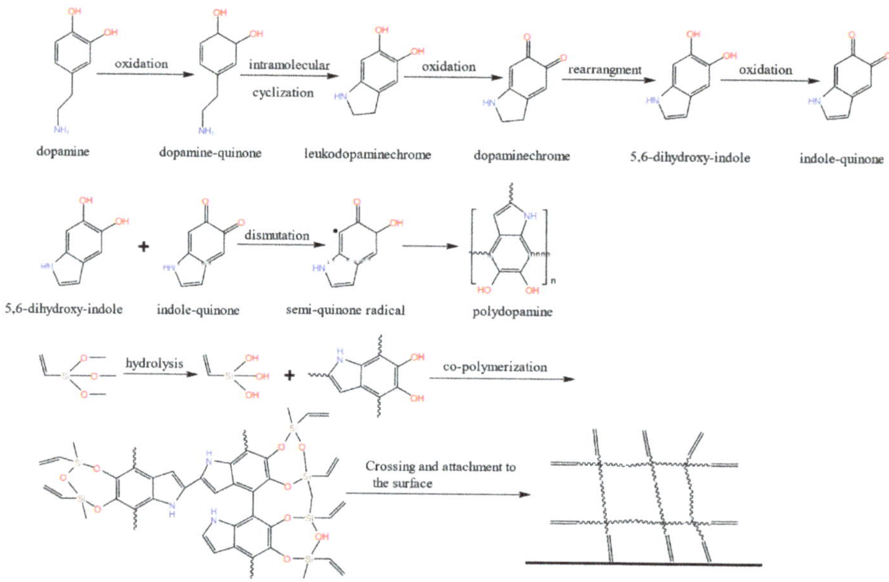

Scheme 1. Schematic illustration of mechanism for polymerization of dopamine and interfacial co-crosslinking mechanism of EPDM/DMMT-10.

Table 1. Formulations of EPDM composites.

Ingredient	Pure EPDM	EPDM/MMT-10	EPDM/DMMT-10
EPDM	100	100	100
VTMS	0	0	2
MMT	0	10	0
DMMT	0	0	10
ZnO	8	8	8
MgO	4	4	4
TAIC	2	2	2
DCP	4.5	4.5	4.5
4010NA	2	2	2

2.4. Characterization and Measurements

Fourier transform infrared spectra (FTIR) of MMT, DMMT and PDA were recorded using a Thermo Scientific Nicolet iS20 spectrometer (Thermo Fisher Scientific Inc., Waltham, MA, USA) with a resolution of 4 cm^{-1}. X-ray photoelectron spectroscopy (XPS) measurements were performed on Thermo Scientific K-Alpha XPS spectrometer (Thermo Fisher Scientific Inc., Waltham, MA, USA) with an Al Kα X-ray source. All the binding energies (B.E.) obtained from the XPS analysis were corrected using the C 1s at 284.6 eV. Thermogravimetric analysis (TG) was conducted using a TGA 550 analyzer (TA Instruments, New Castle, DE, USA) under a nitrogen atmosphere from 45 to 800 °C with a heating rate of 10 °C min^{-1}. X-ray diffraction (XRD) experiments involving EPDM nanocomposites were carried out on a diffractometer (D/Max-III C, Rigaku, Tokyo, Japan) with Cu Kα radiation. The morphological structure of the EPDM composites' tensile-fractured surfaces was observed using scanning electron microscopy (SEM) (FEI Quanta200F, Hitachi Co., Tokyo, Japan) at an accelerating voltage of 5 kV. The tensile properties were assessed by means of a universal test instrument (GT-TC2000, Gotech Testing Machines Inc., Taiwan, China) according to ISO 37-2011 [34]. Dynamic mechanical analysis (DMA) of EPDM composites was performed on a TA Q800 instrument (TA Co., Ltd., Newcastle, DE, USA) under a nitrogen atmosphere. The samples were assessed in tensile mode at 1 Hz and heated at a rate of 2 °C min^{-1}.

3. Results and Discussion

3.1. DMMT Characterization

As illustrated in Figure 1a, the C 1s core-level spectrum of DMMT has three functional peaks, corresponding to C-C species at 284.6 eV, C-N species at 285.8 eV, and C=O species at 287.4 eV. For comparison, the C 1s spectrum of MMT (Figure 1c) only has one peak, corresponding to the C-C species. The N 1s core-level spectrum of DMMT (Figure 1b) can be curved-fitted with two peak components, with B.E. at 401.8 eV for the R–NH2 species and 399.9 eV for the R_1-NH-R_2 species [35]. However, there are no characteristic peaks on the N 1s spectrum of MMT (Figure 1d). The functional groups containing C and N indicate that MMT was successfully modified by PDA. Furthermore, FTIR was used to probe the successful modification of PDA on MMT. As shown in Figure 1e, DMMT exhibited two bands around 3415 cm^{-1} and 1508 cm^{-1}, attributable to catechol O-H and N-H from PDA, respectively [36]. The characteristic bands of DMMT further confirm that PDA was coated on MMT, which is consistent with the XPS results. TGA curves (Figure 1f) show that DMMT has one more weightlessness process than MMT, which is caused by the decomposition of PDA. The content of PDA coating on the surface of MMT was calculated to be ~7.0 wt%. The XRD curves of MMT and DMMT are compared in Figure 2. Diffraction peaks appeared at both 2.6° and 7.2° in DMMT, indicating increased layer spacing, which occurs due to the existence of PDA macromolecules between DMMT layers [37].

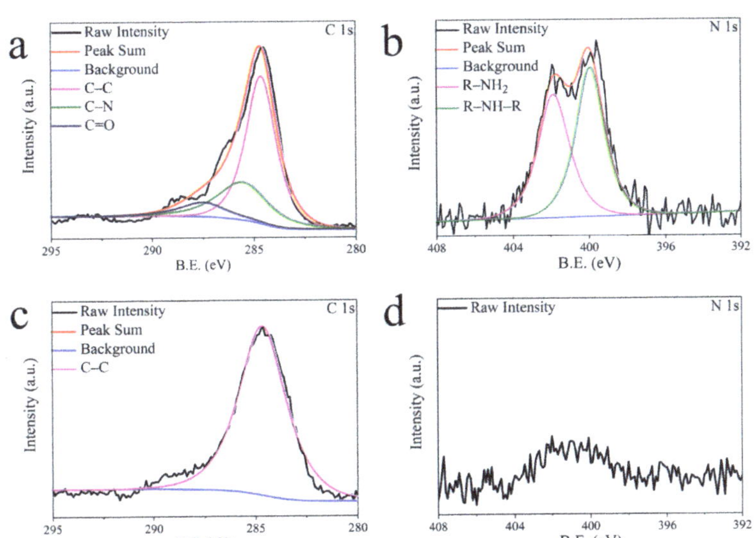

Figure 1. *Cont.*

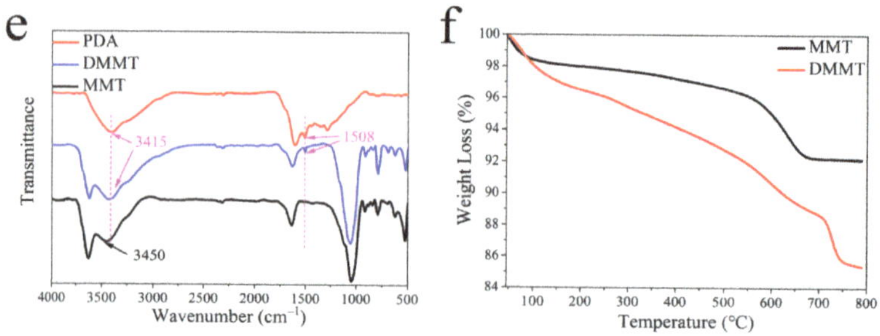

Figure 1. XPS spectra of (**a**,**b**) DMMT and (**c**,**d**) MMT; FTIR spectra (**e**) of PDA, DMMT and MMT; TGA curves (**f**) of MMT and DMMT.

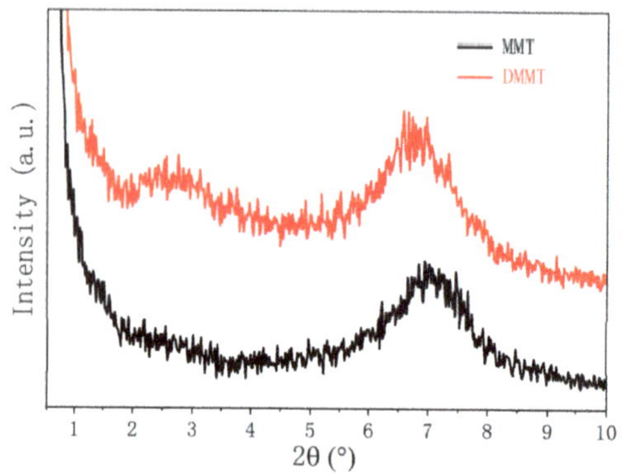

Figure 2. XRD curves of MMT and DMMT.

3.2. Structure and Morphology of EPDM Composites

Figure 3a shows TGA curves. In the figure, we can see that EPDM/MMT-10 and EPDM/DMMT-10 exhibited similar pyrolysis behaviors. The weight loss percentage of EPDM/MMT-10 and EPDM/DMMT-10 was lower than that of pure EPDM rubber. The results showed that MMT and DMMT were well dispersed in rubber and had improved thermal stability [38]. Figure 3b shows DSC curves. As can be seen from the curves, the glass transition temperature of EPDM/DMMT and EPDM/MMT is basically the same as that of EPDM [23].

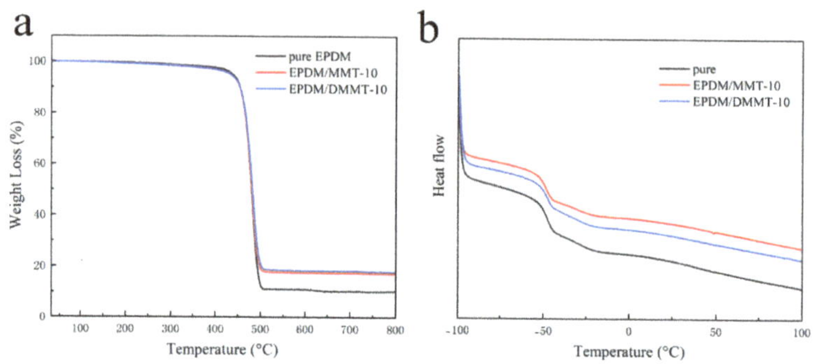

Figure 3. TGA curves (**a**) of pure EPDM, EPDM/MMT-10, and EPDM/DMMT-10; DSC curves (**b**) of pure EPDM, EPDM/MMT-10, and EPDM/DMMT-10.

In Figure 4a, EPDM/MMT-10 shows a diffraction peak at 6.2°, corresponding to a layer spacing of 1.43 nm. Compared to the pristine MMT, the slight expansion of layer spacing is similar to that seen in our previous study [35,39]. This is attributed to the incomplete removal of the water remaining in the MMT layers. When the DMMT is incorporated into EPDM, no diffraction peak is found in the XRD curve. In the stage of EPDM/DMMT-10 nanocomposite preparation, the aggregation of MMT in the hydrophobic matrix is weakened by water and PDA coating, and the compatibility between MMT and EPDM is increased by PDA coating. Even in the vulcanization process with high pressures and high temperatures, the movement of chain segments is enhanced, but the chemical bond is constructed by the PDA, VTMS, and EPDM matrixes on the MMT surface. This restricts the sliding out of macromolecules between layers, and so no diffraction peak exists in the XRD curve.

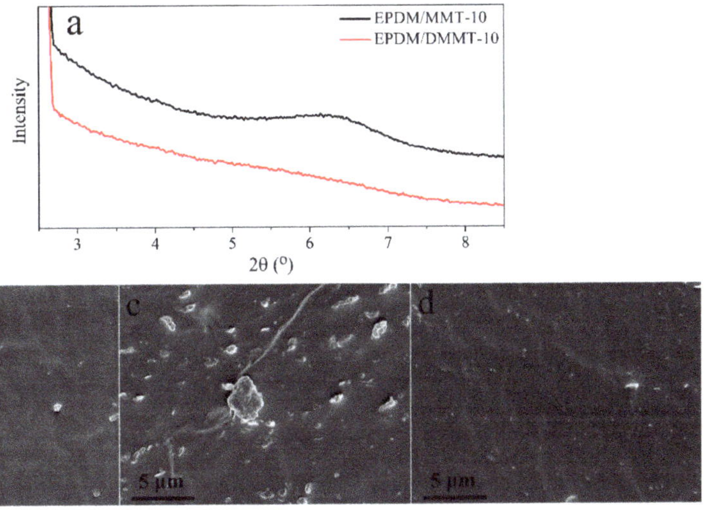

Figure 4. XRD curves (**a**) of EPDM/MMT-10 and EPDM/DMMT-10; SEM images of (**b**) pure EPDM, (**c**) EPDM/MMT-10, and (**d**) EPDM/DMMT-10.

In Figure 4b, the EPDM composite has a dense and smooth surface without defects. In Figure 4c, with the addition of MMT, the size and distribution of MMT aggregates are uneven, and the clear interface of MMT–EPDM indicates that the interface interaction of MMT–EPDM is weak. As shown in Figure 4d, with the addition of DMMT, the aggregates vanish, and the DMMT–EPDM boundaries blur. This comparison shows that DMMT–EPDM has a strong interfacial interaction. The PDA coating on the MMT surface contains a large number of hydroxyl groups, which can be connected to the silane coupling agent (VTMS) by a grafting reaction, while the double bonds contained in VTMS are co-crosslinked with the EPDM matrix during the curing process, which significantly improves the interface interaction [29,40].

3.3. Properties of EPDM Composites

Figure 5a shows the stress–strain curves of EPDM composites. It can be clearly observed that the tensile strength and the elongation at break of all the nanocomposites improved. For example, the tensile strength for pure EPDM is 2.2 MPa; this value increases to 4.7 MPa and 6.5 MPa for EPDM/MMT-10 and EPDM/DMMT-10. However, EPDM/DMMT nanocomposite has a different stress–strain behavior, and the stress increases rapidly when the strain is more than 200%. This can be attributed to the fact that PDA coating on the MMT surface is linked to VTMS by chemical bonding, which successfully introduces a double bond on the MMT surface and causes a connection to the macromolecular chain during curing. In addition, the excellent dispersion of the DMMT the of EPDM matrix also contributes to the improvement of mechanical properties.

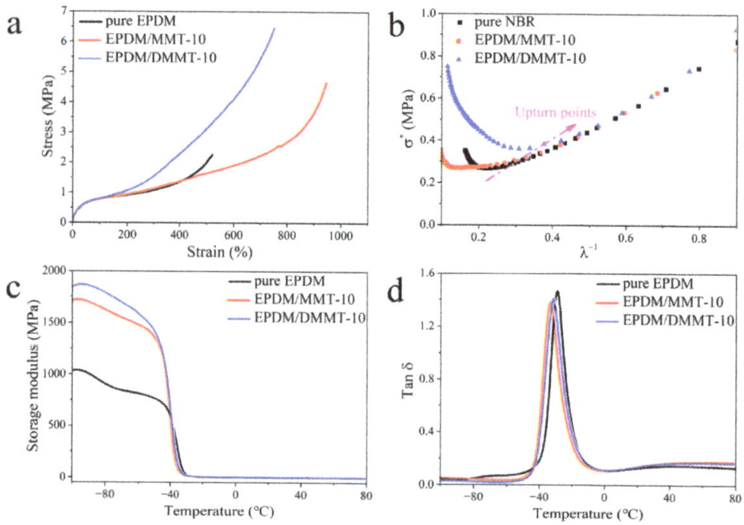

Figure 5. (**a**) Stress–strain curves, (**b**) reduced stress (σ*)–λ^{-1} curves, (**c**) storage modulus–temperature curves, and (**d**) tanδ–temperature curves of EPDM composites.

The Mooney–Rivlin equation was used to evaluate the elastomer network and the interface interaction between EPDM and DMMT as follows:

$$\sigma^* = \frac{\sigma}{\lambda - \lambda^{-2}} = 2C_1 + 2C_2\lambda^{-1}$$

where σ is applied stress, λ is the extension ratio, and C_1 and C_2 are the Mooney–Rivlin constants. As elaborated in Figure 5b, during the process of stretching, the stress decreases first and then increases, and these upturn points are related to the anchoring of the packing to the rubber macromolecular chain. The upturn occurs at smaller deformations for EPDM/DMMT-10 than for EPDM/MMT-10. This phenomenon is ascribed to DMMT serving as the crosslinking center and forming strong chemical interactions, which leads to more rubber chains being introduced into the interface, and these rubber chains are easily oriented during the stretching process.

To investigate the DMMT–EPDM interfacial interaction, dynamic mechanical analysis was carried out. As illustrated in Figure 5c, with the addition of filler into the EPDM matrix, the storage modulus increased regardless of whether the MMT was modified or not. However, the storage modulus of EPDM/DMMT-10 is higher than that of EPDM/MMT-10. The results demonstrated that DMMT has a stronger limiting effect on EPDM molecule under the same curing agent loading, which is attributed to the superior dispersion of DMMT on the matrix and the strong interfacial interaction between DMMT and EPDM. The addition of DMMT or MMT reduces the tanδ value of the composite, which is due to the reduction in the free volume of the rubber molecular chain after the addition of filler, thus limiting the movement of the polymer chain segment. It is worth noting that the tanδ of EPDM/DMMT-10 is 1.41, which is 0.02 higher than that of EPDM/MMT-10. The energy dissipation of PDA macromolecular chains on the MMT surface under small deformation forces and the interaction of hydrogen and covalent bonds at the interface of DMMT–EPDM limit the movement of macromolecular network segments. The combined effect of these two factors ultimately leads to subtle differences in tanδ values. However, the Tg values of EPDM/DMMT-10 and EPDM/MMT-10 nanocomposites were almost constant compared to pure EPDM, which is consistent with other reported results for non-polar rubber/MMT composites [1,41].

Table 2 shows the mechanical properties of various filler-filled EPDM composites in the literature. The reported filler EPDM composites made with carbon black or aluminum hydroxide show a tensile strength of ~7.2 MPa; however, quite a lot of filler was introduced. The EPDM composite filled with 25 phr silica shows a tensile strength of only 1.8 MPa, and the composites filled with 10 phr modified MMT exhibit a tensile strength of ~4.3 MPa. For comparison, the EPDM composite filled with 10 phr DMMT in this study shows a tensile strength of 6.5 MPa, and shows an elongation at break of 720%, which is much higher than that seen in the literature.

Table 2. Comparison of mechanical properties of various filler-filled EPDM composites.

Filler Type	Filler Loading (phr)	Tensile Strength (MPa)	Elongation at Break (%)	References
Carbon black	100	7.0	410	[42]
Silica	25	1.8	290	[42]
Aluminium hydroxide	50	7.2	315	[43]
Dendrimer-modified MMT	10	4.3	135	[44]
Octadecyl ammonium ion-modified MMT	10	4.2	320	[45]
DMMT	10	6.5	720	This work

4. Conclusions

In this study, the surface modification of MMT by PDA and vinyltrimethoxysilane grafting was successfully demonstrated. XRD, XPS, FTIR, and TGA confirmed that the montmorillonite (MMT) surface could be coated with polydopamine (PDA) to obtain DMMT. Subsequently, EPDM composites were prepared by the gel compounding method. Compared with EPDM/MMT-10, EPDM/DMMT-10 showed excellent filler dispersion and strong interfacial interactions. As a result, the mechanical performance of the EPDM/DMMT-10 was greatly enhanced. The highest tensile strength of the composites reached 6.5 MPa with 10 phr DMMT, almost 200% higher than that of pure EPDM. Moreover, the elongation at break of EPDM/DMMT-10 reached 720%, much higher than that in the literature. This study demonstrates the potential of MMT modification and the gel compounding method to prepare high-performance nanocomposites. These methods can improve the dispersion of inorganic fillers in the non-polar rubber matrix and enhance the interfacial strength of fillers and non-polar rubber.

Author Contributions: Data curation, Z.L., Z.C., W.S., Y.L., J.L. and J.W.; Funding acquisition, J.W. and S.H.; Investigation, Z.L.; Methodology, Z.C.; Project administration, S.H.; Resources, J.L. and S.H.; Supervision, X.W. and S.H.; Validation, Z.L.; Visualization, X.W.; Writing—original draft, Z.L. and Z.C.; Writing—review & editing, X.W. and S.H. All authors have read and agreed to the published version of the manuscript.

Funding: This work was financially supported by the National Natural Science Foundation of China (grant numbers: 51973057) and the Fundamental Research Funds for the Central Universities (2023JC005).

Institutional Review Board Statement: Not applicable.

Data Availability Statement: The original contributions presented in the study are included in the article, further inquiries can be directed to the corresponding authors.

Conflicts of Interest: The authors declare that the research was conducted in the absence of any commercial or financial relationships that could be construed as a potential conflict of interest.

References

1. Ahmadi, S.J.; Huang, Y.D.; Li, W. Morphology and Characterization of Clay-reinforced EPDM Nanocomposites. *J. Compos. Mater.* **2016**, *39*, 745–754. [CrossRef]
2. Chiu, C.W.; Huang, T.K.; Wang, Y.C.; Alamani, B.G.; Lin, J.J. Intercalation strategies in clay/polymer hybrids. *Prog. Polym. Sci.* **2014**, *39*, 443–485. [CrossRef]
3. Fei, B.; Qian, B.; Yang, Z.; Wang, R.; Liu, W.C.; Mak, C.L.; Xin, J.H. Coating carbon nanotubes by spontaneous oxidative polymerization of dopamine. *Carbon* **2008**, *46*, 1795–1797. [CrossRef]
4. Feng, J.; Fan, H.; Zha, D.A.; Wang, L.; Jin, Z. Characterizations of the Formation of Polydopamine-Coated Halloysite Nanotubes in Various pH Environments. *Langmuir* **2016**, *32*, 10377–10386. [CrossRef]
5. Fröhlich, J.; Niedermeier, W.; Luginsland, H.D. The effect of filler–filler and filler–elastomer interaction on rubber reinforcement. *Compos. Part A-Appl. Sci. Manuf.* **2005**, *36*, 449–460. [CrossRef]
6. He, S.J.; Wang, J.Q.; Hu, J.B.; Zhou, H.F.; Nguyen, H.; Luo, C.M.; Lin, J. Silicone rubber composites incorporating graphitic carbon nitride and modified by vinyl tri-methoxysilane. *Polym. Test.* **2019**, *79*, 106005–106010. [CrossRef]
7. Jaber, M.; Lambert, J.-F. A New Nanocomposite: L-DOPA/Laponite. *J. Phys. Chem. Lett.* **2009**, *1*, 85–88. [CrossRef]
8. He, S.J.; Bai, F.J.; Liu, S.X.; Ma, H.F.; Hu, J.B.; Chen, L.; Lin, J.; Wei, G.S.; Du, X.Z. Aging properties of styrene-butadiene rubber nanocomposites filled with carbon black and rectorite. *Polym. Test.* **2017**, *64*, 92–100. [CrossRef]
9. Chen, Y.; Wang, K.; Zhang, C.; Yang, W.; Qiao, B.; Yin, L. The Effect of Various Fillers on the Properties of Methyl Vinyl Silicone Rubber. *Polymers* **2023**, *15*, 1584. [CrossRef]
10. Roy, K.; Debnath, S.C.; Potiyaraj, P. A critical review on the utilization of various reinforcement modifiers in filled rubber composites. *J. Elastom. Plast.* **2020**, *52*, 167–193. [CrossRef]
11. Lee, H.; Dellatore, S.M.; Miller, W.M.; Messersmith, P.B. Mussel-inspired surface chemistry for multifunctional coatings. *Science* **2007**, *318*, 426–430. [CrossRef] [PubMed]
12. Chen, Z.L.; Wang, W.C.; Li, J.; Lin, J.; Li, F.Z.; Zhang, L.Q.; He, S.J. Bioinspired design of nitrile-butadiene rubber/montmorillonite nanocomposites with hydrogen bond interactions leading to highly effective reinforcement. *Polymer* **2023**, *277*, 125968. [CrossRef]
13. Li, J.; Liu, X.; Feng, Y.; Yin, J. Recent progress in polymer/two-dimensional nanosheets composites with novel performances. *Prog. Polym. Sci.* **2022**, *126*, 101505. [CrossRef]
14. Zhu, T.T.; Zhou, C.H.; Kabwe, F.B.; Wu, Q.Q.; Li, C.S.; Zhang, J.R. Exfoliation of montmorillonite and related properties of clay/polymer nanocomposites. *Appl. Clay Sci.* **2019**, *169*, 48–66. [CrossRef]
15. Ma, Y.; Wu, Y.P.; Zhang, L.Q.; Li, Q.F. The role of rubber characteristics in preparing rubber/clay nanocomposites by melt compounding. *J. Appl. Polym. Sci.* **2008**, *109*, 1925–1934. [CrossRef]
16. Mohammadpour, Y.; Katbab, A.A. Effects of the ethylene-propylene-diene monomer microstructural parameters and interfacial compatibilizer upon the EPDM/montmorillonite nanocomposites microstructure: Rheology/permeability correlation. *J. Appl. Polym. Sci.* **2007**, *106*, 4209–4218. [CrossRef]
17. Meng, Z.; Li, J.; Zou, Y.; Li, N.; Fu, X.; Zhang, R.; Hu, S.; Liu, Q. Advanced montmorillonite modification by using corrosive microorganisms as an alternative filler to reinforce natural rubber. *Appl. Clay Sci.* **2022**, *225*, 106534. [CrossRef]
18. Phua, S.L.; Yang, L.; Toh, C.L.; Huang, S.; Tsakadze, Z.; Lau, S.K.; Mai, Y.W.; Lu, X. Reinforcement of polyether polyurethane with dopamine-modified clay: The role of interfacial hydrogen bonding. *ACS Appl. Mater. Interfaces* **2012**, *4*, 4571–4578. [CrossRef]
19. Li, P.; Yin, L.; Song, G.; Sun, J.; Wang, L.; Wang, H. High-performance EPDM/organoclay nanocomposites by melt extrusion. *Appl. Clay Sci.* **2008**, *40*, 38–44. [CrossRef]
20. Donchak, V.; Stetsyshyn, Y.; Bratychak, M.; Broza, G.; Harhay, K.; Stepina, N.; Kostenko, M.; Voronov, S. Nanoarchitectonics at surfaces using multifunctional initiators of surface-initiated radical polymerization for fabrication of the nanocomposites. *Appl. Surf. Sci. Adv.* **2021**, *5*, 100104. [CrossRef]
21. Kostenko, M.; Stetsyshyn, Y.; Harhay, K.; Melnyk, Y.; Donchak, V.; Gubriy, Z.; Kracalik, M. Impact of the functionalized clay nanofillers on the properties of the recycled polyethylene terephthalate nanocomposites. *J. Appl. Polym. Sci.* **2024**, *141*, e55543. [CrossRef]
22. Qu, C.; Li, S.; Zhang, Y.; Wang, T.; Wang, Q.; Chen, S. Surface modification of Ti_3C_2-MXene with polydopamine and amino silane for high performance nitrile butadiene rubber composites. *Tribol. Int.* **2021**, *163*, 107150. [CrossRef]
23. Rana, A.S.; Vamshi, M.K.; Naresh, K.; Velmurugan, R.; Sarathi, R. Mechanical, thermal, electrical and crystallographic behaviour of EPDM rubber/clay nanocomposites for out-door insulation applications. *Adv. Mater. Process. Technol.* **2019**, *6*, 54–74. [CrossRef]
24. Sa, R.; Yan, Y.; Wei, Z.; Zhang, L.; Wang, W.; Tian, M. Surface modification of aramid fibers by bio-inspired poly(dopamine) and epoxy functionalized silane grafting. *ACS Appl. Mater. Interfaces* **2014**, *6*, 21730–21738. [CrossRef] [PubMed]
25. Lu, Z.R.; Yuan, X.C.; Jia, X.Y.; Lin, J.; He, S.J. High-performance proton exchange membrane employing water-insoluble hybrid formed by chemically bonding phosphotungstic acid with polydopamine. *Clean Energy Sci. Technol.* **2024**, *2*, 138. [CrossRef]
26. Surya, I.; Muniyadi, M.; Ismail, H. A review on clay-reinforced ethylene propylene diene terpolymer composites. *Polym. Compos.* **2021**, *42*, 1698–1711. [CrossRef]
27. Liao, Y.F.; Weng, Y.X.; Wang, J.Q.; Zhou, H.F.; Lin, J.; He, S.J. Silicone rubber composites with high breakdown strength and low dielectric loss based on polydopamine coated mica. *Polymers* **2019**, *11*, 2030. [CrossRef]
28. Plagge, J.; Lang, A. Filler-polymer interaction investigated using graphitized carbon blacks: Another attempt to explain reinforcement. *Polymer* **2021**, *218*, 123513. [CrossRef]

29. Wang, L.; Hu, L.; Gao, S.; Zhao, D.; Zhang, L.; Wang, W. Bio-inspired polydopamine-coated clay and its thermo-oxidative stabilization mechanism for styrene butadiene rubber. *RSC Adv.* **2015**, *5*, 9314–9324. [CrossRef]
30. Chen, Z.L.; Li, J.; Li, Z.X.; Wang, J.; Li, Q.; Lin, J.; Zhang, L.Q.; He, S.J. Rubber/clay nanocomposites prepared by compounding clay gel with hydrophilically treated styrene-butadiene rubber. *Eur. Polym. J.* **2024**, *213*, 113137. [CrossRef]
31. Chen, Z.L.; Li, J.; Li, Z.X.; Lin, J.; Zhang, L.Q.; He, S.J. A novel strategy to prepare rubber/clay nanocomposites via compounding clay gel into cocoamidopropyl betaine modified styrene butadiene rubber. *Compos. Sci. Technol.* **2024**, *252*, 110602. [CrossRef]
32. He, S.J.; He, T.F.; Wang, J.Q.; Wu, X.H.; Xue, Y.; Zhang, L.Q.; Lin, J. A novel method to prepare acrylonitrile-butadiene rubber/clay nanocomposites by compounding with clay gel. *Compos. Part B-Eng.* **2019**, *167*, 356–361. [CrossRef]
33. Wang, X.; Sinha, T.K.; Sun, J.; Wang, C.; Kim, J.K.; Zong, C. Facile preparation of hydrogenated nitrile butadiene rubber/reduced graphene oxide nanocomposite with one-pot reduction approach via the latex way. *Colloid Polym. Sci.* **2021**, *299*, 1703–1715. [CrossRef]
34. *ISO 37-2011*; Rubber, Vulcanized or Thermoplastic—Determination of Tensile Stress-Strain Properties. International Organization for Standardization: Geneva, Switzerland, 2011.
35. Yang, L.; Phua, S.L.; Teo, J.K.; Toh, C.L.; Lau, S.K.; Ma, J.; Lu, X. A biomimetic approach to enhancing interfacial interactions: Polydopamine-coated clay as reinforcement for epoxy resin. *ACS Appl. Mater. Interfaces* **2011**, *3*, 3026–3032. [CrossRef]
36. Ye, N.; Zheng, J.; Ye, X.; Xue, J.; Han, D.; Xu, H.; Wang, Z.; Zhang, L. Performance enhancement of rubber composites using VOC-Free interfacial silica coupling agent. *Compos. Part B-Eng.* **2020**, *202*, 108301. [CrossRef]
37. Zhou, M.; Liu, Q.; Wu, S.; Gou, Z.; Wu, X.; Xu, D. Starch/chitosan films reinforced with polydopamine modified MMT: Effects of dopamine concentration. *Food Hydrocolloid.* **2016**, *61*, 678–684. [CrossRef]
38. Yang, M.-C.; Kao, B.-J.; Tsou, C.-H.; Suen, M.-C.; Wu, C.-S.; Tsou, C.-Y.; Chu, C.-K.; Yao, W.-H.; Wu, W.-Y.; Hong, W.-S.; et al. The Properties and a New Preparation of Ethylene Propylene Diene Monomer/Montmorillonite Nanocomposites. *Polym. Polym. Compos.* **2015**, *23*, 181–190. [CrossRef]
39. Xu, Z.J.; Song, Y.H.; Zheng, Q. Payne effect of carbon black filled natural rubber compounds and their carbon black gels. *Polymer* **2019**, *185*, 121953. [CrossRef]
40. Li, S.; Li, X.; Shao, M.; Yang, J.; Wang, Q.; Wang, T.; Zhang, X. Regulating interfacial compatibility with amino silane and bio-inspired polydopamine for high-performance epoxy composites. *Tribol. Int.* **2019**, *140*, 105861. [CrossRef]
41. Zheng, H.; Zhang, Y.; Peng, Z.; Zhang, Y. Influence of clay modification on the structure and mechanical properties of EPDM/montmorillonite nanocomposites. *Polym. Test.* **2004**, *23*, 217–223. [CrossRef]
42. Yazici, N.; Kodal, M.; Ozkoc, G. Lab-Scale Twin-Screw Micro-Compounders as a New Rubber-Mixing Tool: 'A Comparison on EPDM/Carbon Black and EPDM/Silica Composites'. *Polymers* **2021**, *13*, 4391. [CrossRef] [PubMed]
43. Yen, Y.Y.; Wang, H.T.; Guo, W.J. Synergistic effect of aluminum hydroxide and nanoclay on flame retardancy and mechanical properties of EPDM composites. *J. Appl. Polym. Sci.* **2013**, *130*, 2042–2048. [CrossRef]
44. Zhang, C.; Wang, J.; Zhao, Y. Effect of dendrimer modified montmorillonite on structure and properties of EPDM nanocomposites. *Polym. Test.* **2017**, *62*, 41–50. [CrossRef]
45. Chang, Y.W.; Yang, Y.; Ryu, S.; Nah, C. Preparation and properties of EPDM/organomontmorillonite hybrid nanocomposites. *Polym. Int.* **2002**, *51*, 319–324. [CrossRef]

polymers

MDPI

Article

Fatigue Performance of 3D-Printed Poly-Lactic-Acid Bone Scaffolds with Triply Periodic Minimal Surface and Voronoi Pore Structures

Hamed Bakhtiari [1], Alireza Nouri [2] and Majid Tolouei-Rad [1,*]

[1] Centre for Advanced Materials and Manufacturing (CAMM), School of Engineering, Edith Cowan University, Joondalup, WA 6027, Australia; h.bakhtiari@ecu.edu.au
[2] School of Engineering, RMIT University, Melbourne, VIC 3001, Australia; alireza.nouri@rmit.edu.au
* Correspondence: m.rad@ecu.edu.au; Tel.: +61-8-6304-5883

Abstract: Bone scaffolds serve a crucial role in tissue engineering, particularly in facilitating bone regeneration where natural repair is insufficient. Despite advancements in the fabrication of polymeric bone scaffolds, the challenge remains to optimize their mechanical resilience. Specifically, research on the fatigue behaviour of polymeric bone scaffolds is scarce. This study investigates the influence of pore architecture on the mechanical performance of poly-lactic-acid (PLA) scaffolds under quasi-static and cyclic compression. PLA scaffolds with a 60% porosity were fabricated using extrusion-based 3D printing in various designs: Gyroid, Lidinoid, Fischer–Koch, IWP, and Voronoi. Results demonstrated that Gyroid scaffolds had the highest compressive strength (6.6 MPa), followed by Lidinoid, Fischer–Koch, IWP, and Voronoi designs. Increased strut thickness was linked to higher compressive strength. However, normalized fatigue resistance showed a different pattern. While scaffolds resisted fatigue cycles at low strain amplitudes, fatigue damage was observed at higher strains. Voronoi structures exhibited the highest normalized fatigue performance, enduring around 58,000 cycles at 85% strain amplitude, followed by Gyroid, Fischer–Koch, Lidinoid, and IWP structures. Enhanced fatigue performance in different topologies correlated with the minimum cross-sectional area of scaffolds. Given the importance of both static and fatigue strength, the Gyroid topology emerges as the superior choice overall.

Keywords: bone scaffold; fatigue; compression; PLA; additive manufacturing

1. Introduction

Bone fractures and defects can result from various factors, including trauma, osteoporosis, overuse, medical conditions, and nutritional deficiencies [1,2]. In 2000, an estimated 9 million bone fractures worldwide were solely due to osteoporosis [1]. The incidence of bone fractures has been found to increase considerably over the years, with 178 million new cases reported around the globe in 2019 [3], underlining the increased burden of bone injuries on global healthcare systems. To address this challenge, effective bone fracture repair techniques have been explored over the past two decades [4]. Conventional techniques such as autograft and allograft often encounter issues such as scarcity, infection risk, and immune rejection [5]. To overcome these challenges, bone tissue engineering (BTE) was developed, beginning in the late 20th century with the use of biodegradable polymeric scaffolds for cell transplantation [6]. The main cornerstones of BTE involve developing biocompatible scaffolds that support cell attachment and bone ingrowth, and which withstand external mechanical loads. These engineered scaffolds aim to replicate the bone's extracellular matrix by providing a three-dimensional (3D) structure that promotes cell growth, differentiation, and vascularization. An ideal scaffold should be biocompatible, biodegradable, and have proper mechanical properties [7].

Recent advancements in 3D printing technology have enabled the development of customized bone tissue scaffolds featuring different materials and complex designs. 3D printing techniques used for bone tissue engineering can be categorized into beam-based 3D printing techniques such as SLS [8] and SLM [9] and beamless techniques such as material jetting [10], binder jetting [11], and FDM [12]. Beam-based techniques (SLS, SLM) offer high precision and superior mechanical properties but are limited by high costs, material restrictions, and low cell viability. Beamless techniques (material jetting, binder jetting, FDM) provide greater material versatility, lower costs, and high cell density but may compromise on resolution and mechanical strength [13].

Bone scaffolds experience various physiological forces, such as tension, compression, and bending, under both static and cyclic conditions throughout their service life. Figure 1 illustrates the static and cyclic loading experienced by bone scaffolds during standing and walking. Loading conditions significantly impact the durability of bone scaffolds. Daily activities like walking, running, cycling, or climbing induce repetitive stresses that can lead to bone stress injury and fatigue failure [14]. Thus, an ideal bone scaffold should withstand both static and fatigue stresses throughout the bone healing period. Despite extensive research on the static performance of polymeric bone scaffolds, their fatigue behaviour remains relatively unexplored. Bakhtiari et al. [15] provided a detailed review of the factors influencing the fatigue resistance of bone scaffolds, highlighting material characteristics, in vivo conditions, topological features, and loading conditions as the most influential parameters. From a material perspective, the fatigue resistance of polymeric bone scaffolds can be improved by using higher-molecular-weight polymers or by reinforcing them with materials such as bioceramics or metals. Conversely, being exposed to body fluids can cause polymer degradation through hydrolysis and oxidation and consequently lead to a reduction in mechanical and fatigue resistance of bone scaffolds. On the other hand, the gradual formation of bone tissue within scaffold pores enhances mechanical support and improves the overall fatigue resistance of bone scaffold [15].

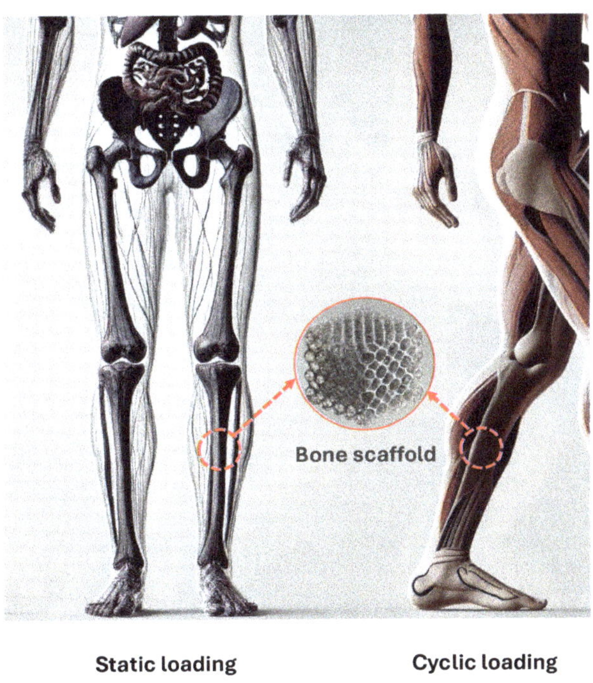

Figure 1. Static and cyclic loading during daily human activities (created by AI).

Topological features such as porosity, pore size, shape, and distribution of pores within a scaffold significantly influence its mechanical and fatigue characteristics [16]. While higher porosity can enhance bone ingrowth [17], it also compromises compressive strength and fatigue resistance of the polymeric scaffold [18,19]. Similar to porosity, size of the pores influences both biological and mechanical characteristics of bone scaffolds [20]. Although pore sizes larger than 300 μm are favourable for bone ingrowth [21], they often reduce the mechanical strength of scaffolds. Research has demonstrated that large pores in scaffolds can expedite crack initiation and propagation, leading to premature fatigue failure. In contrast, smaller pore sizes can significantly improve the scaffold's resistance to fatigue crack propagation [22]. Therefore, a trade-off must be managed between porosity and pore size in scaffold design to optimize both bone ingrowth and mechanical strength [21]. The shape of pores also plays a crucial role in the mechanical behaviour of scaffolds. Pore shapes in scaffolds can vary widely, including both regular geometries (such as circle, triangle, cube [23], tetrahedron and hexagon [24]) and irregular shapes (triply periodic minimal surface (TPMS) [25], metamaterials [26], bioinspired [27], and functionally graded topologies [28,29]). Figure 2 illustrates the variation in pore size, shape, and distribution within the scaffold. The image highlights differences in pore morphology, including both round and irregular shapes, as well as variations in pore size and nonuniform pore distribution across the scaffold.

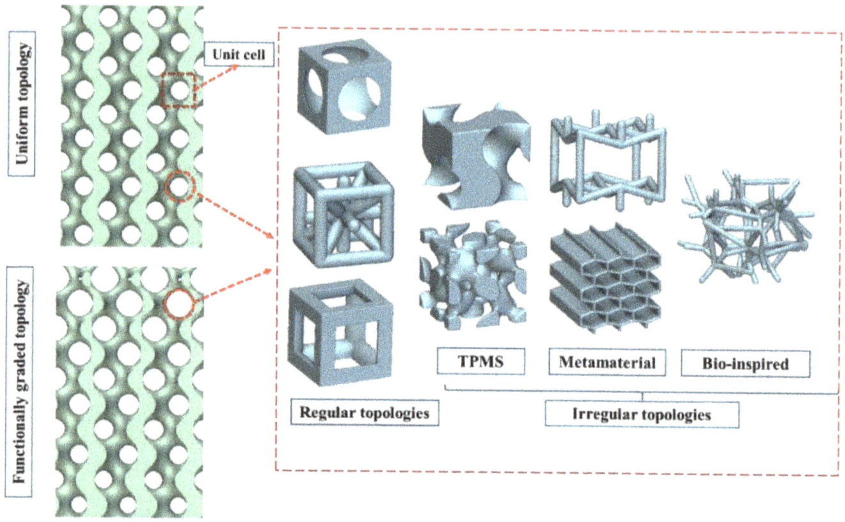

Figure 2. Illustration of different topologies used in designing bone scaffolds.

Few studies have explored the influence of pore architecture on the fatigue behaviour of polymeric bone scaffolds. Orthogonal (0/90°) pores exhibited higher compressive strength [30–32] and superior fatigue performance [18,19] compared to isometric patterns, likely due to increased slippage resistance between layers. In another study, Liang et al. [24] demonstrated the low cycle fatigue behaviour of 3D-printed PLA scaffolds with hexagonal, tetragonal, and wheel-like topologies, all having similar porosities. The results indicated the superiority of tetragonal design under cyclic loading, enduring 4400 fatigue cycles, attributed to its increased number of junction points, which provided enhanced mechanical support. Gong et al. [33] investigated 3D-printed PLA scaffolds with different pore architectures under compression and demonstrated that triangular pores possessed higher strength than circular topology. However, circular pores performed better under dynamic loading.

Complex pore architectures such as TPMS, Voronoi tessellation, functionally graded structures, and auxetic metamaterials in polymers have demonstrated promising load-bearing capacities [34–41]. Notably, the TPMS structure has shown potential in reducing stress concentration due to their smooth surface transitions at strut junctions [42]. TPMS structures feature continuous and smooth surfaces that maintain zero mean curvature at all points of their surfaces. This geometric characteristic is advantageous for higher mechanical and fatigue performance because it facilitates uniform stress distribution across the structure. The absence of sharp angles and discontinuities, which can concentrate stress and initiate cracks, enhances the structural durability under both static and dynamic loads. Voronoi tessellation has also gained attention in the biomedical field for its ability to mimic the natural, irregular patterns of bone microarchitecture. This geometry exhibits load-bearing and permeability characteristics that make it effective for bone scaffold applications [28,43]. However, the fatigue performance of these structures in polymeric bone scaffolds remains poorly understood, as it is underrepresented in the current literature. Without a comprehensive knowledge of how these scaffolds respond to cyclic loading, there is a risk of premature failure, which can compromise the structural integrity and longevity of the scaffold. To address this issue, the present study investigates the fatigue performance of PLA bone scaffolds with various pore architectures under compressive loading at different strain amplitudes. PLA scaffolds with Gyroid, Lidinoid, Fischer–Koch, Voronoi, and IWP pore structures were 3D-printed using an extrusion-based technique, and their static and fatigue strengths under compression were examined. This study underscores the importance of selecting appropriate pore architectures to balance the mechanical strength and fatigue resistance of PLA bone scaffolds, thereby providing valuable insights for optimizing scaffold designs in bone tissue engineering applications.

2. Materials and Methods

2.1. Bone Scaffold Fabrication

PLA was selected as the base material for bone scaffolds. Scaffold samples, featuring Gyroid, Lidinoid, Fischer–Koch, Voronoi, and IWP pore structures, were designed using nTopology 4.11.2 software. To comply with the standard test procedures for strength measurements (ASTM D1621-16 [44]), samples were designed as rectangular prisms with dimensions of 12.7 × 12.7 × 25.4 mm. Table 1 provides schematics of the topologies used in this study, along with the formulations employed to generate each topology. α, β, and γ denote constants related to the unit cell size (L) in x, y, and z, respectively; c is the offset parameter, which equals zero for a single unit cell of the TPMS.

Table 1. Schematic representations of the topological designs utilized in this study, accompanied by the mathematical formulations used to create each structure (X, Y, and Z denote the Cartesian coordinates in 3D space).

Topology	Formulation
 Gyroid	$\sin X \cos Y + \sin Y \cos Z + \sin Z \cos X$
 Lidinoid	$\sin(2X)\cos(Y)\sin(Z) + \sin(2Y)\cos(Z)\sin(X) + \sin(2Z)\cos(X)\sin(Y) - \cos(2X)\cos(2Y) - \cos(2Y)\cos(2Z) - \cos(2Z)\cos(2X)$

Table 1. *Cont.*

Topology	Formulation
 Fischer–Koch	$\cos(2x)\sin(y)\cos(z) + \cos(2y)\sin(z)\cos(x) + \cos(2z)\sin(x)\cos(y)$
 Voronoi	Randomly distributed set of points connected by irregularly shaped struts
 IWP	$2(\cos X\cos Y + \cos Y\cos Z + \cos Z\cos X) - (\cos 2X + \cos 2Y + \cos 2Z)$

To isolate the effect of pore shape, the porosity and unit cell size of all scaffolds were maintained at 60% and 2.5 mm, respectively, creating a 10 × 5 pore network within the scaffold's structure. Unit cell sizes ranging from 1 to 2.5 mm and porosities between 50% and 85% are considered optimal for biomedical implants [45]. Raise3D premium PLA filament (Raise 3D Technologies Inc., Irvine, CA, USA) with a diameter of 1.75 mm was used as the feedstock for the 3D printer. Fused deposition modelling (FDM) was employed to produce the scaffolds using a Raise3D pro2 Plus 3D printer (Raise 3D Technologies Inc., Irvine, CA, USA). The selected printing parameters included a layer thickness of 0.05 mm, an extrusion width of 0.65 mm, a nozzle temperature of 220 °C, and a print speed of 45 mm/s. These parameters were determined based on the findings of a parametric study conducted in our prior research [12,46]. Table 2 provides the physical and mechanical characteristics of the PLA filament used in this study.

Table 2. Physical and mechanical characteristics of the PLA filament [47].

Physical and Mechanical Properties	Value
Density (g/cm^3)	1.2
Glass transition temperature (°C)	62.3
Melting temperature (°C)	150.9
Young's modulus (MPa)	2681 ± 215
Tensile strength (MPa)	40 ± 1
Elongation at break (%)	2.5 ± 0.6
Bending strength (MPa)	68 ± 2

Table 3 summarizes the initial design and 3D-printed bone scaffolds, along with the topological characteristics of each design. SEM images of 3D-printed scaffolds are also provided in Figure 3. As shown in the table, although the porosity of all scaffolds is identical (i.e., 60%), the pore sizes vary between 700 and 900 μm due to the differences in pore shapes across various topologies. This range aligns with biological standards for scaffolds, which suggest that high porosity and interconnected channels with pore diameters ranging from 300 to 900 μm are optimal for bone tissue engineering [48].

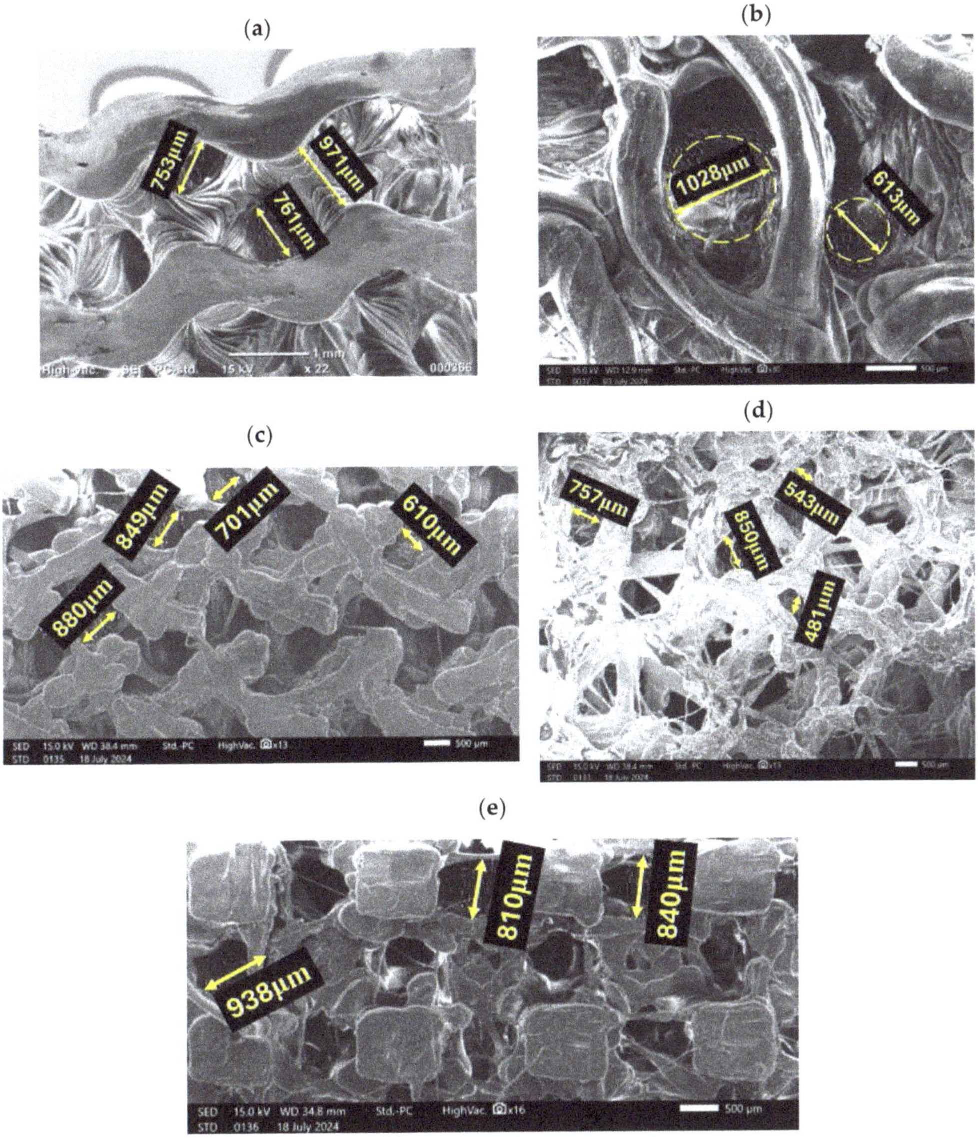

Figure 3. SEM images of 3D-printed (**a**) Gyroid, (**b**) Lidinoid, (**c**) Fischer–Koch, (**d**) Voronoi, and (**e**) IWP scaffolds.

Table 3. Details of the initial design and 3D-printed bone scaffolds, along with the topological characteristics of each design.

TPMS Structure	Designed Scaffold	3D-Printed Scaffold	Porosity (%)	Pore Size (μm)
Gyroid		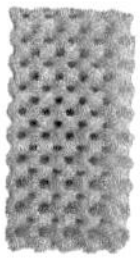	60	~748
Lidinoid			60	~883
Fischer–Koch	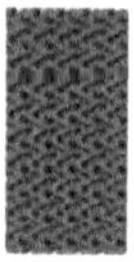		60	~682
Voronoi			60	~708
IWP			60	~936

2.2. *Quasi-Static Compression Test*

Quasi-static compression tests were conducted on scaffold specimens according to ASTM D1621-16 [44], which outlines methods for assessing the compressive properties of rigid cellular plastics. An Instron 8801 machine equipped with a double-acting servo hydraulic actuator was utilized to apply compressive loads. The length, width, and height of the specimens were measured at three points each and rounded to the nearest 0.01 mm. The average of these dimensions was used to calculate the cross-sectional area and volume of the scaffolds. Specimens were subjected to uniaxial loading at a speed of 1.3 mm/min until complete failure. Load-deformation data were recorded at a frequency of 20 Hz, and stress–strain data were derived by dividing the load by the cross-sectional area and the deformation by the specimen's length.

Figure 4 depicts the typical compression behaviour observed in cellular solids such as foams and lattice structures. The linear elastic region is characterised by the linear, reversible deformation of the scaffold under compressive force. Compressive strength and yield strain of the scaffold are determined by continuing the compression until the deformation results in permanent changes to the scaffold. This point is identified as the intersection of the stress–strain curve and a line parallel to the linear region, originating from a 0.2% strain offset [49,50]. After reaching the yield point, the cell walls begin to buckle or collapse, leading to the plateau region. This region is characterised by relatively constant stress because the collapsing pore structure dissipates the applied energy without significantly increasing the overall stress. Further compression moves the scaffold into the densification region, where most pores have collapsed, causing a sharp increase in material density. In this stage, the remaining solid framework bears the load, resulting in a steep rise in stress with additional strain [16,51]. Compressive modulus, compressive strength, plateau stress, and densification strain for each scaffold were derived from the stress–strain graphs. The compressive modulus was calculated by drawing a tangent to the linear region of the stress–strain curve and determining its slope using Equation (1):

$$E = \frac{F \times h_0}{A_0 \times \Delta h} \tag{1}$$

where E, F, h_0, A_0, and Δh denote the modulus of elasticity in compression, compressive load, initial specimen height, initial cross-sectional area, and deformation, respectively. Plateau stress was calculated where the stress–strain curve stabilizes. The densification strain was also measured at the point where the plateau line intersects with the tangent line drawn to the densification region.

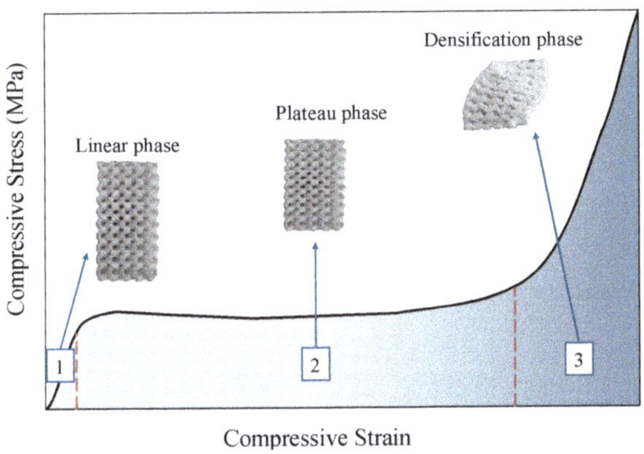

Figure 4. Typical compression behaviour of cellular solids, such as foams and porous scaffolds, is depicted in the stress–strain curve, which shows three phases: linear elastic (reversible deformation with linear stress increase), plateau (constant stress as cell walls collapse), and densification (steep stress rise as pores collapse and material density increases).

2.3. *Fatigue (Cyclic) Test*

Strain-controlled fatigue tests at various strain levels (0.004ε, 0.015ε, 0.025ε, and 0.03ε) were conducted using an Instron 8801 device. Daily activities such as walking and uphill running can induce strains on bones from 0.0004ε to 0.002ε [15,52]. To maintain the loading within the elastic range, the yield strain for each sample was established via compression tests, ensuring that the applied fatigue loads did not exceed the elastic limit. Each scaffold was initially preloaded to approximately 0.2 MPa and then subjected to 100,000 sinusoidal compressive cycles at a frequency of 5 Hz. Force and deflection data were collected, and compressive stress, strain, and modulus were calculated for each cycle. Fatigue damage in polymeric scaffolds manifests as stiffness loss. According to various sources in the literature, fatigue failure in natural bone occurs when the initial stiffness is reduced by 50 to 90% [53–55]. In the present study, the slope of the linear region of the loading force–displacement curve was used to determine the elastic stiffness. The reduction in elastic stiffness is defined by Equation (2):

$$\text{Stiffness reduction } (\%) = \frac{E_0 - E_{cycle}}{E_0} \times 100 \tag{2}$$

where E_0 and E_{cycle} denote the initial elastic stiffness and the elastic stiffness at the current cycle, respectively. The experiment was terminated when either a 60% reduction in the initial scaffold stiffness occurred or at the completion of 100,000 cycles, whichever came first.

3. Results and Discussion

3.1. Geometrical Features of Bone Scaffolds

To correlate the mechanical properties of the scaffolds with their geometries, geometrical features such as strut thickness and cross-sectional area were derived from the designed CAD files. It is important to note that the porosity and pore size of all scaffolds fall within the same range. Strut thickness and cross-sectional area for each scaffold were measured at 10 different sections across the entire height of a repeating unit cell, as illustrated schematically in Figure 5.

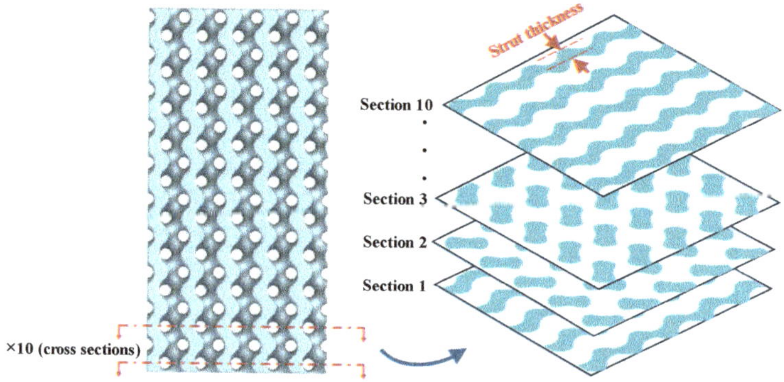

Figure 5. Measurement of strut thickness and cross-sectional area at ten different sections spanning the entire height of a repeating unit cell for each scaffold.

Table 4 summarizes the cross-section, minimum cross-sectional area, and average strut thickness for various pore topology designs, including Gyroid, Lidinoid, Fischer–Koch, IWP, and Voronoi structures. Geometrical data provided in Table 4 indicate no specific relationship between strut thickness and minimum cross-sectional area; a higher strut thickness does not necessarily correlate with a larger cross-sectional area. For example, the minimum cross-sectional area of the Voronoi design is approximately 13.6% greater than that of the Lidinoid, yet its struts are, on average, 27.6% thinner. In fact, these two factors are independent and vary according to the topological design of the scaffolds.

Table 4. Geometric characteristics of various pore topologies.

Geometric Characteristics	Voronoi	IWP	Fischer–Koch	Lidinoid	Gyroid
Cross-section	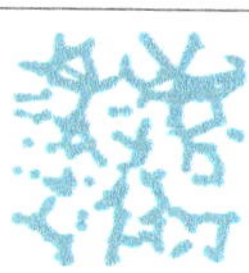			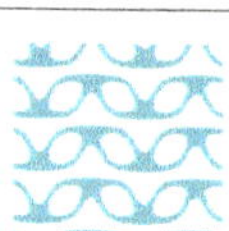	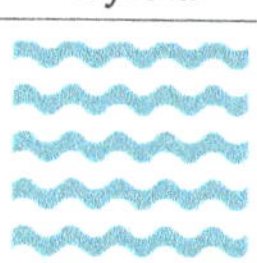
Minimum cross-sectional area (mm^2)	61.8	46.2	58.3	54.4	59.9
Average strut thickness (mm)	0.76	0.80	0.82	0.97	1.1

3.2. Compressive Properties

The stress–strain curves and compressive properties of PLA scaffolds are shown in Figure 6. Gyroid scaffolds exhibited the highest compressive properties, followed by Lidinoid, Fischer–Koch, IWP, and Voronoi topologies. As detailed in Table 5, the Gyroid scaffold demonstrated the highest compressive strength at 6.6 MPa, plateau stress at 7.7 MPa, a compressive modulus of approximately 176 MPa, and a densification strain of 0.26. In contrast, the Voronoi and IWP scaffolds exhibited the lowest compressive strength, around 2 to 3 MPa, and a compressive modulus of approximately 80 to 90 MPa, respectively. Unlike the other structures, the Voronoi scaffolds displayed no plateau region, indicating that densification begins immediately at the onset of plastic deformation.

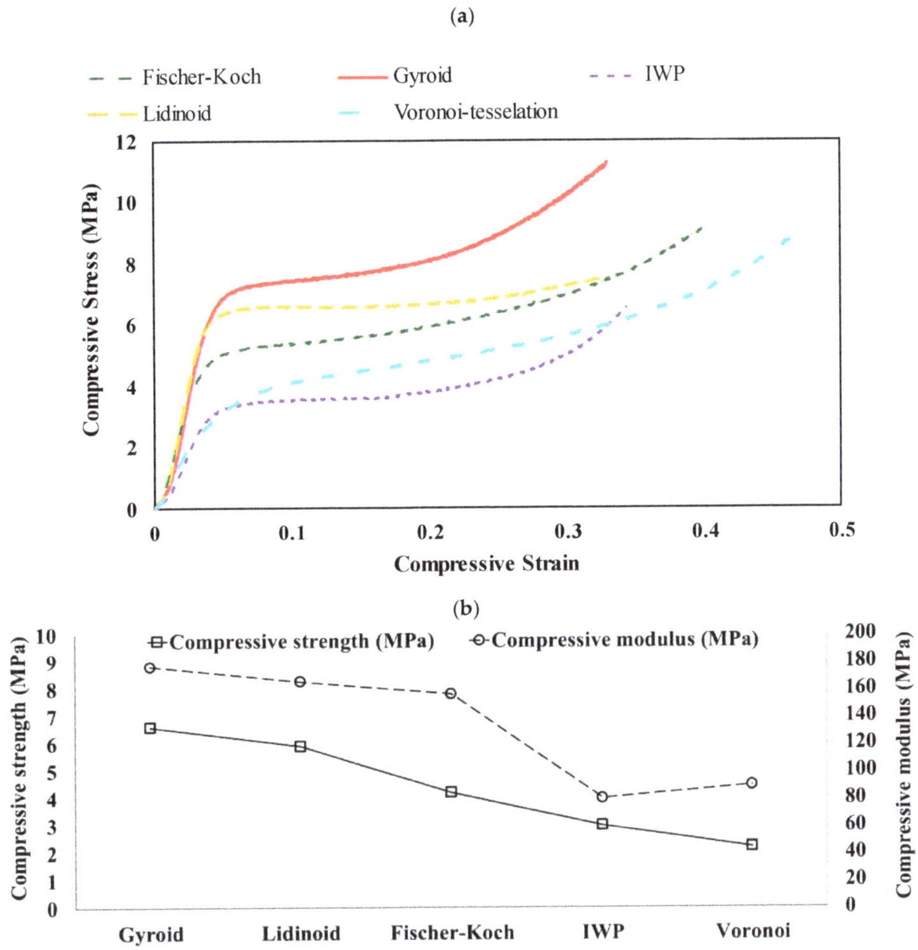

Figure 6. (**a**) Stress–strain curves and (**b**) compressive properties of PLA scaffolds, demonstrating the relationship between applied stress and the resulting strain, as well as the elastic modulus for various scaffold designs.

Table 5. Compressive properties of PLA scaffolds.

TPMS Scaffold	Porosity (%)	Pore Size (µm)	Yield Strain	Compressive Strength (MPa)	Compressive Modulus (MPa)	Plateau Stress (MPa)	Densification Strain
Voronoi	60	710	0.029	2.2	89.3	–	0.05
IWP	60	936	0.043	3.0	79.8	3.6	0.19
Fischer–Koch	60	682	0.032	4.2	156	5.6	0.25
Lidinoid	60	883	0.040	5.9	165	6.6	0.23
Gyroid	60	750	0.047	6.6	176.3	7.7	0.26

The increase in compressive properties (i.e., compressive strength, compressive modulus, plateau stress, and densification strain) correlates with greater strut thickness, as depicted in Figure 7. The Gyroid scaffold exhibits the highest strut thickness, approximately 1.1 mm, followed by the Lidinoid, Fischer–Koch, IWP, and Voronoi scaffolds. The same trend can be seen in their compressive properties. This finding is consistent with the results obtained by Torres et al. [27], who reported that even a small increase in the thickness of rod-like struts (structural elements oriented perpendicular to the applied load) can significantly enhance the structural strength of a polymeric scaffold by 10 to 100 times while having minimal impact on density and stiffness.

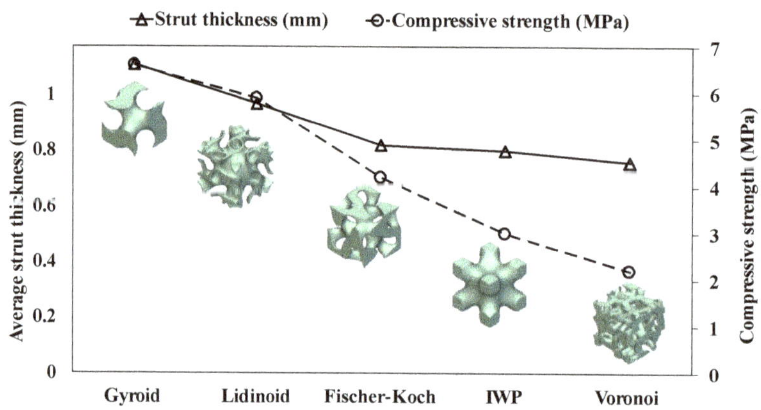

Figure 7. Change in average strut thickness across various topological scaffold structures.

3.3. Fatigue Properties

The fatigue performance of various topologies is shown in Figure 8 as normalised ε–N curves. Normalisation involves scaling the applied strain to the yield strain for each scaffold according to Equation (3):

$$\text{Normalised strain}(\%) = \frac{\text{Applied strain}}{\text{Yield strain}} \times 100 \tag{3}$$

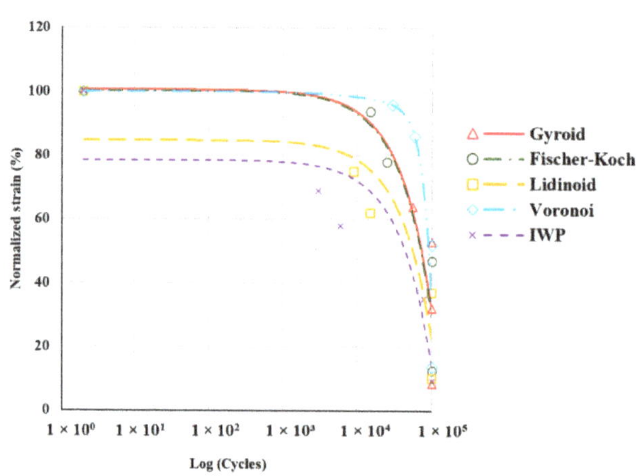

Figure 8. Normalised ε–N curves illustrate the fatigue performance of various topologies. The curves depict the relationship between normalised strain and the number of cycles to failure for each scaffold topology.

In all topologies, increasing the strain amplitude led to a decrease in the number of cycles to failure. This is because higher strain amplitudes impose greater cyclic stress on the material, accelerating fatigue damage accumulation [12]. At low strain amplitudes (<15%), all scaffolds could endure fatigue loading. However, increasing the normalised strain to 35% resulted in the failure of the IWP topology at the 78,000th cycle. At this strain level, all other scaffolds withstood the fatigue cycles. Further increasing the normalised strain to the range of 65–70% caused the failure of all topologies. At this strain level, the IWP, Lidinoid, and Gyroid scaffolds failed at 3000, 15,000, and 55,000 cycles, respectively.

The Voronoi topology exhibited the highest normalised fatigue resistance, followed by the Gyroid, Fischer–Koch, and Lidinoid topologies. In contrast, the IWP topology showed the lowest normalised fatigue performance among the studied designs. These results suggest a significant difference between the fatigue performance of topologies and their behaviour under quasi-static loading. Despite having lower compressive properties, the Voronoi topology outperformed all other topologies in cyclic loading. The observed differences arise from the distinct failure mechanisms under static and dynamic loads [56]. Cubic structures generally show superior strength under static compressive forces [57], whereas structures with circular pores perform better in cyclic loading as a result of their more uniform stress distribution and reduced stress concentration [33]. In a study, Haddock et al. investigated the fatigue strength of vertebral and bovine tibial trabecular bones with varying porosities and found that they exhibited similar fatigue strength despite differences in yield strength [55]. Likewise, a study by Gong et al. on 3D-printed PLA scaffolds revealed that while triangular pore architectures had higher compressive strength, circular pore architectures provided better dynamic stability owing to a more even stress distribution [33].

In the present study, the analysis of the cross-sectional areas of various topologies revealed that the cross-sectional area perpendicular to the applied load plays a significant role in withstanding cyclic loading. These cross-sections provide mechanical support and effectively distribute the applied load across the structure. Since damage typically initiates at the weakest cross-sections, smaller cross-sections are more susceptible to damage under cyclic loading. Consequently, applying the same load to a smaller surface area leads to greater damage and a reduction in the overall stiffness of the structure.

Figure 9 compares the minimum cross-sectional areas of the topologies examined. As shown, scaffolds with smaller cross-sectional areas tend to have lower fatigue resistance, while those with larger cross-sectional areas exhibit higher fatigue resistance. This is consistent with the findings reported in the literature [24], indicating that scaffolds with a higher number of junction points (resulting in a higher cross-sectional area) possess higher fatigue resistance.

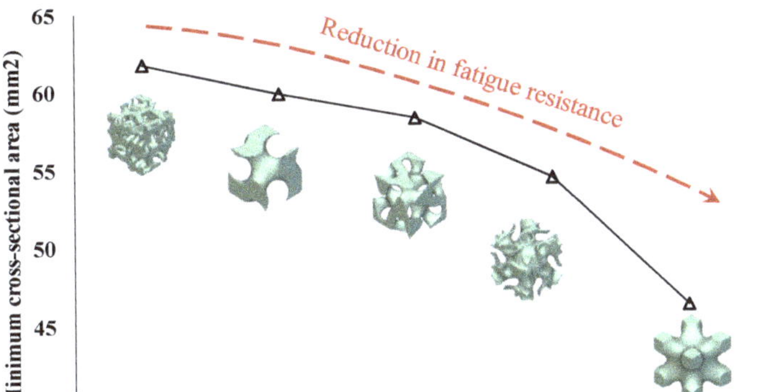

Figure 9. Minimum cross-sectional areas of various topological scaffold structures.

4. Conclusions

PLA is an affordable, accessible, and 3D-printable material that has been approved by the FDA for biomedical use. However, its suitability as a material for biomedical implants is questioned due to its relatively low mechanical strength. While numerous studies have investigated the strength of PLA scaffolds under quasi-static conditions, research on their fatigue properties remains scarce. Given that scaffolds are porous structures, their pore architecture significantly impacts their fatigue durability when implanted in the body. The present study explored the impact of various pore architectures on the mechanical and fatigue performance of PLA bone scaffolds with 60% porosity. An extrusion-based 3D printing technique was used to fabricate five pore structures, including Gyroid, Lidinoid, Fischer–Koch, IWP, and Voronoi. The key conclusions drawn from this study are as follows:

- The Gyroid topology exhibited the highest compressive properties, including compressive strength (6.6 MPa) and modulus (176.3 MPa), making it the most robust structure under static loads.
- The Voronoi topology showed the lowest compressive strength (~2.2 MPa) and modulus (~89.3 MPa) but demonstrated superior normalised fatigue resistance, followed by Gyroid topology.
- Increased strut thickness correlated with higher compressive properties. The Gyroid structure, with the thickest struts, also had the highest compressive performance.
- Scaffolds with larger cross-sectional areas generally exhibited better fatigue resistance. The larger area provided more mechanical support and effective load distribution, reducing localized damage and enhancing longevity under cyclic loads.
- Given that both static and fatigue strength are critical for the efficacy of bone scaffolds, the Gyroid topology emerges as the superior choice overall.

While this study explored the fatigue performance of different topologies, future work can utilize the potential of structural optimization for enhancing the fatigue resistance of polymeric bone scaffolds. Since higher mechanical strength often compromises the biological performance of bone scaffolds due to reducing the pore size and porosity, multiobjective optimization of scaffold structures is necessary. While this paper demonstrates that strut thickness and cross-sectional area play key roles in the mechanical and fatigue strength of PLA bone scaffolds, stress concentration is expected to be among other influencing factors that can be investigated through numerical simulations in future works. With increasing interest in patient-specific and custom-designed polymeric bone scaffolds, these findings offer valuable insights for biomedical engineers and can help expand their application in clinical settings.

Author Contributions: Conceptualization, M.T.-R. and H.B.; methodology, M.T.-R. and H.B.; software, H.B.; validation, H.B. and M.T.-R.; formal analysis, H.B. and M.T.-R., and A.N.; investigation, H.B., M.T.-R. and A.N.; resources, H.B.; data curation, H.B.; writing—original draft preparation, H.B., M.T.-R. and A.N.; writing—review and editing, H.B., M.T.-R. and A.N.; visualization, H.B. and A.N.; supervision, M.T.-R. and A.N.; project administration, M.T.-R. All authors have read and agreed to the published version of the manuscript.

Funding: This research received no external funding.

Institutional Review Board Statement: Not applicable.

Data Availability Statement: The raw data supporting the conclusions of this article will be made available by the authors on request.

Acknowledgments: The first author would like to thank Edith Cowan University for the awarded (ECU-HDR) higher degree research scholarship.

Conflicts of Interest: The authors declare no conflicts of interest.

References

1. Johnell, O.; Kanis, J. An estimate of the worldwide prevalence and disability associated with osteoporotic fractures. *Osteoporos. Int.* **2006**, *17*, 1726–1733. [CrossRef]
2. Nouri, A.; Wang, L.; Li, Y.; Wen, C. Materials and Manufacturing for Ankle–Foot Orthoses: A Review. *Adv. Eng. Mater.* **2023**, *25*, 2300238. [CrossRef]
3. Wu, A.-M.; Bisignano, C.; James, S.L.; Abady, G.G.; Abedi, A.; Abu-Gharbieh, E.; Alhassan, R.K.; Alipour, V.; Arabloo, J.; Asaad, M. Global, regional, and national burden of bone fractures in 204 countries and territories, 1990–2019: A systematic analysis from the Global Burden of Disease Study 2019. *Lancet Healthy Longev.* **2021**, *2*, e580–e592. [CrossRef] [PubMed]
4. Kim, T.; See, C.W.; Li, X.; Zhu, D. Orthopedic implants and devices for bone fractures and defects: Past, present and perspective. *Eng. Regen.* **2020**, *1*, 6–18. [CrossRef]
5. Dimitriou, R.; Jones, E.; McGonagle, D.; Giannoudis, P.V. Bone regeneration: Current concepts and future directions. *BMC Med.* **2011**, *9*, 66. [CrossRef]
6. Vacanti, J.P.; Morse, M.A.; Saltzman, W.M.; Domb, A.J.; Perez-Atayde, A.; Langer, R. Selective cell transplantation using bioabsorbable artificial polymers as matrices. *J. Pediatr. Surg.* **1988**, *23*, 3–9. [CrossRef] [PubMed]
7. O'brien, F.J. Biomaterials & scaffolds for tissue engineering. *Mater. Today* **2011**, *14*, 88–95.
8. Bin, S.; Wang, A.; Guo, W.; Yu, L.; Feng, P. Micro magnetic field produced by Fe_3O_4 nanoparticles in bone scaffold for enhancing cellular activity. *Polymers* **2020**, *12*, 2045. [CrossRef]
9. Attaeyan, A.; Shahgholi, M.; Khandan, A. Fabrication and characterization of novel 3D porous Titanium-6Al-4V scaffold for orthopedic application using selective laser melting technique. *Iran. J. Chem. Chem. Eng.* **2024**, *43*, 66–82.
10. Lee, G.; Carrillo, M.; McKittrick, J.; Martin, D.G.; Olevsky, E.A. Fabrication of ceramic bone scaffolds by solvent jetting 3D printing and sintering: Towards load-bearing applications. *Addit. Manuf.* **2020**, *33*, 101107. [CrossRef]
11. Zafar, M.J.; Zhu, D.; Zhang, Z. 3D printing of bioceramics for bone tissue engineering. *Materials* **2019**, *12*, 3361. [CrossRef] [PubMed]
12. Bakhtiari, H.; Nouri, A.; Tolouei-Rad, M. Impact of 3D printing parameters on static and fatigue properties of polylactic acid (PLA) bone scaffolds. *Int. J. Fatigue* **2024**, *186*, 108420. [CrossRef]
13. Bhushan, S.; Singh, S.; Maiti, T.K.; Sharma, C.; Dutt, D.; Sharma, S.; Li, C.; Tag Eldin, E.M. Scaffold fabrication techniques of biomaterials for bone tissue engineering: A critical review. *Bioengineering* **2022**, *9*, 728. [CrossRef] [PubMed]
14. Hoenig, T.; Ackerman, K.E.; Beck, B.R.; Bouxsein, M.L.; Burr, D.B.; Hollander, K.; Popp, K.L.; Rolvien, T.; Tenforde, A.S.; Warden, S.J. Bone stress injuries. *Nat. Rev. Dis. Primers* **2022**, *8*, 26. [CrossRef] [PubMed]
15. Bakhtiari, H.; Nouri, A.; Khakbiz, M.; Tolouei-Rad, M. Fatigue behaviour of load-bearing polymeric bone scaffolds: A review. *Acta Biomater.* **2023**, *172*, 16–37. [CrossRef]
16. Nouri, A. Novel Metal Structures through Powder Metallurgy for Biomedical Applications. Ph.D. Thesis, Deakin University, Geelong, VIC, Australia, September 2008.
17. Mikos, A.G.; Sarakinos, G.; Lyman, M.D.; Ingber, D.E.; Vacanti, J.P.; Langer, R. Prevascularization of porous biodegradable polymers. *Biotechnol. Bioeng.* **1993**, *42*, 716–723. [CrossRef]
18. Baptista, R.; Guedes, M. Porosity and pore design influence on fatigue behavior of 3D printed scaffolds for trabecular bone replacement. *J. Mech. Behav. Biomed. Mater.* **2021**, *117*, 104378. [CrossRef]
19. Baptista, R.; Guedes, M. Fatigue behavior of different geometry scaffolds for bone replacement. *Procedia Struct. Integr.* **2019**, *17*, 539–546. [CrossRef]
20. Nouri, A. Titanium foam scaffolds for dental applications. In *Metallic Foam Bone*; Elsevier: Amsterdam, The Netherlands, 2017; pp. 131–160.
21. Karageorgiou, V.; Kaplan, D. Porosity of 3D biomaterial scaffolds and osteogenesis. *Biomaterials* **2005**, *26*, 5474–5491. [CrossRef]
22. Wang, Y.; Zhang, D.; Pan, G. Investigating the fatigue behavior of 3D-printed bone scaffolds. *J. Mater. Sci.* **2023**, *58*, 12929–12953. [CrossRef]
23. Wu, J.; Zhang, Y.; Lyu, Y.; Cheng, L. On the various numerical techniques for the optimization of bone scaffold. *Materials* **2023**, *16*, 974. [CrossRef] [PubMed]
24. Liang, X.; Gao, J.; Xu, W.; Wang, X.; Shen, Y.; Tang, J.; Cui, S.; Yang, X.; Liu, Q.; Yu, L. Structural mechanics of 3D-printed poly (lactic acid) scaffolds with tetragonal, hexagonal and wheel-like designs. *Biofabrication* **2019**, *11*, 035009. [CrossRef] [PubMed]
25. Lehder, E.; Ashcroft, I.; Wildman, R.; Ruiz-Cantu, L.; Maskery, I. A multiscale optimisation method for bone growth scaffolds based on triply periodic minimal surfaces. *Biomech. Model. Mechanobiol.* **2021**, *20*, 2085–2096. [CrossRef] [PubMed]
26. Liu, Y.; Hu, H. A review on auxetic structures and polymeric materials. *Sci. Res. Essays* **2010**, *5*, 1052–1063.
27. Torres, A.M.; Trikanad, A.A.; Aubin, C.A.; Lambers, F.M.; Luna, M.; Rimnac, C.M.; Zavattieri, P.; Hernandez, C.J. Bone-inspired microarchitectures achieve enhanced fatigue life. *Proc. Natl. Acad. Sci. USA* **2019**, *116*, 24457–24462. [CrossRef]
28. Cai, S.; Xi, J.; Chua, C.K. A novel bone scaffold design approach based on shape function and all-hexahedral mesh refinement. *Comput.-Aided Tissue Eng.* **2012**, *868*, 45–55.
29. Nazari, K.A.; Hilditch, T.; Dargusch, M.S.; Nouri, A. Functionally graded porous scaffolds made of Ti-based agglomerates. *J. Mech. Behav. Biomed. Mater.* **2016**, *63*, 157–163. [CrossRef] [PubMed]
30. Serra, T.; Planell, J.A.; Navarro, M. High-resolution PLA-based composite scaffolds via 3-D printing technology. *Acta Biomater.* **2013**, *9*, 5521–5530. [CrossRef]

31. Hoque, M.; Hutmacher, D.; Feng, W.; Li, S.; Huang, M.-H.; Vert, M.; Wong, Y. Fabrication using a rapid prototyping system and in vitro characterization of PEG-PCL-PLA scaffolds for tissue engineering. *J. Biomater. Sci. Polym. Ed.* **2005**, *16*, 1595–1610. [CrossRef]
32. Bakhtiari, H.; Aamir, M.; Tolouei-Rad, M. Effect of 3D Printing Parameters on the Fatigue Properties of Parts Manufactured by Fused Filament Fabrication: A Review. *Appl. Sci.* **2023**, *13*, 904. [CrossRef]
33. Gong, B.; Cui, S.; Zhao, Y.; Sun, Y.; Ding, Q. Strain-controlled fatigue behaviors of porous PLA-based scaffolds by 3D-printing technology. *J. Biomater. Sci. Polym. Ed.* **2017**, *28*, 2196–2204. [CrossRef] [PubMed]
34. Spece, H.; Yu, T.; Law, A.; Marcolongo, M.; Kurtz, S. 3D printed porous PEEK created via fused filament fabrication for osteoconductive orthopaedic surfaces. *J. Mech. Behav. Biomed. Mater.* **2020**, *109*, 103850. [CrossRef] [PubMed]
35. Al-Ketan, O.; Abu Al-Rub, R.K. Multifunctional mechanical metamaterials based on triply periodic minimal surface lattices. *Adv. Eng. Mater.* **2019**, *21*, 1900524. [CrossRef]
36. Abou-Ali, A.M.; Al-Ketan, O.; Lee, D.-W.; Rowshan, R.; Al-Rub, R.K.A. Mechanical behavior of polymeric selective laser sintered ligament and sheet based lattices of triply periodic minimal surface architectures. *Mater. Des.* **2020**, *196*, 109100. [CrossRef]
37. Mishra, A.K.; Chavan, H.; Kumar, A. Effect of material variation on the uniaxial compression behavior of FDM manufactured polymeric TPMS lattice materials. *Mater. Today Proc.* **2021**, *46*, 7752–7759. [CrossRef]
38. Wang, S.; Ding, Y.; Yu, F.; Zheng, Z.; Wang, Y. Crushing behavior and deformation mechanism of additively manufactured Voronoi-based random open-cell polymer foams. *Mater. Today Commun.* **2020**, *25*, 101406. [CrossRef]
39. Jin, Y.; Xie, C.; Gao, Q.; Zhou, X.; Li, G.; Du, J.; He, Y. Fabrication of multi-scale and tunable auxetic scaffolds for tissue engineering. *Mater. Des.* **2021**, *197*, 109277. [CrossRef]
40. Callens, S.J.; Tümer, N.; Zadpoor, A.A. Hyperbolic origami-inspired folding of triply periodic minimal surface structures. *Appl. Mater. Today* **2019**, *15*, 453–461. [CrossRef]
41. Al-Ketan, O.; Lee, D.-W.; Rowshan, R.; Al-Rub, R.K.A. Functionally graded and multi-morphology sheet TPMS lattices: Design, manufacturing, and mechanical properties. *J. Mech. Behav. Biomed. Mater.* **2020**, *102*, 103520. [CrossRef]
42. Almeida, H.A.; Bartolo, P.J. Design of tissue engineering scaffolds based on hyperbolic surfaces: Structural numerical evaluation. *Med. Eng. Phys.* **2014**, *36*, 1033–1040. [CrossRef]
43. Zhao, H.; Han, Y.; Pan, C.; Yang, D.; Wang, H.; Wang, T.; Zeng, X.; Su, P. Design and mechanical properties verification of gradient Voronoi scaffold for bone tissue engineering. *Micromachines* **2021**, *12*, 664. [CrossRef] [PubMed]
44. *ASTM D1621-16*; Standard Test Method for Compressive Properties of Rigid Cellular Plastics. ASTM: West Conshohocken, PA, USA, 2016.
45. Ziaie, B.; Velay, X.; Saleem, W. Exploring the Mechanical Properties of the Gyroid Sheet Network for the Additive Manufacturing of Biomedical Structures: A Numerical Analysis Approach. *Eng. Proc.* **2024**, *65*, 7. [CrossRef]
46. Bakhtiari, H.; Nikzad, M.; Tolouei-Rad, M. Influence of three-dimensional printing parameters on compressive properties and surface smoothness of polylactic acid specimens. *Polymers* **2023**, *15*, 3827. [CrossRef] [PubMed]
47. Raise3D Premium PLA Technical Data Sheet. Available online: https://s1.raise3d.com/2023/02/Raise3D-Premium_PLA_TDS_V5.1_EN.pdf (accessed on 10 June 2024).
48. Cheng, C.-H.; Chen, Y.-W.; Kai-Xing Lee, A.; Yao, C.-H.; Shie, M.-Y. Development of mussel-inspired 3D-printed poly (lactic acid) scaffold grafted with bone morphogenetic protein-2 for stimulating osteogenesis. *J. Mater. Sci. Mater. Med.* **2019**, *30*, 1–12. [CrossRef] [PubMed]
49. Hany, R.A. Characterization of scaffolds. *Biomater. J.* **2023**, *2*, 53–67.
50. Will, J.; Detsch, R.; Boccaccini, A.R. Structural and biological characterization of scaffolds. In *Characterization of Biomaterials*; Elsevier: Amsterdam, The Netherlands, 2013; pp. 299–310.
51. Sun, Y.; Li, Q. Dynamic compressive behaviour of cellular materials: A review of phenomenon, mechanism and modelling. *Int. J. Impact Eng.* **2018**, *112*, 74–115. [CrossRef]
52. Cullen, D.M.; Smith, R.; Akhter, M.P. Bone-loading response varies with strain magnitude and cycle number. *J. Appl. Physiol.* **2001**, *91*, 1971–1976. [CrossRef]
53. Michel, M.C.; Guo, X.-D.E.; Gibson, L.J.; McMahon, T.A.; Hayes, W.C. Compressive fatigue behavior of bovine trabecular bone. *J. Biomech.* **1993**, *26*, 453–463. [CrossRef]
54. Bowman, S.; Guo, X.; Cheng, D.; Keaveny, T.; Gibson, L.; Hayes, W.; McMahon, T. Creep contributes to the fatigue behavior of bovine trabecular bone. *J. Biomech. Eng.* **1998**, *120*, 647–654. [CrossRef]
55. Haddock, S.M.; Yeh, O.C.; Mummaneni, P.V.; Rosenberg, W.S.; Keaveny, T.M. Similarity in the fatigue behavior of trabecular bone across site and species. *J. Biomech.* **2004**, *37*, 181–187. [CrossRef]
56. Hoyt, A.J.; Yakacki, C.M.; Fertig III, R.S.; Carpenter, R.D.; Frick, C.P. Monotonic and cyclic loading behavior of porous scaffolds made from poly (para-phenylene) for orthopedic applications. *J. Mech. Behav. Biomed. Mater.* **2015**, *41*, 136–148. [CrossRef]
57. Xu, M.; Zhai, D.; Chang, J.; Wu, C. In vitro assessment of three-dimensionally plotted nagelschmidtite bioceramic scaffolds with varied macropore morphologies. *Acta Biomater.* **2014**, *10*, 463–476. [CrossRef] [PubMed]

polymers

MDPI

Article

Aramid Honeycomb Cores under Constant Pressure: Unveiling the Out-of-Plane Compression Deformation

Xinzheng Huang [1,2,†], Xin Hu [3,†], Jinzhan Guo [3], Dechao Zhang [3], Shunming Yao [1], Lihua Zhan [3,*], Bolin Ma [3,*], Minghui Huang [1] and Lihua Zhang [3]

1 College of Mechanical and Electrical Engineering, Central South University, Changsha 410083, China; 13322995346@163.com (X.H.); yaoshunming@csu.edu.cn (S.Y.); meeh@csu.edu.cn (M.H.)
2 Shenzhen Kuang-Chi Cutting-Edge Technology Co., Ltd., Shenzhen 518000, China
3 Light Alloys Research Institute, Central South University, Changsha 410083, China; ladihu4@gmail.com (X.H.); guruci@163.com (J.G.); 213801004@csu.edu.cn (D.Z.); zhanglihua@csu.edu.cn (L.Z.)
* Correspondence: yjs-cast@csu.edu.cn (L.Z.); mabolin_csu@163.com (B.M.)
† These authors contributed equally to this work.

Abstract: The primary challenge during the secondary bonding process of full-height honeycomb sandwich structures is the aramid honeycomb core's height shrinkage. This paper systematically investigated the height evolution behavior of the honeycomb core by using a creep testing machine. The results showed that the out-of-plane compression deformation curve of aramid honeycomb cores is mainly divided into three stages: the dehumidification stage, the pressurization stage and the creep stage. Under conditions of high temperature and pressure, height shrinkage was attributed to the dehydration caused by moisture infiltration, and the compression creep resulted from the slippage of polymer molecular chains. Dehydration shrinkage is stable, whereas compression creep reflects typical viscoelastic polymer characteristics. By employing the viscoelastic Burgers mechanical model and applying the nonlinear surface fitting method, the total height shrinkage deformation behavior of the aramid honeycomb core during the curing process can be accurately predicted by summing the above three stages. This research contributes valuable insights for the manufacturing process of honeycomb sandwich structures.

Keywords: aramid honeycomb core; out-of-plane compression deformation; moisture infiltration; viscoelasticity; Burgers model

1. Introduction

Composite materials are renowned for their lightweight, high strength and corrosion resistance [1–3]. As a type of composite material, honeycomb sandwich structures primarily consist of top and bottom face sheets and a honeycomb core, and the aramid paper honeycomb core is commonly employed as the core material in this structure. The aramid paper honeycomb core is a bionic honeycomb core material made of aramid paper impregnated with phenolic resin. It has a series of excellent properties, such as its light weight, energy absorption, high specific strength/stiffness, corrosion resistance, radiation resistance, etc., and is widely used in various types of aerospace vehicles [4–6]. According to the type of aramid paper used, there are two common types of aramid paper honeycomb. Among them, those made of poly (m-phenylene isophthalamide) fiber paper are often referred to as NOMEX® honeycombs, and those made of poly (p-phenylene terephthalamide) fiber paper are often referred to as KEVLAR® honeycomb. KEVLAR® honeycomb has more outstanding mechanical and thermal properties than NOMEX® honeycomb. For advanced aircraft, using KEVLAR® honeycomb can achieve a greater weight reduction target [7].

In the edge structures of control surfaces in fighter jets, small unmanned aerial vehicles and other aircraft, a full-height honeycomb integral sandwich construction with upper and lower skins is widely applied. Different from the method applied in the conventional honeycomb sandwich panel structure manufacturing process, honeycomb bonding can significantly decrease structural weight and enhance manufacturing efficiency. During the manufacturing processes of the full-height honeycomb sandwich structure, a secondary bonding process is indispensable. However, because of the high temperature and pressure during the bonding procedure, the honeycomb core experiences shrinkage in the height direction and, consequently, depression appears on the surface of the honeycomb structure. Therefore, clearing the mechanism of this shrinkage process is important for maintaining surface shape and keeping structure size accuracy.

Little research on the shrinkage and creep properties of aramid honeycomb cores under high temperature and pressure has been published, and most work addresses the fundamental mechanical properties of aramid honeycomb sandwich structures, such as tension [8], compression [9–11], shear [12], three-point bending [13] and impact properties [14]. For the creep properties of aramid honeycomb sandwich structures, the published literature often applies three-point bending creep tests. For example, this experimental approach was utilized by Du et al. [15] to investigate the creep behavior of Nomex aramid honeycomb sandwich structures under various stress and humidity conditions, and the time dependent creep strain curve based on the Burgers model was fitted. Meanwhile, Ishak et al. [16] conducted the three-point bending creep performance of honeycomb sandwich structures at different spans, and applied Findley's power law to simulate the viscoelastic response of the cross-arm. These aforementioned studies predominantly focused on the creep formability of honeycomb sandwich structures by conducting three-point bending at room temperature, and factors such as temperature and pressure were ignored.

Furthermore, there is also limited research on the compressive creep behavior of aramid honeycomb cores in the in-plane direction. Gibson and Ashby [17] calculated the overall elastic moduli of porous structures in both the in-plane and out-of-plane directions from a mechanical perspective based on material properties, and validated the in-plane viscoelastic and creep deformations by experiment data. However, the creep characteristics in the out-of-plane direction of the porous structures were not investigated. Lin and Huang [18] utilized a representative volume element (RVE) to analyze the in-plane creep characteristics of hexagonal honeycombs with non-uniform thickness cell edges. However, their investigation did not extend to examining the out-of-plane creep behavior, leaving a potential avenue for further research in this domain. Balkan and Mecitoglu [19] utilized a dynamic mechanical analyzer (DMA) to identify the viscoelasticity of aramid sandwich samples and carbon/epoxy resin samples. They investigated the nonlinear dynamic response of viscoelastic sandwich plates under explosion load. However, they did not explore the creep properties of aramid honeycomb cores, despite their viscoelastic characteristics. It can be found from the above works that the existing body of research has not fully revealed the creep characteristics of honeycomb structures in different directions.

Apart from the creep attribute under high pressure or stress, aramid honeycomb core experiences shrinkage deformation due to dehydration at high temperatures. The primary cause of this shrinkage is the hygroscopic expansion of the aramid fibers and phenolic resin that occurred before the test. There is a relatively larger body of research concerning the hygroscopicity of aramid honeycomb cores [20,21]. For example, Sala [22] investigated the infiltration levels and performance impacts of different liquids (water, fuel and dichloromethane) on aramid fibers and carbon fibers by utilizing an optical microscopy, and significantly demonstrated the swelling of aramid polyamide fibers after the wet treatment, confirming the moisture absorption swelling effect at the microstructural level. Guo et al. [23] provided a microscopic explanation for the moisture-induced expansion mechanism in composite materials, i.e., the formation of one or more hydrogen bonds between bound water and the molecular chains of fibers and resins.

Based on the observed out-of-plane shrinkage phenomenon under high temperature and pressure condition, this study systematically investigated the out-of-plane compressive creep behavior of aramid honeycomb cores. In this study, a creep test machine was employed to simulate the common forming process environment (temperature and pressure parameters), and the compressive creep phenomenon of aramid honeycomb cores was discussed in detail for the first time. After the conclusion of this experiment, the viscoelastic Burgers mechanical model was further applied to fit the time–compressive creep strain curve of the aramid honeycomb core in the creep stage. Then, the shrinkage deformation curve equation of the aramid honeycomb core was obtained by superimposing the deformations of the dehumidification, pressurization and creep stages. The research results of this paper have great guiding significance for the manufacturing process of aramid honeycomb sandwich structure.

2. Materials and Methods

2.1. Experimental Materials

Supplied by Shenzhen Kuang-chi Cutting-edge Technology Co., Ltd. (Shenzhen, China), aramid honeycomb core with a density of 48 kg/m^3 was utilized in this work. The aramid honeycomb core has a repetitive hexagonal unit structure, and different numbers of complete honeycomb lattices may have appeared under the same sample size. To ensure consistency, the W direction of the sample included 9 complete honeycomb lattices, and the L direction included 11 complete honeycomb lattices, for a total of 99 complete honeycomb cores. The specific sample dimensions and honeycomb lattice dimensions are illustrated in Figure 1.

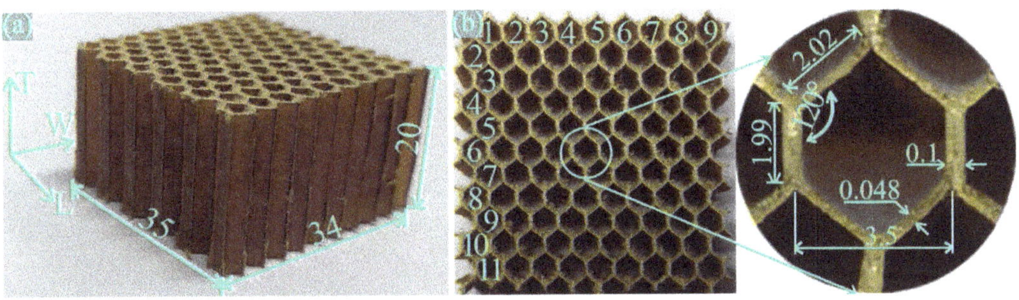

Figure 1. Detailed dimensions of experimental samples and honeycomb lattice structure. (**a**) Sample dimensions and (**b**) honeycomb lattice dimensions.

2.2. Experimental Methods

This experiment was conducted using the SUST creep testing machine, which has a maximum control temperature of 350 °C and a force accuracy of 0.01 N. In the experiment, extensometers manufactured by Weihai Dikong Electronic Technology Co., Ltd. (Weihai, China), with the model number D-KB10-05-T4 and a displacement accuracy of 0.0001 mm, were employed. To adapt to the needs of out-of-plane compression creep, the experimental fixture was self-modified, as shown in Figure 2. In the mechanism of the out-of-plane compression creep instrument, the plane compression platens primarily convert the pressure provided by the creep testing machine into a planar pressure. Extension rods were employed to transmit the deformation of the sample height to the extensometers. Simultaneously, the environmental chamber controlled the ambient temperature, while extensometers recorded the sample's height shrinkage in real-time. The specific working mechanism of the apparatus is illustrated in Figure 3.

In this study, the selected temperature conditions were chosen as they are commonly used in honeycomb structure manufacturing processes: a medium curing temperature of 120 °C and a high curing temperature of 180 °C, with set pressure conditions of 0.2 MPa and 0.3 MPa, respectively. Each experimental condition employed five identical aramid honeycomb cores of the same material, with the same dimensions and lattice counts. Initially, a force of 10 N was applied to the specimen to ensure intimate contact between the plane compression platens and the surfaces of the aramid honeycomb core, thereby enhancing measurement accuracy. Upon reaching the processing temperature, no additional pressure was directly applied; instead, the specimen was maintained at this temperature for 30 min to eliminate moisture from within the aramid honeycomb core and mitigate the influence of humidity on its creep behavior. In industrial settings, subjecting the honeycomb core to temperatures exceeding 100 °C for 30 min effectively removes internal moisture. Consequently, whether baked at 120 °C or 180 °C for 30 min, water molecules within the honeycomb core are efficiently expelled. It is worth noting that the selected test samples have a buckling strength of 1 MPa. Therefore, under the experimental pressure conditions (0.2 MPa, 0.3 MPa), no structural buckling in the aramid honeycomb cores occurred. The specific experimental procedures are illustrated in Figure 4.

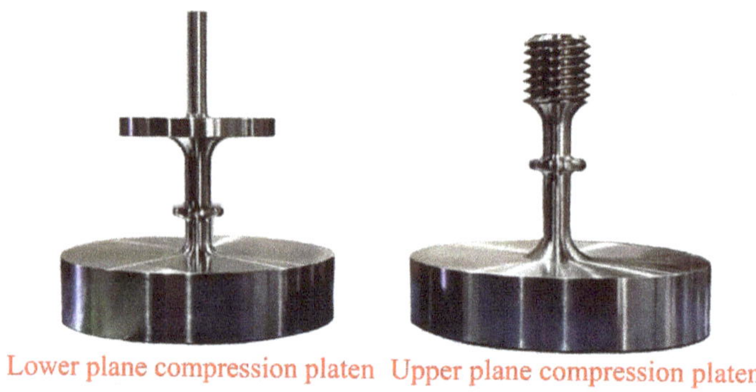

Figure 2. Upper and lower plane compression platens.

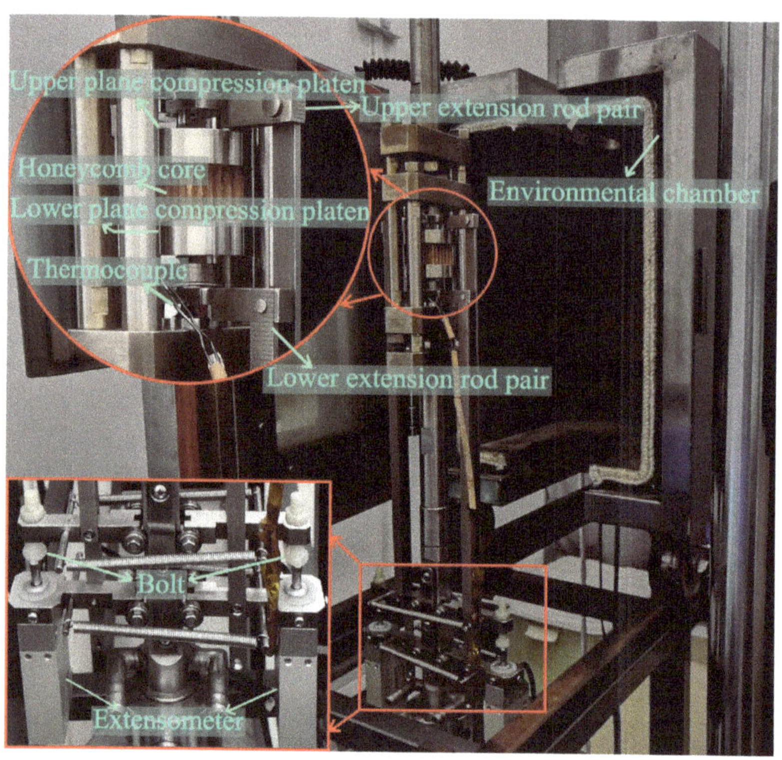

Figure 3. Out-of-plane compression creep experimental instrument.

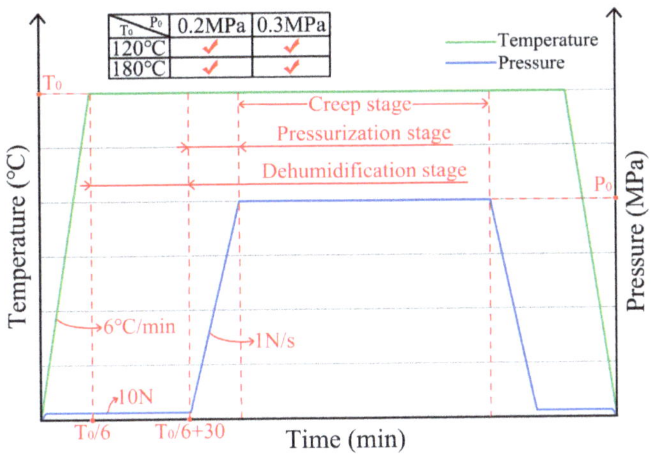

Figure 4. Specific experimental procedures.

3. Results

Under each experimental condition, a dataset exhibiting relative stability and minimal fluctuations was selected. Subsequently, sampling was applied at every fifth data point to plot the corresponding time–compression deformation curves, as depicted in Figure 5. It can be observed that the aramid honeycomb cores exhibited a consistent trend during the out-of-plane compression creep test, characterized by compression–expansion–platform–pressurization creep.

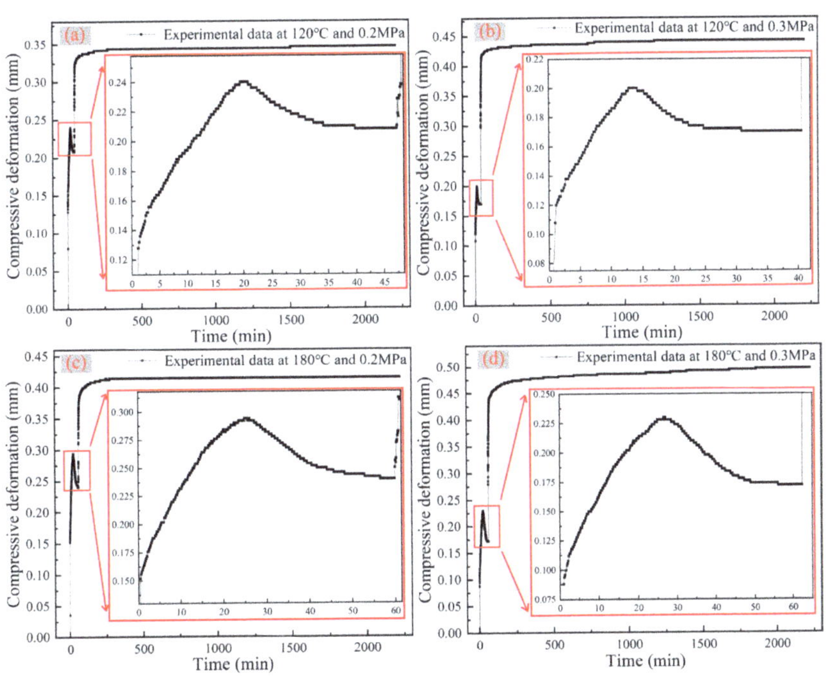

Figure 5. Time–compression deformation curves under various experimental conditions: (**a**) experimental data at 120 °C and 0.2 MPa, (**b**) experimental data at 120 °C and 0.3 MPa, (**c**) experimental data at 180 °C and 0.2 Mpa and (**d**) experimental data at 180 °C and 0.3 MPa.

Based on this trend, a typical results trend curve was plotted, which delineated the consistent variation pattern of the honeycomb core's height direction under high temperature and pressure condition, as shown in Figure 6. Among them, point O represents the initial point of the experiment, point A is the first sampling point beyond 10 s, point B indicates the peak shrinkage of the aramid honeycomb core, point C is the expansion peak, point D denotes the end of the dehumidification stage and the beginning of the pressurization stage, and point E marks the end of the pressurization stage and the beginning of the creep stage. In Figure 6, three distinct deformation stages are observed: the OD stage, the DE stage and the EG stage. The OD stage is primarily attributed to the volume contraction caused by the dehydration of the honeycomb at high temperatures, thus named the dehumidification stage. The DE stage exhibited obvious elastic characteristics, representing the elastic deformation of the honeycomb under out-of-plane compression, named the pressurization stage. The deformation observed in the EG stage is the deformation attained under the conditions of the creep test, and therefore, the EG stage is designated as the creep stage.

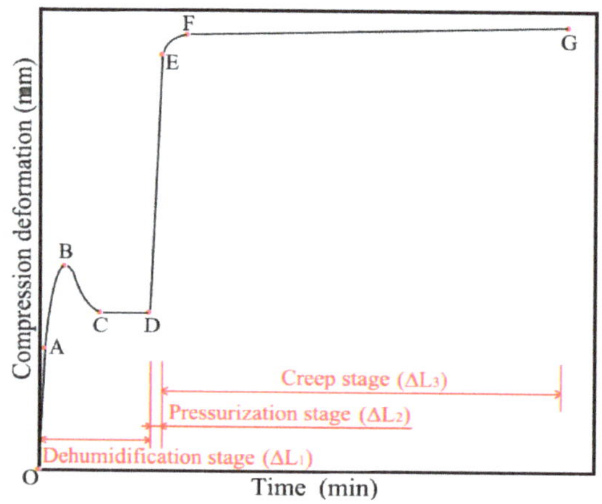

Figure 6. Typical trend curve of creep experiment.

Define total deformation ΔL, where $\Delta L = \Delta L_1 + \Delta L_2 + \Delta L_3$ and $\Delta L_1 = \Delta L_{11} + \Delta L_{12}$. The deformation ΔL_1 during the dehumidification stage mainly consists of two parts: the elastic deformation ΔL_{11} in the OA stage and the humidity deformation ΔL_{12} in the AD stage. The deformations during the pressurization and creep stages are represented by ΔL_2 and ΔL_3, respectively.

3.1. Dehumidification Stage

The deformation ΔL_1 in the dehumidification stage is mainly composed of two parts: the elastic deformation ΔL_{11} in the OA stage and the humidity deformation ΔL_{12} in the AD stage. For each experimental condition with five data sets, the highest and lowest values were excluded, and then the three central data sets for ΔL_{11} and ΔL_{12} were selected to be presented in Table 1.

Table 1. Deformations of ΔL_{11} and ΔL_{12} under each experimental condition.

Experimental Conditions	ΔL_{11} (mm)	ΔL_{12} (mm)
120 °C, 0.2 MPa	0.112, 0.118, 0.132	0.060, 0.062, 0.066
120 °C, 0.3 MPa	0.088, 0.104, 0.110	0.046, 0.050, 0.058
180 °C, 0.2 MPa	0.100, 0.100, 0.130	0.050, 0.058, 0.066
180 °C, 0.3 MPa	0.106, 0.106, 0.116	0.054, 0.064, 0.064
Overall mean value	0.110	0.058

ΔL_{11} represents the deformation resulting from the application of a 10 N force at room temperature, while ΔL_{12} is attributed to moisture dehydration. Given the consistent experimental site and a dehumidification temperature exceeding 100 °C, it is theoretically expected that ΔL_{11} and ΔL_{12} under the four experimental conditions would be fixed at specific values. Consequently, the average values of all ΔL_{11} and ΔL_{12} were calculated separately and are presented in Table 1. The errors in ΔL_{11} and ΔL_{12} arise from the discrepancies in the experiment's starting time.

To investigate whether the deformation ΔL_{12} of the 20 mm aramid honeycomb core remains consistently around 0.058 mm during the dehumidification stage, two sets of aramid honeycomb cores with identical dimensions were measured for height in the T direction before and after drying, as shown in Table 2. By summing up the height differences of the four points before and after drying and taking the average, the moisture-induced deformation of the honeycomb was measured to be 0.055 mm, which closely matches the data obtained from the creep testing machine.

Table 2. Height of aramid honeycomb core before and after drying.

Sample	Before Drying		After Drying	
	Height 1 (mm)	Height 2 (mm)	Height 1 (mm)	Height 2 (mm)
Sample 1	19.95	19.83	19.89	19.77
Sample 2	20.00	20.02	19.95	19.97

Based on a detailed observation of Figure 6, it is further revealed that the aramid honeycomb core does not remain in a continuous state of contraction during the dehumidification phase. Specifically, after reaching the shrinkage peak point B, the aramid honeycomb core exhibits an unexpected expansion trend (as seen in the BD segment of Figure 6). That is because, prior to the initiation of the experiment, hydrogen bonds had formed between the water molecules and the polar groups within the aromatic polyamide fiber/phenolic resin, consequently resulting in the expansion of the aramid honeycomb core [23,24]. In order to confirm the hygroscopic expansion of the aramid honeycomb samples prior to experimentation, this study utilized a Thermo Fisher Nicolet iS20 infrared spectrometer to measure the functional groups of pure aramid paper before and after drying. As shown in Figure 7, the infrared spectrum revealed hydroxyl groups exhibiting stretching vibrations around 3300–3400 cm^{-1} and bending vibrations around 1500 cm^{-1}. These characteristic peaks indicate that prior to the experiment, the aramid fibers had already formed hydrogen bonds with external water molecules, thereby also validating the mechanism that aramid honeycomb structures expand due to moisture absorption under humid conditions.

Since the aramid honeycomb core had already experienced hygroscopic swelling before the experiment began, at the initial stages of the experiment, the increase in environmental temperature disrupted the hydrogen bonds formed between water and the polar groups, causing water evaporation and resulting in a contraction phenomenon (as seen in the AB segment of Figure 6). As the temperature further increased to the conditioned level, the enhanced thermal motion of molecules weakened the intermolecular interactions, leading to an increase in the polymer's free volume and resulting in the overall expansion of the aramid honeycomb core (as seen in the BC segment of Figure 6). After the expansion phenomenon and before the pressurization stage, there was a relatively long duration of plateau (as seen in the CD segment of Figure 6). This plateau stage marks the end of the dehumidification stage and signifies the complete removal of moisture from the aramid honeycomb core.

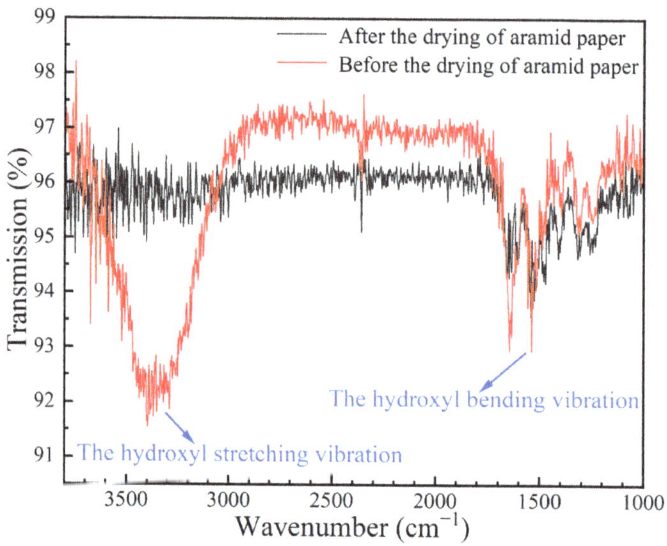

Figure 7. The infrared spectrum of aramid paper before and after drying.

3.2. Pressurization Stage

The averages of ΔL_2 and ΔL_3 under each experimental condition are presented in Table 3. According to Table 3, it can be observed that the deformation ΔL_2 during the pressurization stage occupies a significant proportion of the total deformation ΔL_2, approximately around 55%. The magnitude of ΔL_2 is primarily influenced by the applied force, meaning that as the load increases, the corresponding ΔL_2 also increases. It can also be found that under the condition of keeping the temperature constant, when the pressure is expanded to 1.5 times, the deformation ΔL_2 does not increase in the same proportion, revealing a nonlinear relationship. It shows that under high temperature and pressure conditions, the aramid paper honeycomb core impregnated with phenolic resin is not a pure linear elastomer, nor is it a pure viscous fluid, and it is more likely to be a viscoelastic material between elasticity and viscosity. It is worth noting that this finding is consistent with the results obtained by Balkan and Mecitoglu using a dynamic mechanical analyzer (DMA) to measure the viscoelasticity of aramid honeycomb cores [19].

Table 3. Deformation of each stage.

Experimental Conditions	ΔL (mm)	ΔL_1 (mm)		ΔL_2 (mm)	ΔL_3 (mm)
		ΔL_{11} (mm)	ΔL_{12} (mm)		
120 °C, 0.2 MPa	0.392	0.110	0.058	0.198	0.026
120 °C, 0.3 MPa	0.442			0.242	0.032
180 °C, 0.2 MPa	0.430			0.224	0.038
180 °C, 0.3 MPa	0.492			0.260	0.064

3.3. Creep Stage

3.3.1. Explanation of Creep Stage

Aramid fibers belong to the category of aromatic polyamide organic polymers, characterized by their molecular structure consisting of rigid main chains aligned along the fiber axis, composed of aromatic rings and amide bonds. These main chains interact through strong hydrogen bonding, collectively forming a rod-like fiber structure [25,26]. The manufacturing process of aramid fibers employs the dry-jet wet spinning technique, yielding thermoplastic polymer fibers with a distinct skin-core structure. Within this structure, the molecular chains at the fiber's core exhibit segments that are not parallel to the fiber axis, which leads to a certain degree of creep behavior under stress. Additionally, under experimental conditions, an increase in temperature causes the softening of the phenolic resin matrix, increasing the likelihood of fiber slippage within the resin, ultimately resulting in the creep phenomenon observed in aramid honeycomb structures.

It is evident from Table 3 that the aramid honeycomb core manifests creep phenomena. The magnitude of creep deformation, denoted as ΔL_3, constitutes a minor fraction, not surpassing 10% of the total deformation. However, ΔL_3 is positively correlated with temperature, time and pressure. With the increase in temperature, time and pressure, the deformation of ΔL_3 is more significant. The experimental results indicate that the honeycomb composite material (consisting of thermoplastic poly (p-phenylene terephthalamide) fibers and thermosetting phenolic resin) exhibits the creep phenomenon characteristic of viscoelastic materials under high temperature and pressure. The heights of the specimens were measured before and after the experiment (once moisture equilibrium was achieved), indicating that the plastic deformation and the deformation attributed to creep (ΔL_3) of the honeycomb were found to be nearly consistent.

3.3.2. Burgers Model Introduction and Fitting Results

Under the influence of specific temperature and small constant stress, viscoelastic polymer will exhibit the creep phenomenon. In order to characterize the viscoelastic properties of polymers, researchers often use simple mechanical elements such as Hooke's spring and Newton's dashpot. In the related research, mechanical models capable of describing the creep phenomenon of polymeric materials have been developed through series, parallel and other combinations of Hooke's spring and Newton's dashpot [27]. Since the creep deformation of the viscoelastic honeycomb core during the experiment depends simultaneously on time, temperature and pressure, its creep deformation is fitted individually. Furthermore, since the four-element Burgers model can encompass various scenarios described by both the two-element (Maxwell model and Kelvin model) and three-element models, Burgers model was directly chosen to fit the time–creep strain curves under each experimental temperature condition. The Burgers model's constitutive equation and model are presented in Equation (1) and depicted in Figure 8, respectively.

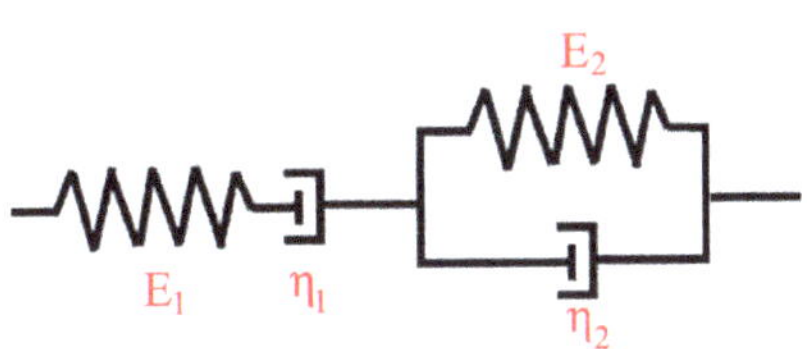

Figure 8. Burgers mechanical constitutive model.

The Burgers constitutive equation is as follows:

$$\varepsilon(t) = \frac{\sigma_0}{E_1} + \frac{\sigma_0}{\eta_1}t + \frac{\sigma_0}{E_2}(1 - \exp(-E_2 t/\eta_2)) \quad (1)$$

where $\varepsilon(t)$ represents the strain; σ_0 indicates the initial stress; E_1 and E_2 denote the elastic modulus of the left and right springs, respectively; and η_1 and η_2 are the viscosity of the left and right dashpot, respectively.

The creep experimental data of aramid honeycomb under different experimental conditions are shown in Figure 9. Through the analysis of the experimental conditions and the creep strain, it was found that an increase in temperature or pressure leads to the increase in creep amount. Moreover, an increase in pressure delayed the time it took for the aramid honeycomb core to reach a steady creep state. Under the experimental conditions of 120 °C, 0.2 MPa and 180 °C, 0.3 MPa, the honeycomb did not show a distinct steady creep state. This could potentially be attributed to the damage sustained by the aramid fiber/phenolic resin interface or the continuous molecular chain slippage of the aramid fiber under high temperatures and pressure. Subsequently, MATLAB R2020a software was employed to fit the creep experimental data at 120 °C and 180 °C using the nonlinear surface fitting method, treating time and stress as independent variables. The fitting parameters and errors of the Burgers model are presented in Table 4. Ultimately, the Burgers' constitutive equations of creep strain at 120 °C and 180 °C are given by Equations (2) and (3), respectively.

$$\varepsilon(t) = \frac{\sigma_0}{4.616} + \frac{\sigma_0}{25580}t + \frac{\sigma_0}{3.571}(1-\exp(-3.571t/484.1)) \quad (2)$$

$$\varepsilon(t) = \frac{\sigma_0}{3.481} + \frac{\sigma_0}{8231}t + \frac{\sigma_0}{1.862}(1-\exp(-1.862t/229)) \quad (3)$$

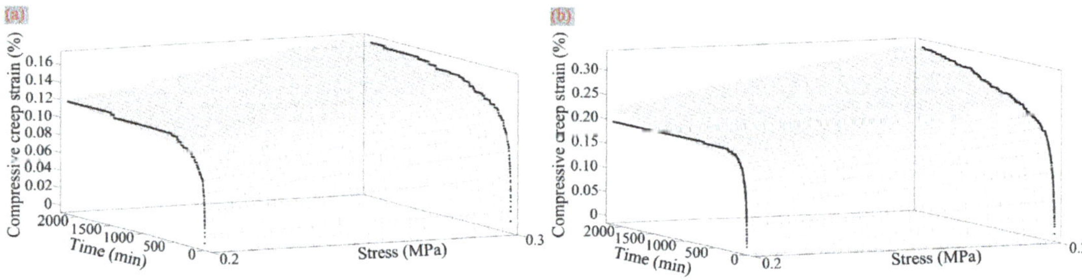

Figure 9. Time–creep strain curves and fitting results under various experimental conditions: (**a**) the creep curves and nonlinear surface fitting result at 120 °C and (**b**) the creep curves and nonlinear surface fitting result at 180 °C.

Table 4. Summary of the fitted parameters for creep strain tested under experimental conditions based on the Burgers model.

Parameters	Experimental Conditions			
	120 °C	Error Range	180 °C	Error Range
E_1	4.616	(−0.088, +0.087)	3.481	(−0.038, +0.039)
η_1	25,580	(−800, +800)	8231	(−59, +58)
E_2	3.571	(−0.051, +0.051)	1.862	(−0.011, +0.01)
η_2	484.1	(−17.2, +17.2)	229	(−3.3, +3.2)
Fitting degrees	0.9226		0.9428	

According to the Burgers constitutive equation, it is evident that as $t \rightarrow 0$, the strain $\varepsilon(t)$ approaches $\frac{\sigma_0}{E_1}$. Simultaneously, as $t \rightarrow \infty$, the strain $\varepsilon(t)$ approaches $\frac{\sigma_0}{E_1} + \frac{\sigma_0}{\eta_1}t + \frac{\sigma_0}{E_2}$, and the strain rate $\dot{\varepsilon}(t)$ approaches $\frac{\sigma_0}{\eta_1}$. Furthermore, based on Equations (2) and (3), it was observed that under consistent nominal stress, at $t \rightarrow 0$, higher temperatures correspond to smaller E_1, indicating greater creep deformation at elevated temperatures. At $t \rightarrow \infty$, higher temperatures correspond to smaller η_1, indicating higher creep deformation rates under elevated temperatures. In addition to the above, E_1, E_2 and η_1 collectively dictate the long-term creep deformation $\varepsilon(\infty)$, whereas E_2 and η_2 jointly influence the time required to achieve stable creep deformation.

It can also be observed from Table 4 that the application of nonlinear surface fitting methods to fit the Burgers model with creep test data yields a relatively ideal fitting effect. Furthermore, the fitting results exhibit good representativeness and are capable of predicting the creep curves under different stress levels at the same temperature conditions, which holds significant practical implications for engineering applications.

4. Conclusions

As the most widely used sandwich material in aerospace applications, the deformation of aramid honeycomb cores in the secondary bonding process has a great influence on the quality of the manufacturing process. In this paper, the compression phenomenon of para-aramid honeycomb cores under high temperatures and high pressure was systematically discussed. The main conclusions are as follows:

1. The out-of-plane compression performance of the aramid honeycomb core was systematically investigated for the first time. Based on the compression deformation curve, the compression behavior can be divided into three stages: the dehumidification stage, the pressurization stage and the creep stage. These stages contribute approximately 35%, 55% and 10%, respectively, to the total deformation.
2. During the dehumidification stage, the aramid honeycomb core undergoes a process of compression, expansion and plateau. The primary reason for the phenomenon of honeycomb cores is the combined effect of thermal expansion and dehydration shrinkage on aramid honeycomb cores. Prior to the experiment, hydrogen bonds formed between water molecules and the polar groups in the aramid honeycomb core, causing expansion. As the temperature increased, the hydrogen bonds were disrupted, leading to water evaporation and the contraction of the honeycomb core. Further temperature increases weakened intermolecular interactions, increasing the polymer's free volume and resulting in the overall expansion of the aramid honeycomb core.
3. In the pressurization stage, the deformation ΔL_2 and pressure exhibit a nonlinear relationship, indicating that the aramid honeycomb core impregnated with phenolic resin is a not purely linear elastomer. Simultaneously, in the creep stage, the aramid honeycomb core exhibits compression creep phenomena related to time, temperature and pressure under high temperature and pressure conditions. Taking into account the aforementioned phenomena, the aramid honeycomb core under high temperature and pressure conditions can be characterized as a viscoelastic material.
4. After the out-of-plane compression creep experiment, the nonlinear surface fitting method was employed to fit the creep deformation equations at various stress levels under a constant temperature, resulting in a high degree of fitting accuracy. This method is also capable of predicting the creep deformation behavior under the influence of varying stress levels. Finally, the total height deformation of the aramid honeycomb core under typical manufacturing processes can be obtained by integrating the dehumidification (ΔL_1), pressurization (ΔL_2) and creep (ΔL_3) stages.

Author Contributions: X.H. (Xinzheng Huang): conceptualization, methodology and validation; X.H. (Xin Hu): conceptualization, methodology, software, validation and writing—original draft; J.G., D.Z. and S.Y.: methodology and software; L.Z. (Lihua Zhan), B.M., M.H. and L.Z. (Lihua Zhang): supervision, editing and review. All authors have read and agreed to the published version of the manuscript.

Funding: This research was funded by the Scientific and Technological Innovation Project of Shenzhen, grant number CJGJZD20200617102400001.

Institutional Review Board Statement: Not applicable.

Data Availability Statement: The original contributions presented in the study are included in the article, further inquiries can be directed to the corresponding author.

Conflicts of Interest: Author Xinzheng Huang was employed by the company Shenzhen Kuang-Chi Cutting-Edge Technology Co., Ltd. The remaining authors declare that the research was conducted in the absence of any commercial or financial relationships that could be construed as a potential conflict of interest.

References

1. Feng, J.P.; Zhan, L.H.; Ma, B.L.; Zhou, H.; Xiong, B.; Guo, J.Z.; Xia, Y.N.; Hui, S.M. Metal-Metal Bonding Process Research Based on Xgboost Machine Learning Algorithm. *Polymers* **2023**, *15*, 4085. [CrossRef] [PubMed]
2. Liu, S.; Zhan, L.H.; Ma, B.L.; Guan, C.L.; Yang, X.B. Simulation and Experimental Study on the Internal Leak Behavior in Carbon Fiber Reinforced Composite Components. *Polymers* **2023**, *15*, 2758. [CrossRef] [PubMed]
3. Zhang, D.C.; Zhan, L.H.; Guan, C.L.; Guo, J.Z.; Ma, B.L.; Dai, G.M.; Yao, S.M. Optimization of Vibration Pretreatment Microwave Curing in Composite Laminate Molding Process. *Polymers* **2023**, *15*, 296. [CrossRef] [PubMed]
4. Rathod, S.; Tiwari, G. Structural analysis of cylindrical composite structures for high velocity applications: A parametric optimization study. *Thin Walled Struct.* **2023**, *192*, 111178. [CrossRef]
5. Shan, J.F.; Xu, S.L.; Zhou, L.J.; Wang, D.R.; Liu, Y.G.; Zhang, M.; Wang, P.F. Dynamic fracture of aramid paper honeycomb subjected to impact loading. *Compos. Struct.* **2019**, *223*, 110962. [CrossRef]
6. Zhao, Y.Y.; Sun, Y.X.; Li, R.Y.; Sun, Q.R.; Feng, J.T. Response of aramid honeycomb sandwich panels subjected to intense impulse loading by Mylar flyer. *Int. J. Impact Eng.* **2017**, *104*, 75–84. [CrossRef]
7. Kumar, A.; Angra, S.; Chanda, A.K. Fabrication and mechanical characterisation of composite sandwich structure having kevlar honeycomb core. *Adv. Mater. Process. Technol.* **2023**, 1–18. [CrossRef]
8. Liu, L.Q.; Wang, H.; Guan, Z.W. Experimental and numerical study on the mechanical response Of Nomex honeycomb core under transverse loading. *Compos. Struct.* **2015**, *121*, 304–314. [CrossRef]
9. Zhang, Y.W.; Liu, T.; Tizani, W. Experimental and numerical analysis of dynamic compressive response of Nomex honeycombs. *Compos. Part B Eng.* **2018**, *148*, 27–39. [CrossRef]
10. Liu, L.Q.; Meng, P.; Wang, H.; Guan, Z.W. The flatwise compressive properties of Nomex honeycomb core with debonding imperfections in the double cell wall. *Compos. Part B Eng.* **2015**, *76*, 122–132. [CrossRef]
11. Asprone, D.; Auricchio, F.; Menna, C.; Morganti, S.; Prota, A.; Reali, A. Statistical finite element analysis of the buckling behavior of honeycomb structures. *Compos. Struct.* **2013**, *105*, 240–255. [CrossRef]
12. Yang, W.C.; Zhang, X.F.; Yang, K.J.; Pan, B.; Fei, B.J.; Yi, X.S.; Chen, Y.L. Shear property characterization of aramid paper and its application to the prediction of honeycomb behaviors. *Compos. Struct.* **2020**, *254*, 112800. [CrossRef]
13. Naufal, A.M.; Prabowo, A.R.; Muttaqie, T.; Hidayat, A.; Purwono, J.; Adiputra, R.; Akbar, H.I.; Smaradhana, D.F. Characterization of sandwich materials–Nomex-Aramid carbon fiber performances under mechanical loadings: Nonlinear FE and convergence studies. *Rev. Adv. Mater. Sci.* **2024**, *63*, 20230177. [CrossRef]
14. Kolopp, A.; Rivallant, S.; Bouvet, C. Experimental study of sandwich structures as armour against medium-velocity impacts. *Int. J. Impact Eng.* **2013**, *61*, 24–35. [CrossRef]
15. Du, Y.; Yan, N.; Kortschot, M.T. An experimental study of creep behavior of lightweight natural fiber-reinforced polymer composite/honeycomb core sandwich panels. *Compos. Struct.* **2013**, *106*, 160–166. [CrossRef]
16. Ishak, M.R.; Yidris, N.; Zuhri, M.Y.M. Flexural creep response of honeycomb sandwich pultruded GFRP composite cross-arm: Obtaining full-scale viscoelastic moduli and creep coefficients. *J. Mater. Res. Technol.* **2024**, *29*, 225–241.
17. Gibson, L.J.; Ashby, M.F. *Cellular Solids: Structure and Properties*; Cambridge University Press: Cambridge, UK, 1997.
18. Lin, J.Y.; Huang, J.S. Creep of hexagonal honeycombs with Plateau borders. *Compos. Struct.* **2005**, *67*, 477–484. [CrossRef]
19. Balkan, D.; Mecitoglu, Z. Nonlinear dynamic behavior of viscoelastic sandwich composite plates under non-uniform blast load: Theory and experiment. *Int. J. Impact Eng.* **2014**, *72*, 85–104. [CrossRef]
20. Birman, V.; Kardomateas, G.A. Review of current trends in research and applications of sandwich structures. *Compos. Part B Eng.* **2018**, *142*, 221–240. [CrossRef]
21. Castanie, B.; Bouvet, C.; Ginota, M. Review of composite sandwich structure in aeronautic applications. *Compos. Part C Open Access* **2020**, *1*, 100004. [CrossRef]
22. Sala, G. Composite degradation due to fluid absorption. *Compos. Part B Eng.* **2000**, *31*, 357–373. [CrossRef]
23. Guo, J.Z.; Zhan, L.H.; Ma, B.L.; Zhang, D.C.; Fan, Y.S.; Yao, S.M.; Feng, J.P. A review on failure mechanism and mechanical performance improvement of FRP-metal adhesive joints under different temperature-humidity. *Thin Walled Struct.* **2023**, *188*, 110788. [CrossRef]
24. Viana, G.; Costa, M.; Banea, M.D.; da Silva, L.F.M. A review on the temperature and moisture degradation of adhesive joints. *Proc. Inst. Mech. Eng. Part L J. Mater.-Des. Appl.* **2017**, *231*, 488–501. [CrossRef]
25. Torki, A.M.; Stojanovic, D.B.; Zivkovic, I.D.; Marinkovic, A.; Skapin, S.D.; Uskokovic, P.S.; Aleksic, R.R. The Viscoelastic Properties of Modified Thermoplastic Impregnated Multiaxial Aramid Fabrics. *Polym. Compos.* **2012**, *33*, 158–168. [CrossRef]
26. Obradovic, V.; Simic, D.; Sejkot, P.; Machalicka, K.V.; Vokác, M. Moisture absorption characteristics and effects on mechanical properties of Kolon/epoxy composites. *Curr. Appl. Phys.* **2021**, *26*, 16–23. [CrossRef]
27. Rubinstein, M.; Colby, R.H. *Polymer Physics*; Oxford University Press: Northamtonshire, UK, 2003.

polymers

MDPI

Review

Exploring the Origins of Low-Temperature Thermochromism in Polydiacetylenes

Magdalena Wilk-Kozubek [1], Bartłomiej Potaniec [1], Patrycja Gazińska [2] and Joanna Cybińska [1,3,*]

1 Materials Science and Engineering Center, Łukasiewicz Research Network—PORT Polish Center for Technology Development, 147 Stabłowicka Street, 54-066 Wrocław, Poland; magdalena.wilk-kozubek@port.lukasiewicz.gov.pl (M.W.-K.); bartlomiej.potaniec@port.lukasiewicz.gov.pl (B.P.)
2 Center for Population Diagnostics, Łukasiewicz Research Network—PORT Polish Center for Technology Development, 147 Stabłowicka Street, 54-066 Wrocław, Poland; patrycja.gazinska@port.lukasiewicz.gov.pl
3 Faculty of Chemistry, University of Wrocław, 14 F. Joliot-Curie Street, 50-383 Wrocław, Poland
* Correspondence: joanna.cybinska@port.lukasiewicz.gov.pl

Abstract: This review article delves into the intriguing phenomenon of low-temperature thermochromism, whereby materials change color in response to temperature variations, with a particular focus on its applications in temperature-sensitive fields like medical storage. By closely examining thermochromic materials, this article highlights their potential to offer innovative solutions for monitoring and preserving thermolabile products that require strict temperature control. This leads to a special emphasis on polydiacetylenes (PDAs), a class of conjugated polymers with unique low-temperature thermochromic properties, positioning them as promising candidates for reliable temperature indicators. This article then explores the underlying mechanisms for fine-tuning the thermochromic behavior of PDAs, particularly discussing recent advancements in PDA design, such as structural alterations of monomers to achieve low-temperature thermochromism. These modifications, influenced by factors like side-chain length, hydrogen-bonding interactions, and the use of copolymers, are intended to result in irreversible color transitions at specific low temperatures, which is crucial to maintaining the integrity of thermally sensitive products. Finally, this article discusses the potential applications of PDAs as thermochromic sensors in tissue biobanking, where their ability to provide visual indications of temperature fluctuations could significantly enhance the monitoring and management of biological samples.

Keywords: low-temperature thermochromism; irreversible thermochromism; temperature indicators; thermochromic sensors; polydiacetylenes; side-chain length; hydrogen-bonding interactions; copolymers; biobanking; freshness sensors

1. Introduction

The phenomenon of thermochromism, characterized by the alteration in substances' color in response to temperature changes, has fascinated scientists for centuries [1]. This process, which can be either reversible or irreversible, is driven by various mechanisms, such as molecular (conformational) and crystal structure (symmetry) transformations, alterations in coordination geometry and number, band gap energy shifts, and phase transitions (Figure 1) [2]. These mechanisms are observed in a wide range of materials, including inorganic [3], organic [4], organic–inorganic hybrid [5], and polymeric [6] ones. Due to their thermochromic behavior, these compounds hold significant potential for practical applications, inspiring innovative and environmentally friendly technologies. They form the basis of intriguing products, like invisible ink [7], heat-sensitive fax paper [8], and the iconic Hypercolor T-shirts [9]. Primarily, however, thermochromic materials that demonstrate color changes above ambient temperature serve as temperature indicators, providing crucial information about temperature variations [10]. For instance, they can measure temperature distribution in industrial heating devices [11], indicate temperature fluctuations in fire-resistant coatings [12], and monitor body temperature [13] or blood circulation [14] in medical applications (Table 1).

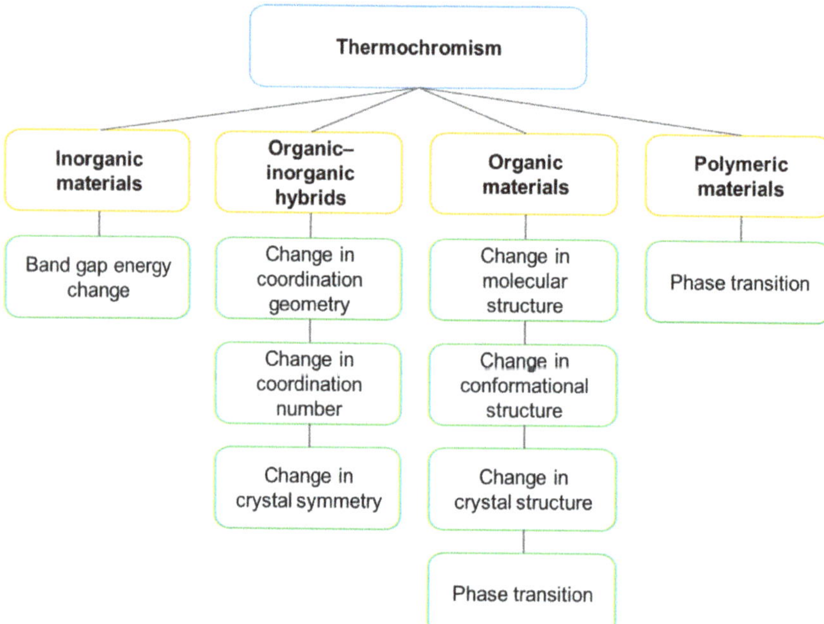

Figure 1. Mechanisms of action of thermochromic materials.

Table 1. Summary of applications, types of thermochromic materials, and their mechanisms of action.

Application	Thermochromic Material	Mechanism	Reference
Invisible ink	Organic	Change in molecular structure	[7]
Heat-sensitive fax paper	Organic	Change in molecular structure	[8]
Hypercolor T-shirt	Organic	Change in molecular structure	[9]
Industrial heating device	Organic	Phase transition	[11]
Fire-resistant coating	Organic–inorganic	Change in coordination geometry and number	[12]
Monitoring of body temperature	Organic	Phase transition	[13]
Monitoring of blood circulation	Organic	Phase transition	[14]

Recently, there has been a significant surge in interest in compounds exhibiting color alterations at sub-ambient temperature. This trend is primarily fueled by the increasing demand for advanced materials suitable for temperature sensors, which are essential to monitoring temperature-sensitive products like food, pharmaceuticals, and biological materials. These goods require strict temperature control during storage and transportation, as any deviations from optimal storage conditions can reduce their effectiveness [15,16]. For example, dairy products must be kept at 4 °C [17], while meat, fish, fruits, and vegetables need to be stored at −18 °C [18–20], and ice cream and confectionery products should be maintained at −29 °C [21]. In contrast, inactivated vaccines, dispersed cells, and polypeptide-based drugs such as insulin and antibiotics demand storage within a temperature range of 2 to 8 °C for short-term preservation [22–24]. Certain pharmaceuticals, like botulinum toxin A and methyl carprost, must be stored below −5 °C [25], while DNA and urine samples need to be kept at −20 °C for long-term preservation [26,27]. Moreover, live attenuated viral vaccines and red blood cells require storage for extended periods at even lower temperature, typically −70 and −80 °C, respectively [28,29].

Traditionally, the temperature monitoring of these thermolabile products has relied on temperature data loggers. However, the use of disposable models proves economically inefficient, while the retrieval and reuse of these devices present logistical challenges. Consequently, sensors employing thermochromic compounds emerge as a promising alternative, offering potential advancements in both cost effectiveness and logistical practicality for the monitoring of temperature-sensitive pharmaceuticals and biological materials [30]. In particular, compounds that undergo an irreversible color shift are more desirable for this application, as a color transition that cannot be reversed upon cooling provides a permanent record, indicating whether the products were exposed to excessive temperature at any point in the past. Moreover, the exact temperature at which the color change occurs is crucial, as it must align precisely with the temperature thresholds relevant to the stability of thermally sensitive products. While various materials with thermochromic properties can manifest irreversible color alterations, the greatest challenge is finding those that exhibit such transitions at specific low temperatures. For instance, inorganic and hybrid compounds generally do not reveal color shifts at sub-ambient temperature; instead, they change color at higher temperatures, ranging from 70 to 500 °C [31]. In contrast, organic materials, like leuco dye–developer systems, can alter color even at temperatures as low as −100 °C [32]. However, leuco dyes are typically sensitive to external factors such as moisture, oxygen, and pH changes, so they require protection [33]. This is achieved through encapsulation, which involves enclosing the dyes in protective coatings [34]. This method ensures better control over the conditions to which these organic compounds are exposed, regulates their release within strictly defined temperature ranges, and thus enables precise management of color shifts when mixed with developers [35]. Unlike organic materials, which commonly need a combination of active ingredients to attain thermochromic effects, polymeric materials inherently have the ability to undergo color transitions in response to temperature variations without requiring additional compounds [36]. This intrinsic capability arises from the unique structural properties of polymers, which can be engineered to exhibit thermochromic behavior over a broad temperature range, including sub-ambient levels, through direct modifications of their monomers. Consequently, polymeric materials offer a reliable and straightforward solution for managing temperature-induced changes, making them particularly valuable in applications necessitating accurate temperature monitoring.

This review, therefore, focuses on the use of thermochromic materials, including polymeric ones, in the medical field, with special emphasis on biobanking. In this area, precise temperature control of storage conditions is crucial to preventing the degradation of thermolabile biological samples and ensuring their integrity for future research and medical applications. To meet these requirements, the compounds used need to demonstrate irreversible color-changing properties at sub-ambient temperature. Accordingly, this review highlights recent advancements in the design of polydiacetylenes, a subgroup of polymeric materials known for their low-temperature thermochromism. It explores innovative strategies involving the manipulation of monomer chemical structures, the interactions between monomers, and the integration of monomers into nanoscale architectures to fine-tune the thermochromic behavior of polydiacetylenes (Scheme 1).

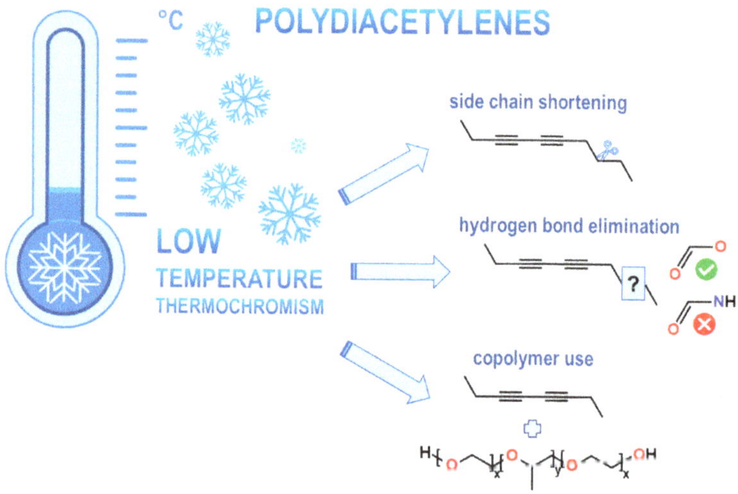

Scheme 1. A general scheme showing the influence of the structural modifications of diacetylene monomers on the low-temperature thermochromism of the corresponding polydiacetylenes.

2. Selection Criteria for Polydiacetylenes

Polydiacetylenes (PDAs) have been selected for low-temperature thermochromic applications, such as temperature indicators, due to unique properties that distinguish them from other materials. One of their key advantages is their ability, as single-component systems, to demonstrate a distinct color change—typically from blue to red—in response to temperature fluctuations [37]. This inherent characteristic allows PDAs to function effectively without the additional components, such as developers, required in systems based on leuco dyes. Consequently, the absence of these additives simplifies the design process and enhances the convenience of PDAs for practical applications. Additionally, the color alteration in PDAs is often more noticeable compared with leuco dye-based systems, where the change can be less pronounced and more gradual, depending on the strength of the developer. In contrast, color shifts in liquid crystals are angle-dependent, complicating interpretation and making them less effective for visual monitoring. Thus, the vivid color transition observed in PDAs makes them particularly suitable for visual temperature monitoring. Their high sensitivity to even slight temperature variations further underscores their appropriateness for precise temperature sensing. While leuco dyes generally exhibit lower sensitivity and are better suited for applications requiring a wider temperature range, they are not ideal for precise measurements. Although liquid crystals show high sensitivity, their optical changes are dependent on the angle of observation, limiting their usability in precision applications.

Moreover, PDAs can be easily chemically modified, allowing researchers to tailor their thermochromic properties to meet specific requirements. By introducing different functional groups, it is possible to fine-tune the temperature sensitivity, the range over which the color change occurs, and the intensity of this change, thereby customizing PDAs for various applications. In contrast, the ability to modify thermochromic properties in systems containing leuco dyes and liquid crystals is limited [38]. PDAs also offer the advantage of easy integration into various matrices, such as polymer films or membranes, thereby broadening their potential applications. Additionally, some PDAs can be designed to be more environmentally friendly than other thermochromic materials. For instance, organic–inorganic hybrids often contain hazardous heavy metals like lead [39], while leuco dye-based systems may include harmful components like bisphenol A [40]. In contrast, PDAs provide a safer alternative, making them an attractive option for applications where sustainability and environmental safety are critical considerations.

3. PDAs

PDAs are a class of conjugated polymers characterized by alternating double and triple bonds along their backbone. This ene-yne arrangement endows PDAs with thermochromic properties, making them a significant interest for various advanced applications, including temperature sensors [41], smart coatings [42], and biomedical devices [43]. First prepared by Wegner in 1969 [44], they are typically synthesized through a topochemical polymerization process. This solid-state method allows for precise control over the polymer's molecular structure and eliminates the need for solvents, initiators, or catalysts [45]. Consequently, it minimizes side reactions and by-products, leading to the formation of high-purity PDAs without extensive post-synthesis purification steps [46].

Efficient polymerization, however, requires diacetylene (DA) monomers to be pre-organized into highly ordered structures. For amphiphilic DA molecules, this organization usually occurs through self-assembly processes. Depending on specific conditions and the nature of the DA derivatives, self-assembly can lead to the formation of various architectures, such as micelles, vesicles, films, or monolayers [47]. In these self-assembled structures, the DA monomers are aligned approximately 5 Å apart, with each monomer oriented at about a 45° angle relative to the stacking axis of the assembly [48]. Upon exposure to ultraviolet (UV) light at a wavelength of 254 nm, the well-ordered DA derivatives undergo 1,4-addition polymerization (Scheme 2). This photochemical reaction leads to the formation of long conjugated polymer chains, which are characteristic of PDAs. PDAs typically undergo a gradual color change from blue to red upon heating. However, the detailed mechanism of the thermochromic transitions in PDAs is still under discussion. Earlier reports suggested a well-ordered planar structure for the blue state and a disordered non-planar structure with a reduced main-chain conjugation length for the red state. Recent reports, however, indicate that the thermochromic effect may originate from a transition between two different chain conformations, both of which may be perfectly ordered [49].

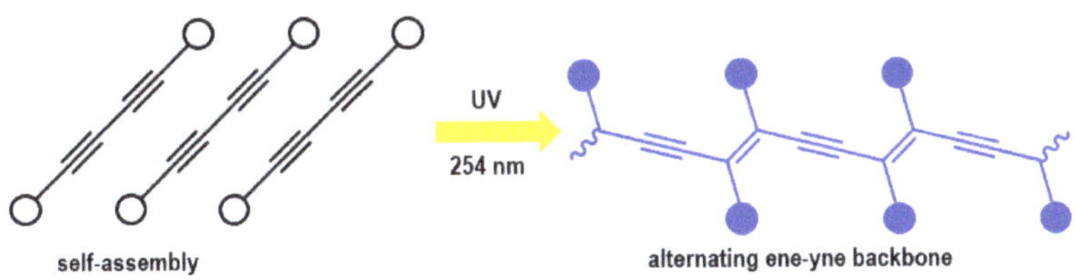

Scheme 2. Topochemical photo-polymerization of DA monomer.

For better insight into the structural evolution of diacetylenes during photo-polymerization and subsequent heat treatment, a possible mechanism is proposed. Generally, DA monomers generate blue-colored polydiacetylene upon UV irradiation, which transforms to red when exposed to heat. The planar "blue" form has unbroken π-electron overlap and a full conjugation length. Upon heat treatment, the accumulated stress during photo-polymerization is released by the partial disordering of long alkyl chains, which eventually twist the π orbitals into non-planar states, reducing the effective conjugation length to that of the red form.

Temperature-induced conformational changes within the polymer chain result in shifts in the energy of the frontier orbitals, which are observed as visible color alterations and variations in the UV–vis spectra. Specifically, these changes lead to a blue shift (hypsochromic effect) in the absorption bands of the UV–vis spectra. The blue PDA phase typically shows an absorption band centered around ~640 nm, due to π-electron delocalization along the conjugated polymer backbone. As the temperature increases, structural alterations shorten the conjugation length of the polymer chain and increase the energy gap. This results in a shift of the absorption band to ~540 nm, indicating the transition to the red PDA phase [46]. In addition to the commonly observed blue and red phases, other chromatic states, like purple and orange, can also occur, with absorption bands at ~590 and ~530 nm, respectively. The presence of these states allows for the observation of a purple-to-orange phase transition upon heating. Further temperature increases can enhance the hypsochromic effect, leading to the yellow PDA phase, characterized by an absorption band around ~470 nm [50].

The color changes discussed above can be either reversible or irreversible, depending on the chemical structure of the PDA and the nature of the interactions between its chains. Specifically, π–π stacking interactions, hydrogen bonding [51], and other non-covalent interactions between the chains play a crucial role in determining this behavior. These interactions cause the chains to arrange closely, which influences the conjugation of the π-electron system along the chains. Consequently, when PDAs with strong intermolecular interactions are exposed to elevated temperatures, their chains undergo minor conformational changes that alter the torsion angles without permanently breaking chemical bonds. These slight modifications in chain arrangement affect the length of the conjugated system, leading to a transition in the observed color. Since these alterations are minor, the PDA can revert to its original conformation and color once the temperature is lowered, demonstrating the influence of chemical interactions on thermal response and color reversibility. Conversely, when the intermolecular interactions are weak or absent, significant structural changes and alterations in the conjugated system occur during heating, resulting in irreversible color shifts [52]. In such cases, the PDA does not return to its original state upon cooling, as the new conformation represents a permanent alteration that is not easily reversed. Thus, the nature of the intermolecular interactions between the PDA chains and the extent of structural modifications in response to temperature fluctuations significantly impact whether color changes are reversible or permanent.

Understanding the aspects that influence whether color changes in PDAs are reversible or irreversible is crucial, as is determining the temperature at which these changes occur. This temperature, also known as the thermochromic transition temperature, is influenced by the chemical structure of the PDA, including chain length, types of functional groups, and strength of non-covalent interactions between PDA chains [53,54]. These factors closely resemble those affecting the melting point of DA derivatives. As a result, the thermochromic transition temperature of PDAs often aligns closely with the melting point of the corresponding DA monomers [53]. This similarity arises because both thermochromism and melting are associated with overcoming energy barriers resulting from similar intermolecular interactions, thus requiring comparable temperature to induce these processes. Consequently, different strategies aimed at lowering the temperature needed for thermochromic transition involve making structural modifications to the DA monomer to effectively reduce its melting point. By doing so, the energy barrier related to these processes is lowered, allowing the color changes to occur at the temperature required for a specific application.

4. Influence of Side-Chain Length

Achieving low-temperature thermochromism in PDAs primarily involves designing DA monomers with short alkyl side chains. Such derivatives exhibit a notably lower melting point, consequently leading to a reduced thermochromic transition temperature compared with counterparts with longer alkyl side chains [55]. The phenomenon of decreasing melting point with shortening chain length, observed in DAs with alkyl side chains, resembles the trend seen in linear alkanes [56]. This similarity can be attributed to the reduction in chain length, which diminishes the energy associated with attractive van der Waals forces, thereby requiring less energy to overcome them. Therefore, this results in both a lower melting point and a reduced thermochromic temperature [57].

Rougeau et al. demonstrated the efficacy of this approach by investigating a series of symmetric diynes and diyne esters containing alkyl groups [58]. In their study, diynes with longer alkyl chains, comprising 11 and 9 methylene units, exhibited the color transition at 35 and 14 °C, respectively. In contrast, diynes with shorter alkyl chains, featuring seven and five methylene groups, displayed the color shift at the significantly lower temperatures of −3 and −50 °C, respectively (Scheme 3a). These results clearly indicate that the thermochromic transition temperature decreases with a reduction in the number of carbon atoms in the side chains. However, it is worth noting that relying solely on chain length to predict the accurate temperature at which the color change will take place presents challenges, as there is not a strong correlation between these variables.

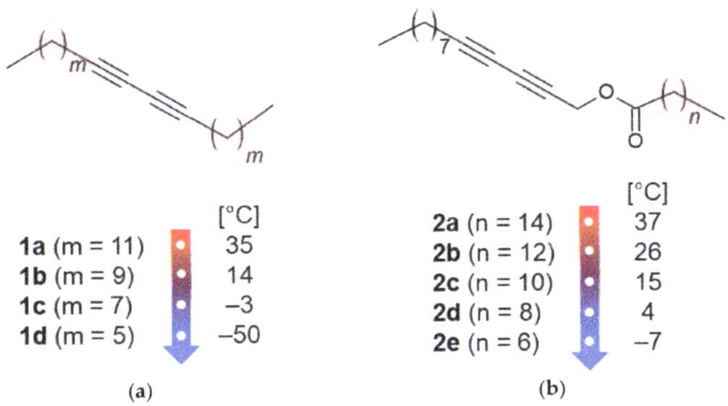

Scheme 3. (**a**) Chemical structures of diyne monomers **1a–1d** and the thermochromic temperature of their polymers; (**b**) chemical structures of ester monomers **2a–2e** and the thermochromic temperature of their polymers.

The situation becomes more straightforward when considering the thermochromic behavior of esters derived from trideca-2,4-diyn-1-ol and linear carboxylic acids [58]. In these esters, which contain 14, 12, 10, 8, and 6 methylene groups in the carboxylate moiety, the color change manifests at temperatures of 37, 26, 15, 4, and −7 °C, respectively (Scheme 3b). This observation suggests a more discernible relationship between chain length and thermochromic temperature. Specifically, for every decrease of two carbon atoms in the chain, the temperature drops by approximately 10 °C.

5. Influence of Methylene Groups Close to DA Moiety

An alternative method for inducing low-temperature thermochromism in PDAs involves obtaining DA derivatives with the minimum number of methylene groups directly attached to the DA unit. The length of the linker connecting the DA portion to any functional group, such as an ester group, exerts a comparable influence on the temperature at which the color change appears, as does the length of the side chain. Rougeau et al. observed that introducing a three-methylene linker in the ester of pentadeca-4,6-diyn-1-ol results in the color transition at 7 °C (Scheme 4). Conversely, for the ester of trideca-2,4-diyn-1-ol, incorporating a methylene linker causes the color shift to occur at a lower temperature of −7 °C [58]. This finding suggests that reducing the length of the methylene linker contributes to a decrease in the thermochromic transition temperature.

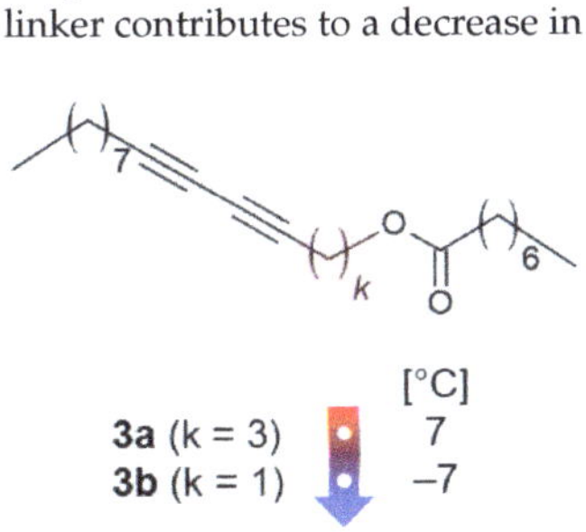

Scheme 4. Chemical structures of ester monomers **3a–3b** and the thermochromic temperature of their polymers.

6. Influence of Hydrogen-Bonding Interactions

A distinct approach to promote low-temperature thermochromism in PDAs involves designing DA monomers with functional groups that hinder the formation of strong interactions, particularly hydrogen bonds, among the polymer's side chains. The presence of such functional groups, similarly to other organic compounds, leads to a reduction in attractive van der Waals forces, consequently lowering the melting point [59]. This reduction, in turn, contributes to a decrease in the thermochromic transition temperature of DA derivatives.

The literature provides several examples that elucidate this concept, one of which involves the thermochromic behavior of poly-pentacosa-10,12-diynoic acid (poly-PCDA) and its ester derivatives. Park et al. observed that poly-PCDA demonstrates a color change at 65 °C [55] (Scheme 5a). Conversely, its methyl derivative (poly-PCDA-Me) exhibits a thermochromic transition at a lower temperature, approximately 29 °C [55] (Scheme 5b). For its esters with ethylene glycol monomethyl ether (EGME), diethylene glycol monomethyl ether (DGME), and triethylene glycol monomethyl ether (TGME), incorporated into a poly(vinyl alcohol) (PVA) polymer matrix, the color shift occurs at even lower temperatures. Specifically, poly(PCDA-EGME)/PVA, poly(PCDA-DGME)/PVA, and poly(PCDA-TGME)/PVA films manifest a color change at temperatures above 15, 10, and 5 °C, respectively [60] (Scheme 5c). The higher thermochromic temperature of poly-PCDA compared with its ester derivatives is ascribed to the presence of carboxyl groups, which facilitate the formation of hydrogen bonds. The replacement of acidic head groups with ester functional groups results in a lower transition temperature for the ester compounds because the potential hydrogen bonding between ester groups is either weak or absent. The notably lower thermochromic temperature of glycolic esters compared with poly-PCDA-Me is likely due to weaker attractive forces between ethylenoxy groups. Additionally, the reduction in transition temperature observed for glycolic esters with a higher number of ethylenoxy units may be attributed to increased strain in the side chains.

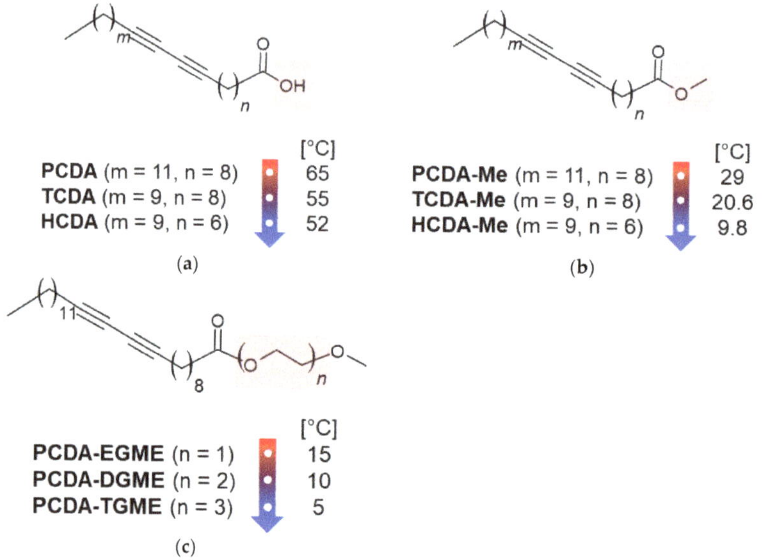

Scheme 5. (**a**) Chemical structures of PCDA, TCDA, and HCDA monomers, along with the thermochromic temperature of their polymers; (**b**) chemical structures of PCDA-Me, TCDA-Me, and HCDA-Me monomers, along with the thermochromic temperature of their polymers; (**c**) chemical structures of PCDA-EGME, PCDA-DGME, and PCDA-TGME monomers, along with the thermochromic temperature of their polymers in PVA films.

In their next study, Park et al. transformed the PCDA monomer into its isocyanate derivative, 1-isocyanatotetracosa-9,11-diyne (TCD-NCO), using a two-step process. Analysis of the thermochromic properties revealed that poly-TCD-NCO shows a color shift near 11 °C [61].

It can be noticed that the isocyanate derivative, similar to ester derivatives, exhibits a lower thermochromic temperature compared with poly-PCDA. It occurs because an isocyanate head group also lacks strong interaction capabilities.

Another instance illustrating this strategy relates to the thermochromic behavior of diamide and diester derivatives of poly-docosa-10,12-diynedioic acid (poly-DCDA). Wrackmeyer et al. demonstrated that its diamides with *n*-pentyl and *n*-hexyl substituents on the termini undergo a color transition at 145 and 130 °C, respectively [62] (Scheme 6a). Meanwhile, Mergu and Son synthesized its diesters with EGME, DGME, and TGME, which manifest alteration in color at 2, −10, and −16 °C, respectively [50] (Scheme 6b). The thermochromic properties of poly-DCDA used to obtain these derivatives are not known. However, according to the generally accepted principle, it can be expected that poly-DCDA will reveal a color change at a temperature close to the melting point of the DCDA monomer (CAS: 28393-02-4), which is 110–112 °C. It is noteworthy that symmetric amides capable of forming hydrogen bonds show transition temperatures that are as high as that of poly-DCDA. They are significantly higher compared with those of symmetric esters, which lack this ability. Furthermore, despite the substantial difference in transition temperature between diamides and diesters, it is evident that in both cases, the temperature decreases with an increase in the number of methylene and ethylenoxy units in the side chains. This decrease may result from increased strain in the side chains.

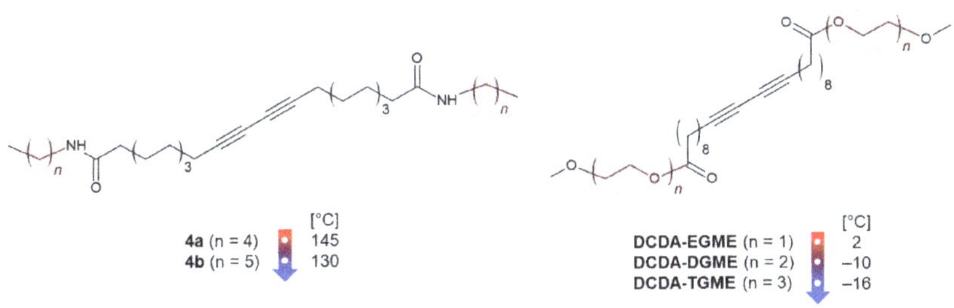

(a) (b)

Scheme 6. (**a**) Chemical structures of amide monomers **4a–4b** and the thermochromic temperature of their polymers; (**b**) chemical structures of DCDA-EGME, DCDA-DGME, and DCDA-TGME monomers, along with the thermochromic temperature of their polymers.

7. Influence of Copolymer

Exploring low-temperature thermochromism in PDAs takes a completely different direction with the application of copolymers. This approach involves blending DA monomers with copolymers, resulting in nanoblends that effectively lower the chromatic transition temperature of PDAs while preserving the DA structure. This reduction in transition temperature is achieved by fine-tuning the hydrophobic/hydrophilic balance, average molecular weight (M_n), and concentration of the copolymer.

A significant advantage of this method lies in its ability to utilize known DA derivatives, as presented in a study conducted by Ferreira et al., which examined the thermochromic properties of poly-PCDA and poly-tricosa-10,12-dyinoic acid (poly-TCDA) [63]. For comparison, the polymerization process was carried out by using classic vesicles, as well as special amphiphilic reactors formed by various triblock copolymers (TCs), including L35, L64, P123, F68, and F127. These copolymers, composed of ethylene oxide (EO) and propylene oxide (PO) segments arranged in an alternating EO-PO-EO linear fashion, vary in the number of EO and PO segments, as well as M_n [64].

The study revealed that while poly-PCDA in vesicles exhibits a color change at 60 °C, incorporation of copolymers into nanoblends generally decreases the temperature of the chromatic transition. Additionally, increasing the concentration of the copolymer leads to a further reduction in the transition temperature. Notably, among the tested copolymers, the sequence L35 < F68 < L64 < F127 < P123, with an increasing number of PO segments (17 < 30 < 31 < 64 < 69), induces the most significant decrease in transition temperature (Scheme 7).

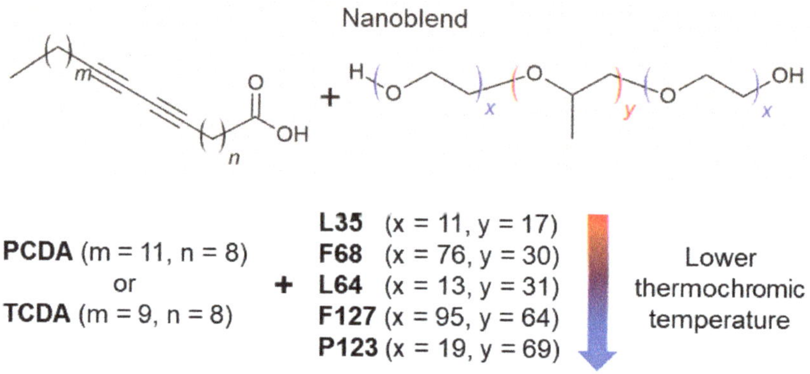

Scheme 7. Chemical structures of PCDA and TCDA monomers and the thermochromic temperature of nanoblends of their polymers with the corresponding copolymers.

Of particular interest is the use of the P123 copolymer in poly-PCDA nanoblends, which triggers a color shift within a temperature range of 12 to 18 °C. This range is over 40 °C lower than that observed with poly-PCDA in vesicles. This observation suggests distinct intermolecular interactions of poly-PCDA in amphiphilic reactors compared with vesicular forms. Specifically, in nanoblends, poly-PCDA can engage in both poly-PCDA–poly-PCDA and poly-PCDA–P123 interactions. The latter competes with the interactions between adjacent PCDA monomers. It is noteworthy that the P123 copolymer, being the most hydrophobic among the investigated copolymers, forms stronger interactions with PCDA monomers and promotes a more hydrophobic environment for the insertion of PCDA monomers. This leads to increased distance between neighboring PCDA monomers and weakened PCDA–PCDA interactions. Consequently, diminished van der Waals forces between proximate carbon chains reduce the thermal energy required to disrupt poly-PCDA–poly-PCDA interactions and overcome the poly-PCDA rotational energy barrier, thereby facilitating the chromatic transition in nanoblends compared with vesicles.

Between PCDA and TCDA monomers, one might expect the poly-TCDA–P123 nanoblend to demonstrate a lower transition temperature compared with the poly-PCDA–P123 nanoblend. This expectation arises from the shorter carbon chain in poly-TCDA, which would result in fewer van der Waals interactions between the poly-TCDA tail and the hydrophobic region of the P123 copolymer, as well as less intense poly-TCDA–poly-TCDA. However, experimental observations do not support this expectation. This discrepancy suggests that hydrophobicity alone does not solely dictate the chromatic transition temperature of the nanostructures. Other factors, such as hydrophilicity and M_n of the copolymer, should also be taken into account. Consequently, it is proposed that the poly-PCDA–P123 nanoblend exhibit a lower color transition temperature due to the weakened poly-PCDA–poly-PCDA interactions. This weakening is attributed to the increased number of poly-PCDA–P123 interactions formed in the hydrophobic environment.

8. Applications of PDAs

PDAs exhibit a unique ability to change color in response to various external factors, such as temperature, pH, and the presence of specific chemical compounds. Notably, when exposed to rising temperature, they can act as reliable indicators, signaling potential breaches in temperature-sensitive environments like the cold supply chain. This capability is particularly critical in fields where temperature control directly influences product quality and safety, such as the handling of food, pharmaceuticals, and vaccines, which may degrade or become harmful under excessive heat.

Given this property, PDAs could also be effectively utilized in thermochromic indicators for tissue biobanking, which involves the collection, analysis, storage, and distribution of biospecimens for basic, translational, and clinical research [65]. By providing visual indications of temperature fluctuations in each stage, these indicators could enhance the monitoring and management of biological samples. Consequently, this would ensure their quality, integrity, safety, and viability, which are essential to obtaining reliable and reproducible results in genomic, transcriptomic, and proteomic assays [66,67].

However, one significant challenge during sample collection is the unrecorded time delay between patient excision and subsequent pathological or molecular analysis. This delay can lead to substantial alterations in protein signaling pathways within just 30 min of excision, triggering a stress response in the tissue [68]. During this response, cytokines are released due to wounding, resulting in hypoxia-related changes that can compromise biospecimen quality [69]. Both warm and cold ischemia can exacerbate these effects, negatively impacting the sample's integrity and hindering subsequent analysis [70]. Moreover, prolonged hypoxia further deteriorates biospecimen quality by increasing tissue lactate levels and lowering pH. Microarray studies suggest that the extent of acidosis during the agonal state could be a key factor influencing RNA variation across biospecimens [71–73].

Once autopsy samples are procured, it becomes imperative to keep them on ice, and all handling procedures must be carried out promptly to avoid degradation. PDA-based thermochromic sensors could play a pivotal role in this process by providing immediate visual alerts when temperature or time thresholds are exceeded. This feature helps prevent mishandling and ensures that samples remain intact prior to freezing and storage. For long-term preservation, tissues, blood, urine, and other biospecimens that require high nucleic acid or protein integrity must be maintained at ultralow temperature, typically −80 °C or lower [72,74–76]. While maintaining high-quality biosamples is crucial for biomedical research, repeated freeze–thaw cycles during storage can degrade biomolecule stability. In fact, research by Ji et al. demonstrated that RNA integrity is better preserved when thawed on ice rather than at room temperature [77].

Integrating these thermochromic sensors into storage protocols could facilitate real-time temperature monitoring, provide immediate visual alerts for any deviations, and simplify compliance with audit requirements. Furthermore, during the transportation of biobanked samples, maintaining controlled conditions is critical. Here, these sensors could offer a reliable method to track temperature variations, ensuring the reliability of the biobank.

In addition to temperature fluctuations, PDAs can also alter color in response to shifts in pH levels that occur during spoilage, as well as to the volatile chemical compounds released during this process [78]. This versatility allows PDAs not only to serve as temperature indicators but also as freshness sensors for food products such as milk, meat, and fish. Some of these sensors can even function at sub-ambient temperature, which is particularly beneficial for monitoring the spoilage of temperature-sensitive goods. Therefore, this review specifically focuses on them.

For instance, Weston et al. created a PDA/ZnO/agarose disc designed to detect milk quality from 4 °C to room temperature [79]. As milk spoils, bacteria convert lactose into lactic acid, resulting in a pH reduction from 6.8 to 4.0. The increasing acidity leads to ZnO dissociation, disrupting interactions with PDA and causing a visible color change. By selecting the appropriate DA monomer and adjusting lipid doping, the system's sensitivity to lactic acid can be finely tuned. Consequently, it can distinguish between fresh (pH 6.8–6.0), spoiling (pH 6.0–4.5), and spoiled milk (pH 4.5–4.0), which is indicated by a corresponding blue-to-purple-to-red color shift.

In another example, Nguyen et al. developed a PDA/cellulose nanocrystal/chitosan film tailored for detecting ammonia emissions from spoiling meat, operating from −20 °C to room temperature [80]. As meat spoils, proteolysis leads to an increase in ammonia levels, which this innovative sensor can monitor with high sensitivity. Notably, the film demonstrates a visible color change with 500 ppm ammonia under sub-zero conditions and detects 100 ppm at room temperature, providing a reliable indicator of product quality. This capability allows for the identification of spoilage before it becomes visually apparent.

In addition to meat, Li et al. fabricated a PDA/DMPC/cellulose nitrate membrane to detect fish spoilage between −20 °C and room temperature by measuring histamine levels, which increase as fish ages [81]. The sensor is stable for two weeks at room temperature and exhibits a linear response to histamine concentrations ranging from 70 to 2240 ppm, with minimal interference from gases such as nitrogen, oxygen, carbon dioxide, argon, and hydrogen sulfide. This specificity makes it a highly reliable tool for monitoring fish freshness over extended periods.

By incorporating these PDA-based indicators into food packaging, stakeholders can obtain immediate visual feedback on the freshness and safety of perishable goods. This proactive approach not only ensures that products meet quality standards before consumption but also significantly enhances food safety while reducing waste.

9. Future Directions

PDAs, which exhibit irreversible color changes at sub-ambient temperature, present exciting opportunities for precise temperature monitoring in various industries. The future directions outlined here pave the way for a deeper understanding and broader exploration of these materials.

One particularly promising avenue to advance this understanding is the use of computational modeling and simulation techniques. The application of advanced methods, such as molecular dynamics simulations and density functional theory (DFT), provides an opportunity to predict the behavior of novel DA monomers. This predictive capability enables researchers to investigate the relationships between molecular structures and the thermochromic properties of PDAs. By applying this knowledge, suitable materials for low-temperature applications can be identified more efficiently while simultaneously establishing a theoretical framework that will better guide future experimental efforts.

Building on these computational insights, the next significant step naturally shifts toward the strategic synthesis of DA derivatives. By designing side chains and functional groups that modulate intermolecular interactions between DA monomers, researchers can further refine the properties of PDAs. This systematic exploration will not lead to the development of materials with precisely controlled thermochromic transition temperature but also deepen our understanding of how specific molecular modifications influence thermochromic behavior. With this knowledge, predictive models can be constructed to tailor PDAs to specific temperature ranges in a targeted manner.

As research progresses, blending synthesized DA derivatives with copolymers represents another promising direction. This strategy could yield materials that can be rapidly evaluated by using high-throughput screening platforms. Advanced analytical methods, such as ultraviolet differential scanning calorimetry (UV-DSC) and temperature-dependent UV–vis spectroscopy, will play a crucial role in assessing these materials. Such rapid assessments will accelerate the identification of compositions with desired thermochromic properties, which, in turn, will facilitate the development of PDAs designed for specific industrial applications.

In practical terms, PDAs exhibiting thermochromic behavior at low temperature could serve an important role in industries where precise temperature control is required. They could be applied in the transportation of pharmaceuticals (Figure 2), the preservation of thermolabile medical products, and the monitoring of biobanking samples. By integrating PDAs into temperature-monitoring systems, the real-time tracking of temperature fluctuations becomes possible, ensuring that sensitive products remain within their optimal temperature ranges and thus maintaining their effectiveness and safety. To bring these innovations to the market, collaboration with industry partners will be essential. Such partnerships will help define specific functional requirements and support the advancement of PDAs for real-world applications.

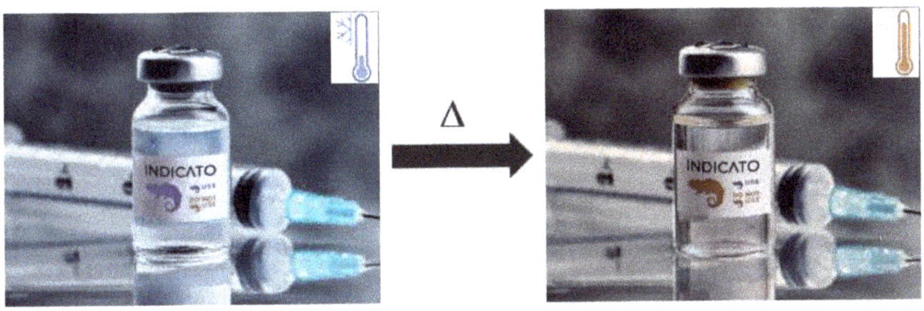

Figure 2. Example of using a PDA-based temperature indicator on a vaccine vial label (the idea of the figure was taken from [82]).

However, to ensure the long-term success of PDAs in these applications, their environmental impact must be carefully considered. The sustainable development of these materials will require future research to prioritize the creation of non-toxic and environmentally friendly DA monomers. Comprehensive studies of cytotoxicity and migration will be necessary to confirm that PDAs are safe for human health and the environment. Moreover, evaluating the long-term stability and degradation behavior of these materials under various conditions will be crucial to verifying their sustainable in practical use.

In summary, the future directions provided here offer a roadmap for advancing the field of PDAs showing low-temperature thermochromism. The insights gained from the proposed approaches will not only enhance our understanding of these materials but also contribute to the development of innovative solutions with far-reaching impacts across different industries. By focusing on these research directions, PDAs can continue to evolve, opening new opportunities for temperature-sensitive applications.

10. Conclusions

This review highlights the burgeoning field of low-temperature thermochromism, emphasizing its importance and potential in diverse applications, particularly in the biobanking sector. Thermochromic materials, especially polydiacetylenes (PDAs), have shown significant promise due to their ability to provide visible, irreversible color changes at sub-ambient temperature. These materials are particularly advantageous for monitoring temperature-sensitive biological samples, ensuring their integrity throughout storage and transportation.

PDAs, with their unique ene-yne backbone structure, exhibit a range of color changes in response to temperature variations. Recent advances have demonstrated that manipulating the chemical structure of DA monomers, such as adjusting side-chain lengths and functional groups, can effectively tune the thermochromic transition temperature of PDAs (Table 2). Innovations like the use of copolymers further expand the potential of PDAs by enabling precise control over their thermal response and improving their applicability in real-world scenarios.

Table 2. The influence of structural modifications of DA monomers on the thermochromic temperature of the corresponding PDAs.

Structural Parameter	Structural Modification	Thermochromic Temperature
Number of methylene units in side chain	Decrease	Decrease
Number of methylene units between DA moiety and any functional group	Decrease	Decrease
Number of hydrogen-bonding interactions	Decrease	Decrease
Number of PO segments in copolymer	Increase	Decrease

The integration of PDAs into biobanking practices represents a significant step forward in maintaining the quality of biological specimens. By offering real-time visual indications of temperature fluctuations, these thermochromic materials enhance the monitoring and management of biosamples, addressing critical issues such as temperature excursions during storage and transportation. This capability is essential to preserving the integrity of sensitive materials like biological samples, but also vaccines and pharmaceuticals, which are pivotal for research and medical applications.

Future research should focus on refining the synthesis of DA derivatives to achieve even lower thermochromic transition temperatures and better control over their properties. Computational modeling and high-throughput screening techniques will play crucial roles in accelerating the development of novel PDAs tailored for specific applications. Collaborative efforts between researchers and industry will be instrumental in translating these advances into practical solutions for temperature monitoring.

In summary, the ongoing developments in low-temperature thermochromism offer exciting opportunities to advance temperature-sensitive technologies. The ability to monitor temperature changes with high precision and reliability holds promise for improving the management of valuable biological materials and other thermolabile products, ultimately contributing to better outcomes in research, medicine, and beyond.

Author Contributions: Conceptualization, J.C.; writing—original draft preparation, M.W.-K. and P.G.; writing—review and editing, B.P. and J.C.; visualization, M.W.-K. and B.P.; supervision, J.C.; funding acquisition, M.W.-K. All authors have read and agreed to the published version of the manuscript.

Funding: This work was financed by the National Centre for Research and Development under the "Lider XI" program, in accordance with grant agreement No. LIDER/38/0135/L-11/NCBR/2020; the Ministry of Science and Education in Poland under grant agreement No. NdS-II/SN/0278/2024/01 (Science for Society II program); and the Medical Research Agency in Poland under grant agreements No. 2023/ABM/02/00004-00 and 2023/ABM/02/00005-00.

Conflicts of Interest: The authors declare no conflicts of interest.

References

1. Houston, E.J. On the Change of Color Produced in Certain Chemical Compounds by Heat. *J. Frankl. Inst.* **1871**, *92*, 115–127. [CrossRef]
2. Hakami, A.; Srinivasan, S.S.; Biswas, P.K.; Krishnegowda, A.; Wallen, S.L.; Stefanakos, E.K. Review on Thermochromic Materials: Development, Characterization, and Applications. *J. Coat. Technol. Res.* **2022**, *19*, 377–402. [CrossRef]
3. Sone, K.; Fukuda, Y. *Inorganic Thermochromism*; Inorganic Chemistry Concepts; Springer: Berlin/Heidelberg, Germany, 1987; Volume 10, ISBN 978-3-642-51019-9.
4. Crano, J.C.; Guglielmetti, R.J. (Eds.) *Organic Photochromic and Thermochromic Compounds: Volume 2: Physicochemical Studies, Biological Applications, and Thermochromism*, 1st ed.; Topics in Applied Chemistry; Kluwer Academic Publishers: Boston, UK, 2002; ISBN 0-306-46912-X.
5. Jia, Q.-Q.; Luo, Q.-F.; Ni, H.-F.; Su, C.-Y.; Fu, D.-W.; Xie, L.-Y.; Lu, H.-F. High-Sensitivity Organic–Inorganic Hybrid Materials with Reversible Thermochromic Property and Dielectric Switching. *J. Phys. Chem. C* **2022**, *126*, 1552–1557. [CrossRef]
6. Seeboth, A.; Lötzsch, D.; Ruhmann, R.; Muehling, O. Thermochromic Polymers—Function by Design. *Chem. Rev.* **2014**, *114*, 3037–3068. [CrossRef] [PubMed]
7. Strahl, W. Sympathetic Ink. U.S. Patent 3926645A, 16 December 1975.

8. Aitken, D.; Burkinshaw, S.M.; Griffiths, J.; Towns, A.D. Textile Applications of Thennochromic Systems. *Rev. Prog. Color. Relat. Top.* **1996**, *26*, 1–8. [CrossRef]
9. Bide, M. Hot Shirts. *ChemMatters* **1992**, 8–11.
10. Supian, A.B.M.; Asyraf, M.R.M.; Syamsir, A.; Najeeb, M.I.; Alhayek, A.; Al-Dala'ien, R.N.; Manar, G.; Atiqah, A. Thermochromic Polymer Nanocomposites for the Heat Detection System: Recent Progress on Properties, Applications, and Challenges. *Polymers* **2024**, *16*, 1545. [CrossRef]
11. Kakade, V.U.; Lock, G.D.; Wilson, M.; Owen, J.M.; Mayhew, J.E. Accurate Heat Transfer Measurements Using Thermochromic Liquid Crystal. Part 1: Calibration and Characteristics of Crystals. *Int. J. Heat. Fluid. Flow.* **2009**, *30*, 939–949. [CrossRef]
12. Lv, L.-Y.; Cao, C.-F.; Qu, Y.-X.; Zhang, G.-D.; Zhao, L.; Cao, K.; Song, P.; Tang, L.-C. Smart Fire-Warning Materials and Sensors: Design Principle, Performances, and Applications. *Mater. Sci. Eng. R. Rep.* **2022**, *150*, 100690. [CrossRef]
13. Guk, K.; Han, G.; Lim, J.; Jeong, K.; Kang, T.; Lim, E.-K.; Jung, J. Evolution of Wearable Devices with Real-Time Disease Monitoring for Personalized Healthcare. *Nanomaterials* **2019**, *9*, 813. [CrossRef]
14. Stasiek, J.; Jewartowski, M.; Aleksander Kowalewski, T. The Use of Liquid Crystal Thermography in Selected Technical and Medical Applications—Recent Development. *J. Cryst. Process Technol.* **2014**, *4*, 46–59. [CrossRef]
15. Guo, S.; Zhang, A. Review of Different Temperatures for Biopreservation. *Int. J. Refrig.* **2024**, *157*, 53–59. [CrossRef]
16. Ahmad, S.A.H.; Ab Rahman, M.N.; Muhamed, A.A. Optimal Temperature in Cold Storage for Perishable Foods. In Proceedings of the 5th NA International Conference on Industrial Engineering and Operations Management, Detroit, MI, USA, 10–14 August 2020; IEOM Society International: Southfield, MI, USA, 2020; pp. 3449–3456.
17. Paludetti, L.F.; Kelly, A.L.; O'Brien, B.; Jordan, K.; Gleeson, D. The Effect of Different Precooling Rates and Cold Storage on Milk Microbiological Quality and Composition. *J. Dairy Sci.* **2018**, *101*, 1921–1929. [CrossRef] [PubMed]
18. Rojas, H.E.; Contreras, R.L.; Schnake, F.G.; Melín, P.; Marín, P.G.L.; Castillo, P.B. Factors Determining Meat Quality and Cold Preservation Methods to Extend Shelf Life. *Open Access J. Biomed. Sci.* **2022**, *4*, 1459–1470. [CrossRef]
19. Suh, S.; Kim, Y.E.; Shin, D.; Ko, S. Effect of Frozen-Storage Period on Quality of American Sirloin and Mackerel (*Scomber japonicus*). *Food Sci. Biotechnol.* **2017**, *26*, 1077–1084. [CrossRef]
20. Alsailawi, H.A.; Mudhafar, M.; Abdulrasool, M.M. Effect of Frozen Storage on the Quality of Frozen Foods—A Review. *J. Chem. Chem. Eng.* **2020**, *14*, 86–89. [CrossRef]
21. Buyck, J.R.; Baer, R.J.; Choi, J. Effect of Storage Temperature on Quality of Light and Full-Fat Ice Cream. *J. Dairy Sci.* **2011**, *94*, 2213–2219. [CrossRef]
22. Geeraedts, F.; Saluja, V.; ter Veer, W.; Amorij, J.-P.; Frijlink, H.W.; Wilschut, J.; Hinrichs, W.L.J.; Huckriede, A. Preservation of the Immunogenicity of Dry-Powder Influenza H5N1 Whole Inactivated Virus Vaccine at Elevated Storage Temperatures. *AAPS J.* **2010**, *12*, 215–222. [CrossRef]
23. Currie, L.M.; Harper, J.R.; Allan, H.; Connor, J. Inhibition of Cytokine Accumulation and Bacterial Growth during Storage of Platelet Concentrates at 4 °C with Retention of in Vitro Functional Activity. *Transfusion* **1997**, *37*, 18–24. [CrossRef]
24. Khurana, G.; Gupta, V. Effect on Insulin upon Storage in Extreme Climatic Conditions (Temperature and Pressure) and Their Preventive Measures. *J. Soc. Health Diabetes* **2019**, *07*, 006–010. [CrossRef]
25. Li, X.; An, J.; Dong, Z.J. Stability and Storage Conditions of Refrigerated and Frozen Drugs Commonly Used in Hospitals. *China Pharm.* **2016**, *27*, 983–986. [CrossRef]
26. Cordsmeier, L.; Hahn, M.B. DNA Stability in Biodosimetry, Pharmacy and DNA Based Data-Storage: Optimal Storage and Handling Conditions. *ChemBioChem* **2022**, *23*, e202200391. [CrossRef] [PubMed]
27. Ng, H.H.; Ang, H.C.; Hoe, S.Y.; Lim, M.-L.; Tai, H.E.; Soh, R.C.H.; Syn, C.K.-C. Simple DNA Extraction of Urine Samples: Effects of Storage Temperature and Storage Time. *Forensic Sci. Int.* **2018**, *287*, 36–39. [CrossRef] [PubMed]
28. Shokri, S.; Shahkarami, M.K.; Shafyi, A.; Mohammadi, A.; Esna-ashari, F.; Hamta, A. Evaluation of the Thermal Stability of Live-Attenuated Rubella Vaccine (Takahashi strain) Formulated and Lyophilized in Different Stabilizers. *J. Virol. Methods* **2019**, *264*, 18–22. [CrossRef] [PubMed]
29. Li, T.H.; Long, X.F.; Liang, J.Y. Cryopreservation of Rare Blood Group Erythrocytes and Detection of Resuscitation Function. *China Med. Eng.* **2014**, *22*, 70–71.
30. Sadoh, A.; Hossain, S.; Ravindra, N.M. Thermochromic Polymeric Films for Applications in Active Intelligent Packaging—An Overview. *Micromachines* **2021**, *12*, 1193. [CrossRef]
31. Hossain, S.; Sadoh, A.; Ravindra, N.M. Principles, Properties and Preparation of Thermochromic Materials. *Mater. Sci. Eng. Int. J.* **2023**, *7*, 146–156. [CrossRef]
32. White, M.A.; LeBlanc, M. Thermochromism in Commercial Products. *J. Chem. Educ.* **1999**, *76*, 1201. [CrossRef]
33. Groeneveld, I.; Kanelli, M.; Ariese, F.; van Bommel, M.R. Parameters That Affect the Photodegradation of Dyes and Pigments in Solution and on Substrate—An Overview. *Dyes Pigment.* **2023**, *210*, 110999. [CrossRef]
34. Sharifi, M.; Sharifi, A.; Abaee, M.S.; Mirzaei, M. A Review on Fluoran Compounds as Widely Used Leuco Dyes. *Dyes Pigment.* **2024**, *221*, 111783. [CrossRef]
35. Li, Y.; Wang, Q.; Zheng, X.; Li, Y.; Luan, J. Microcapsule Encapsulated with Leuco Dye as a Visual Sensor for Concrete Damage Indication via Color Variation. *RSC Adv.* **2020**, *10*, 1226–1231. [CrossRef] [PubMed]
36. Wen, J.T.; Roper, J.M.; Tsutsui, H. Polydiacetylene Supramolecules: Synthesis, Characterization, and Emerging Applications. *Ind. Eng. Chem. Res.* **2018**, *57*, 9037–9053. [CrossRef]

37. Jelinek, R.; Ritenberg, M. Polydiacetylenes—Recent Molecular Advances and Applications. *RSC Adv.* **2013**, *3*, 21192. [CrossRef]
38. Luo, C.-C.; Wang, X.-J.; Han, L.-J.; Jia, Y.-G.; Ying, S.-M.; Wang, J.-W. Preparation, Structure and Optical Properties of Thermochromic Liquid Crystal Compounds Containing (−)-Menthyl with Selective Reflection. *J. Mol. Liq.* **2019**, *275*, 241–250. [CrossRef]
39. Yu, H.; Wei, Z.; Hao, Y.; Liang, Z.; Fu, Z.; Cai, H. Reversible Solid-State Thermochromism of a 2D Organic–Inorganic Hybrid Perovskite Structure Based on Iodoplumbate and 2-Aminomethyl-Pyridine. *New J. Chem.* **2017**, *41*, 9586–9589. [CrossRef]
40. Zhu, X.; Liu, Y.; Dong, N.; Li, Z. Fabrication and Characterization of Reversible Thermochromic Wood Veneers. *Sci. Rep.* **2017**, *7*, 16933. [CrossRef]
41. Saenjaiban, A.; Singtisan, T.; Suppakul, P.; Jantanasakulwong, K.; Punyodom, W.; Rachtanapun, P. Novel Color Change Film as a Time–Temperature Indicator Using Polydiacetylene/Silver Nanoparticles Embedded in Carboxymethyl Cellulose. *Polymers* **2020**, *12*, 2306. [CrossRef]
42. Kingchok, S.; Nontasorn, P.; Laohhasurayotin, K.; Traiphol, N.; Traiphol, R. Reversible Thermochromic Polydiacetylene/Zinc-Aluminium Layered Double Hydroxides Nanocomposites for Smart Paints and Colorimetric Sensors: The Crucial Role of Zinc Ions. *Colloids Surf. A Physicochem. Eng. Asp.* **2021**, *610*, 125733. [CrossRef]
43. Fang, F.; Meng, F.; Luo, L. Recent Advances on Polydiacetylene-Based Smart Materials for Biomedical Applications. *Mater. Chem. Front.* **2020**, *4*, 1089–1104. [CrossRef]
44. Wegner, G. Topochemische Reaktionen von Monomeren Mit Konjugierten Dreifachbindungen/Tochemical Reactions of Monomers with Conjugated Triple Bonds. *Z. Naturforsch. B* **1969**, *24*, 824–832. [CrossRef]
45. Hema, K.; Ravi, A.; Raju, C.; Sureshan, K.M. Polymers with Advanced Structural and Supramolecular Features Synthesized through Topochemical Polymerization. *Chem. Sci.* **2021**, *12*, 5361–5380. [CrossRef] [PubMed]
46. Qian, X.; Städler, B. Recent Developments in Polydiacetylene-Based Sensors. *Chem. Mater.* **2019**, *31*, 1196–1222. [CrossRef]
47. Huang, Q.; Wu, W.; Ai, K.; Liu, J. Highly Sensitive Polydiacetylene Ensembles for Biosensing and Bioimaging. *Front. Chem.* **2020**, *8*, 565782. [CrossRef] [PubMed]
48. Cho, E.; Jung, S. Biomolecule-Functionalized Smart Polydiacetylene for Biomedical and Environmental Sensing. *Molecules* **2018**, *23*, 107. [CrossRef]
49. Seeboth, A.; Lötzsch, D. *Thermochromic and Thermotropic Materials*, 1st ed.; Seeboth, A., Lötzsch, D., Eds.; Pan Stanford Publishing Pte. Ltd.: Singapore, 2013; Volume 114, ISBN 978-981-4411-02-8.
50. Mergu, N.; Son, Y.-A. Design and Synthesis of Polydiacetylenes, and Their Low Temperature Irreversible Thermochromic Properties. *Dyes Pigment.* **2021**, *184*, 108839. [CrossRef]
51. Dong, W.; Zhang, R.; Lin, G. Molecular-Level Design of Excellent Reversible Thermochromic Polydiacetylene Materials with the Simultaneous Enhancement of Multiple Performances. *Mater. Chem. Front.* **2021**, *5*, 7041–7050. [CrossRef]
52. Ahn, D.J.; Lee, S.; Kim, J. Rational Design of Conjugated Polymer Supramolecules with Tunable Colorimetric Responses. *Adv. Funct. Mater.* **2009**, *19*, 1483–1496. [CrossRef]
53. Foley, J.L.; Li, L.; Sandman, D.J. Polydiacetylenes with Long Wavelength Absorption. *Chem. Mater.* **1998**, *10*, 3984–3990. [CrossRef]
54. Mino, N.; Tamura, H.; Ogawa, K. Analysis of Color Transitions and Changes on Langmuir-Blodgett Films of a Polydiacetylene Derivative. *Langmuir* **1991**, *7*, 2336–2341. [CrossRef]
55. Park, I.S.; Park, H.J.; Jeong, W.; Nam, J.; Kang, Y.; Shin, K.; Chung, H.; Kim, J.-M. Low Temperature Thermochromic Polydiacetylenes: Design, Colorimetric Properties, and Nanofiber Formation. *Macromolecules* **2016**, *49*, 1270–1278. [CrossRef]
56. Ogden, S.; Klintberg, L.; Thornell, G.; Hjort, K.; Bodén, R. Review on Miniaturized Paraffin Phase Change Actuators, Valves, and Pumps. *Microfluid. Nanofluidics* **2014**, *17*, 53–71. [CrossRef]
57. Slovokhotov, Y.L.; Batsanov, A.S.; Howard, J.A.K. Molecular van Der Waals Symmetry Affecting Bulk Properties of Condensed Phases: Melting and Boiling Points. *Struct. Chem.* **2007**, *18*, 477–491. [CrossRef]
58. Rougeau, L.; Picq, D.; Rastello, M.; Frantz, Y. New Irreversible Thermochromic Polydiacetylenes. *Tetrahedron* **2008**, *64*, 9430–9436. [CrossRef]
59. Yalkowsky, S.H.; Alantary, D. Estimation of Melting Points of Organics. *J. Pharm. Sci.* **2018**, *107*, 1211–1227. [CrossRef]
60. Goyal, S.; Sharma, D.; Kumar, K. Synthesis of Low-Temperature Irreversible Thermochromic Indicator Based on Functional Polydiacetylene for Food Storage Applications. *J. Mater. Sci.* **2024**, *59*, 7561–7573. [CrossRef]
61. Park, I.S.; Park, H.J.; Kim, J.-M. A Soluble, Low-Temperature Thermochromic and Chemically Reactive Polydiacetylene. *ACS Appl. Mater. Interfaces* **2013**, *5*, 8805–8812. [CrossRef]
62. Wrackmeyer, M.; O'Rourke, A.P.; Pugh, T.; Turner, M.L.; Webb, S.J. Effect of Varying Substituent on the Colour Change Transitions of Diacetylene Pigments. *Dyes Pigment.* **2021**, *192*, 109397. [CrossRef]
63. Ferreira, G.M.D.; Ferreira, G.M.D.; do Carmo Hespanhol, M.; de Paula Rezende, J.; dos Santos Pires, A.C.; Ortega, P.F.R.; da Silva, L.H.M. A Simple and Inexpensive Thermal Optic Nanosensor Formed by Triblock Copolymer and Polydiacetylene Mixture. *Food Chem.* **2018**, *241*, 358–363. [CrossRef]
64. Yu, J.; Qiu, H.; Yin, S.; Wang, H.; Li, Y. Polymeric Drug Delivery System Based on Pluronics for Cancer Treatment. *Molecules* **2021**, *26*, 3610. [CrossRef]
65. Watson, P.H.; Wilson-McManus, J.E.; Barnes, R.O.; Giesz, S.C.; Png, A.; Hegele, R.G.; Brinkman, J.N.; Mackenzie, I.R.; Huntsman, D.G.; Junker, A.; et al. Evolutionary Concepts in Biobanking—The BC BioLibrary. *J. Transl. Med.* **2009**, *7*, 95. [CrossRef]

66. Avella-Oliver, M.; Morais, S.; Puchades, R.; Maquieira, Á. Towards Photochromic and Thermochromic Biosensing. *TrAC Trends Anal. Chem.* **2016**, *79*, 37–45. [CrossRef]
67. Powell, S.; Molinolo, A.; Masmila, E.; Kaushal, S. Real-Time Temperature Mapping in Ultra-Low Freezers as a Standard Quality Assessment. *Biopreservation Biobanking* **2019**, *17*, 139–142. [CrossRef] [PubMed]
68. Espina, V.; Mueller, C.; Edmiston, K.; Sciro, M.; Petricoin, E.F.; Liotta, L.A. Tissue Is Alive: New Technologies Are Needed to Address the Problems of Protein Biomarker Pre-analytical Variability. *Proteom. Clin. Appl.* **2009**, *3*, 874–882. [CrossRef] [PubMed]
69. Li, J.; Gould, T.D.; Yuan, P.; Manji, H.K.; Chen, G. Post-Mortem Interval Effects on the Phosphorylation of Signaling Proteins. *Neuropsychopharmacology* **2003**, *28*, 1017–1025. [CrossRef] [PubMed]
70. Blondal, T.; Jensby Nielsen, S.; Baker, A.; Andreasen, D.; Mouritzen, P.; Wrang Teilum, M.; Dahlsveen, I.K. Assessing Sample and MiRNA Profile Quality in Serum and Plasma or Other Biofluids. *Methods* **2013**, *59*, S1–S6. [CrossRef]
71. Li, J.Z.; Vawter, M.P.; Walsh, D.M.; Tomita, H.; Evans, S.J.; Choudary, P.V.; Lopez, J.F.; Avelar, A.; Shokoohi, V.; Chung, T.; et al. Systematic Changes in Gene Expression in Postmortem Human Brains Associated with Tissue pH and Terminal Medical Conditions. *Hum. Mol. Genet.* **2004**, *13*, 609–616. [CrossRef]
72. Liu, N.; Luo, Y.; Zhu, Y.; Peng, H.; Zou, C.; Zhou, Z.; Chen, W.; Wang, H.; Liu, H.; Hu, Y.; et al. Effects of Warm Ischemia Time, Cryopreservation, and Grinding Methods on RNA Quality of Mouse Kidney Tissues. *Biopreservation Biobanking* **2021**, *19*, 306–311. [CrossRef]
73. Sheedy, D.; Say, M.; Stevens, J.; Harper, C.G.; Kril, J.J. Influence of Liver Pathology on Markers of Postmortem Brain Tissue Quality. *Alcohol. Clin. Exp. Res.* **2012**, *36*, 55–60. [CrossRef]
74. Tashjian, R.S.; Williams, R.R.; Vinters, H.V.; Yong, W.H. Autopsy Biobanking: Biospecimen Procurement, Integrity, Storage, and Utilization. In *Biobanking: Methods and Protocols*; Yong, W.H., Ed.; Springer: New York, NY, USA, 2019; pp. 77–87.
75. Tas, R.P.; Sampaio-Pinto, V.; Wennekes, T.; van Laake, L.W.; Voets, I.K. From the Freezer to the Clinic. *EMBO Rep.* **2021**, *22*, e52162. [CrossRef]
76. Shabihkhani, M.; Lucey, G.M.; Wei, B.; Mareninov, S.; Lou, J.J.; Vinters, H.V.; Singer, E.J.; Cloughesy, T.F.; Yong, W.H. The Procurement, Storage, and Quality Assurance of Frozen Blood and Tissue Biospecimens in Pathology, Biorepository, and Biobank Settings. *Clin. Biochem.* **2014**, *47*, 258–266. [CrossRef]
77. Ji, X.; Wang, M.; Li, L.; Chen, F.; Zhang, Y.; Li, Q.; Zhou, J. The Impact of Repeated Freeze-Thaw Cycles on the Quality of Biomolecules in Four Different Tissues. *Biopreservation Biobanking* **2017**, *15*, 475–483. [CrossRef] [PubMed]
78. Weston, M.; Pham, A.-H.; Tubman, J.; Gao, Y.; Tjandra, A.D.; Chandrawati, R. Polydiacetylene-Based Sensors for Food Applications. *Mater. Adv.* **2022**, *3*, 4088–4102. [CrossRef]
79. Weston, M.; Kuchel, R.P.; Ciftci, M.; Boyer, C.; Chandrawati, R. A Polydiacetylene-Based Colorimetric Sensor as an Active Use-by Date Indicator for Milk. *J. Colloid. Interface Sci.* **2020**, *572*, 31–38. [CrossRef] [PubMed]
80. Nguyen, L.H.; Naficy, S.; McConchie, R.; Dehghani, F.; Chandrawati, R. Polydiacetylene-Based Sensors to Detect Food Spoilage at Low Temperatures. *J. Mater. Chem. C* **2019**, *7*, 1919–1926. [CrossRef]
81. Li, Q.; Ren, S.; Peng, Y.; Lv, Y.; Wang, W.; Wang, Z.; Gao, Z. A Colorimetric Strip for Rapid Detection and Real-Time Monitoring of Histamine in Fish Based on Self-Assembled Polydiacetylene Vesicles. *Anal. Chem.* **2020**, *92*, 1611–1617. [CrossRef]
82. Wilk-Kozubek, M.; Potaniec, B.; Cybińska, J. Diacetylene Derivatives, Thermochromic Temperature Change Indicator Comprising these Derivatives, Method of Obtaining Thereof and Use Thereof. Patent Application P.445107, Poland, 2023.

polymers

MDPI

Review

Eco-Friendly Polymer Nanocomposite Coatings for Next-Generation Fire Retardants for Building Materials

Haradhan Kolya and Chun-Won Kang *

Department of Housing Environmental Design, Research Institute of Human Ecology, College of Human Ecology, Jeonbuk National University, Jeonju 54896, Jeonbuk, Republic of Korea; hdk@jbnu.ac.kr
* Correspondence: kcwon@jbnu.ac.kr

Abstract: The increasing global commitment to carbon neutrality has propelled a heightened focus on sustainable construction materials, with wood emerging as pivotal due to its environmental benefits. This review explores the development and application of eco-friendly polymer nanocomposite coatings to enhance wood's fire resistance, addressing a critical limitation in its widespread adoption. These nanocomposites demonstrate improved thermal stability and char formation properties by integrating nanoparticles, such as nano-clays, graphene oxide, and metal oxides, into biopolymer matrices. This significantly mitigates the flammability of wood substrates, creating a robust barrier against heat and oxygen. The review provides a comprehensive examination of these advanced coatings' synthesis, characterization, and performance. By emphasizing recent innovations and outlining future research directions, this review underscores the potential of eco-friendly polymer nanocomposite coatings as next-generation fire retardants. This advancement supports the expanded utilization of wood in sustainable construction practices and aligns with global initiatives toward achieving carbon neutrality.

Keywords: carbon neutrality; wood; eco-friendly; polymer coatings; fire resistance

1. Introduction

The international community, including key organizations such as the United Nations Framework Convention on Climate Change (UNFCCC), has increasingly prioritized policies aimed at reducing carbon emissions to combat the climate crisis [1]. In this context, the issue of fire safety, particularly in structures made of wood, takes on a new urgency. One significant milestone in the global effort to combat climate change is the widespread declaration of carbon neutrality, or net-zero emissions, by around 130 countries, including the EU, the United States, and Japan, with some nations already enacting legislation to achieve this goal by 2050 [2,3]. Central to these initiatives is the promotion of green building materials, particularly wood, which is recognized for its renewability and environmental benefits [4,5]. However, the inherent combustibility of wood poses significant challenges in ensuring the safety and resilience of structures against fire hazards [6]. Traditional fire retardants, often based on halogenated compounds, have been effective in enhancing fire resistance but pose significant environmental and health risks [7]. These conventional fire retardants release toxic gases and persistent organic pollutants during combustion, leading to severe environmental and human health issues [8]. Consequently, there is a pressing need to develop safer, more sustainable alternatives that can provide effective fire protection without compromising environmental integrity [9].

Considering the pressing need for safer fire retardants, the focus has shifted to eco-friendly alternatives. The application methods for these retardants on wood-based materials include impregnation [10] and surface coatings [11]. Impregnation, although effective in penetrating deep into the wood, is often expensive and leads to substantial chemical and solvent wastage, raising environmental and economic concerns. In contrast, coatings offer a more efficient and less wasteful alternative, especially when using adhesive polymers to ensure durability and effectiveness. These eco-friendly coatings not only enhance fire resistance but also contribute to a more sustainable and environmentally conscious approach to fire safety.

Recently, eco-friendly polymer nanocomposite coatings have garnered significant attention from researchers [12]. Bio-composites made from renewable polymers, such as poly (lactic acid) [13], furfuryl alcohol [14], gluten [15], soy flour [15], and starch [16], combined with naturally available fibers, have been attracting significant interest due to their environmentally friendly characteristics [12]. These coatings incorporate nanoparticles, such as graphene [17], layered double hydroxides (LDH) [18], natural clay minerals [19], MXene [20], nano-metal oxides [21], polyphosphazene [22], cellulose nanofibrils (CNF) [23], biochar [24], and responsive color-changing materials. Notably, graphene oxide (GO), a two-dimensional carbon-based nanomaterial [25], and chitin composites have garnered significant attention in recent decades [26]. Natural clay minerals, including montmorillonite, kaolin, and bentonite, are frequently utilized in the composites industry due to their widespread availability, recyclability, and non-toxic nature [27]. The resulting nanocomposites provide enhanced fire resistance and align with green chemistry principles due to their low toxicity and biodegradability. Therefore, a review focusing on the development of novel eco-friendly polymer nanocomposite coatings for fire retardancy could provide valuable insights for researchers.

This review provides an overview of the current state of eco-friendly polymer nanocomposite coatings for fire retardancy. It examines fire-retardant treatment methods, including impregnation and coatings, and explores various nanofillers, such as layered silicates, carbon-based nanomaterials, and bio-based nanoparticles. The review covers these nanocomposites' synthesis, fabrication, and characterization and the mechanisms underlying their fire-retardant properties. Environmental and health considerations, recent advances, case studies, and prospects are also discussed. This comprehensive analysis offers valuable insights for researchers and industry professionals in developing safer and more sustainable fire-retardant systems.

2. Fire-Retardant Chemicals

Fire-retardant chemicals are substances used to reduce the flammability of materials and delay their combustion. Traditional fire retardants include halogenated compounds, ammonium polyphosphate, borates, and intumescent systems [28]. Halogenated compounds, such as brominated and chlorinated flame retardants, are highly effective but pose significant environmental and health risks due to releasing toxic gases during combustion [8,29]. Ammonium polyphosphate and borates work by promoting char formation and suppressing flame propagation, making them safer alternatives [30]. Intumescent systems form a protective char layer during heat exposure, providing an insulating barrier that slows down combustion [31]. Despite their effectiveness, many conventional fire retardants face scrutiny for their potential toxicity and environmental persistence, leading to a growing interest in developing eco-friendly alternatives [32]. A list of fire-retardant chemicals with their chemical formulas and melting points (MP) or boiling points (BP) are shown in Table 1.

Table 1. List of fire-retardant chemicals, along with their chemical formulas and melting or boiling points (MP/BP), sourced from the internet (Wikipedia).

Fire-Retardant Chemicals	Chemical Formula	MP/BP (°C)	Ref.
Aluminum hydroxide	$Al(OH)_3$	300	[33]
Aluminum phosphate	$AlPO_4$	1800	[34]
Ammonium polyphosphate	$(NH_4)_3PO_4$	260–320	[35]
Antimony trioxide	Sb_2O_3	656	[36]
Borax	$Na_2B_4O_7 \cdot 10H_2O$	743	[37]
Boric acid	H_3BO_3	170	[38]
Brominated flame retardants	Varies	280–345	[39]
Chlorinated flame retardants	Varies	Varies	[39]
Diethyl ethyl phosphonate	$C_6H_{15}O_3P$	198	[40]
Dimethyl methyl phosphonate	$C_3H_9O_3P$	181	[41]
Graphene and graphene oxide	$C, C_xH_yO_z$	3000	[42]
Graphite	C	3650	[43]
Hydrated lime	$Ca(OH)_2$	580	[44]
Layered double hydroxides	Varies	>900	[45]
Magnesium hydroxide	$Mg(OH)_2$	350	[46]
Melamine	$C_3H_6N_6$	345	[46]
Melamine cyanurate	$C_6H_9N_9O_3$	350	[47]
Melamine phosphate	$C_3H_9N_6O_4P$	120–122	[48]
Melamine polyphosphate	$C_3H_9N_6O_4P$	>350	[49]
Red phosphorus	P_4	590	[50]
Silicon dioxide (silica)	SiO_2	1610	[51]
Trimethylphosphate	$C_3H_9O_4P$	156	[52]
Triphenyl phosphate	$C_{18}H_{15}O_4P$	50	[53]
Tris(1,3-dichloro-2-propyl) phosphate	$C_9H_{15}C_{13}O_4P$	315	[54]
Zinc borate	$Zn_2B_6O_{11} \cdot 3.5H_2O$	980	[55]
Zinc oxide	ZnO	1975	[56]
Vanadium oxide	V_2O_5	1967	[57]
Tin oxide	SnO, SnO_2	1630	[58]

2.1. Treatment Methods

Fire retardants can be applied to materials using several methods, such as impregnation and coating, each with its distinct advantages and drawbacks.

2.1.1. Impregnation

The impregnation process for fire retardants involves several meticulous steps to ensure deep penetration and effectiveness in enhancing the fire resistance of wood. First, the wood is prepared by drying it to a specific moisture content, as the moisture level affects the uptake of the fire-retardant solution. Second, fire-retardant chemicals are dissolved in suitable solvents to create a homogeneous solution. Common fire retardants include ammonium polyphosphate, borates, and other inorganic salts, with the choice of solvent depending on these chemicals' solubility and ability to penetrate the wood structure effectively. Water is the most common solvent, but organic solvents can also be used for specific formulations. Third, the wood is placed in a vacuum chamber, and a vacuum is applied to remove air from the wood's pores, facilitating deeper penetration of the fire-retardant solution. Fourth, after the vacuum treatment, the fire-retardant solution is introduced into the chamber, and the pressure is increased to force the solution into the wood's cellular structure. This pressure treatment ensures the chemicals reach deep into the wood, providing comprehensive fire protection. Fifth, the impregnated wood is removed from the chamber and dried to remove excess solvent and stabilize the wood, which can be performed using kilns or air drying, depending on the desired final moisture content and the type of wood. A schematic of the vacuum pressure impregnation chamber is shown in Figure 1a.

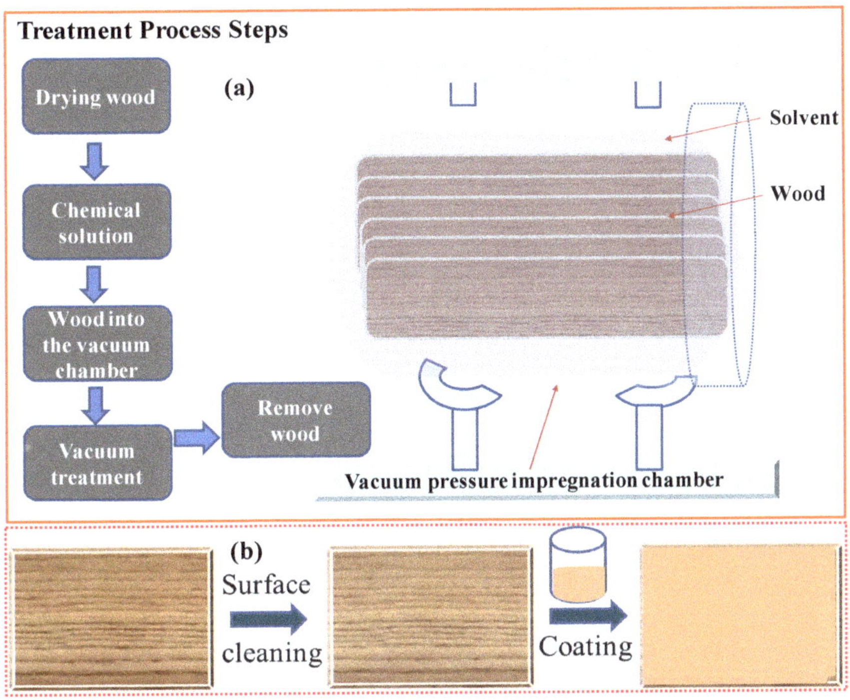

Figure 1. Schematic of (**a**) the vacuum pressure impregnation process and (**b**) surface coating.

The impregnation process for fire retardants is a highly effective method to enhance the fire resistance of wood by introducing protective chemicals deep into its structure [59]. This method involves sophisticated equipment and precise control over various factors, including solvent selection, chemical solubility, and wood pore structure. Impregnation with fire-retardant chemicals can lead to color changes in the wood due to the interaction between the chemicals and the natural components of the wood, such as tannins and lignin [60,61]. Depending on the fire-retardant formulation, the treated timber may darken or take on a different hue. While this color change can sometimes be mitigated through post-treatment processes or additives that stabilize the wood's appearance, it is often an unavoidable aspect of chemical impregnation. Throughout this process, considerations such as the solubility of chemicals, dispersion within the wood's cell wall, and the wood's pore structure are paramount, influencing the efficacy and uniformity of the treatment. Impregnation, although effective in penetrating deep into the wood, is often expensive and leads to substantial chemical and solvent wastage, raising environmental and economic concerns [62].

2.1.2. Coatings

Coating methods offer an alternative approach to applying fire retardants to materials, including wood, offering distinct advantages over impregnation [63]. Below, we delve into the details of coating methods for fire retardants, exploring the procedures, types of coatings, application techniques, and considerations involved. Prior to coating, meticulous surface preparation is imperative. This involves cleaning and smoothing the material's surface to ensure optimal adhesion and uniform coverage of the fire-retardant coating (as shown in Figure 1b). The formulation process entails blending fire-retardant chemicals with appropriate binders and additives to create a homogeneous mixture. Standard binders encompass adhesive polymers, such as acrylics, epoxies, and polyurethanes. A list of polymers with their chemical formulas and melting or boiling points is noted in Table 2.

Table 2. List of polymers used for making fire-retardant composite polymers, along with their melting or boiling points (MP/BP), sourced from the internet (Wikipedia).

Polymer or Adhesive Polymer	Chemical Formula	MP/BP (°C)	Ref.
Polyurethane (PU)	$(C_3H_8N_2O)_n$	136	[64]
Acrylic resins	$(C_5H_8O_2)_n$	160	[65]
Epoxy resins	$(C_{21}H_{25}ClO_5)n$	120	[66]
Polyvinyl chloride (PVC)	$(C_2H_3Cl)_n$	100	[67]
Polyethylene (PE)	$(C_2H_4)_n$	115–135	[68]
Polypropylene (PP)	$(C_3H_6)_n$	130–170	[69]
Polyester resins	$(C_{14}H_{22}O_6)_n$	170–172	[69]
Silicone polymers	$(SiO_2)_n$	1414	[70]
Ethylene-vinyl acetate (EVA) copolymers	$(C_2H_4)_n(C_4H_6O_2)_m$	90	[71]
Polyvinyl alcohol (PVA)	$(C_2H_4O)_n$	200	[72]
Ethylene propylene diene monomer (EPDM)	$(C_8H_{16})_n$	100–160	[73]
Polyvinyl acetate (PVAc)	$(C_4H_6O_2)_n$	60	[74]
Polycarbonate (PC)	$(C_{16}H_{14}O_3)_n$	220–230	[75]
Acrylonitrile butadiene styrene (ABS)	$(C_8H_8{\cdot}C_4H_6{\cdot}C_3H_3N)_n$	105	[76]
Polyimides	$(C_{22}H_{10}O_4)_n$	360	[77]
Polyamide (Nylon)	$(C_{12}H_{22}N_2O_2)_n$	220	[78]
Phenolic resins	$C_8H_6O_2$	90–150	[79]
Urea-formaldehyde resins	$C_2H_6N_2O_2$	130	[80]
Melamine-formaldehyde resins	$C_4H_8N_6O$	354	[81]
Polylactic acid (PLA)	$(C_3H_4O_2)n$	150–160	[82]
Polyhydroxy alkanoates (PHA)	$(C_6H_{10}O_2)_n$	170	[83]
Starch-based polymers	$(C_6H_{10}O_5)_n$	200–220	[84]
Cellulose acetate	$(C_{10}H_{16}O_8)_n$	230	[85]
Lignin-based polymers	Varies	108–150	[86]

Different coating processes include spraying, brushing, rolling, dipping, and electrostatic coating. After application, the coated material can cure or dry, depending on the type of coating used [87]. Depending on the coating formulation, curing may involve air-drying, heat-curing in ovens, or exposure to ultraviolet (UV) light. Coatings offer a cost-effective, flexible, and environmentally friendly alternative to impregnation, though they may require more frequent maintenance and provide primarily surface-level protection.

3. Synthesis and Characterization of Fire-Retardant Polymers

Historical records show the use of various materials to develop flame-retardant properties. Over time, alum, ferrous sulfate, stannic oxide, borax, and ammonium phosphates were used to absorb heat and prevent fire spread. The infusion of financial resources and advancements in polymeric materials has led to the introduction of hybrid materials for fire safety. Generally, solution-mixing, in situ, and ex situ methods are employed to create flame-retardant polymer nanocomposites by optimizing the composition and conditions. In the in situ method, nanomaterials are dispersed in a liquid monomer, with surfactants controlling filler agglomeration and geometry during polymerization. This method enhances porosity, making the composite lighter.

The solution-mixing method involves dissolving one or more components in a solvent to create a solution, which is then mixed thoroughly to ensure uniform distribution of the components. After mixing, the solvent may be evaporated or removed to yield the final product. The ex situ methods often involve preparing or modifying materials outside their intended operational environment. For example, nanoparticles can be synthesized separately and incorporated into a composite material. A schematic of each process is shown in Figure 2.

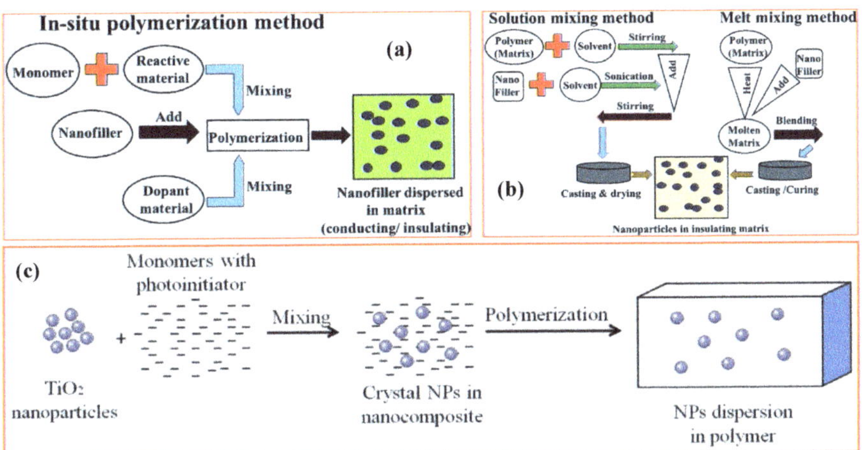

Figure 2. A schematic illustration of the preparation of fire-retardant polymer nanocomposites using (**a**) the in situ method and (**b**) the solution-mixing method [88]. Copyright 2019, reproduced with permission from the authors, RSC. (**c**) Ex situ method [89]. Copyright 2014, reproduced with permission from the authors, MDPI, Basel.

Generally, the characterization of prepared materials involves thermal analysis (DSC, DTA, TGA, TMA, and DMA) [90], microscopy (TEM, SEM, and AFM) [91], spectroscopy (UV-Visible, FTIR, NMR, and Raman) [92], tribological properties [93], and X-ray diffraction techniques, which are employed for chemical characterization of polymer nanocomposites [94]. Mechanical properties for structural stability are assessed using universal testing machines, dynamic mechanical analyzers, and impact and surface analyzers [95]. Weathering effects and electrical properties are measured with resistivity meters and dielectric strength analyzers [96]. Chakraborty et al. provided detailed insights into microscopy and analytical techniques for cellulose morphological, structural, chemical, and thermal characterization [97].

Flame retardancy assessment for fire-retardant polymer nanocomposites uses cone calorimetry and UL-94 tests [98,99]. UL-94, a standard preliminary test (as shown in Figure 3a), categorizes polymers based on vertical and horizontal burn testing and thin films, grading them into V_0, V_1, and V_2 based on the burning time, rate, and dripping behavior (details are shown in Table 3) [100]. Limiting of oxygen index (LOI) testing (ISO 4589-2 [101]) measures the minimum oxygen concentration needed to sustain combustion in a closed atmosphere, with better flame-retardant materials requiring higher oxygen concentrations (as shown in Figure 3b).

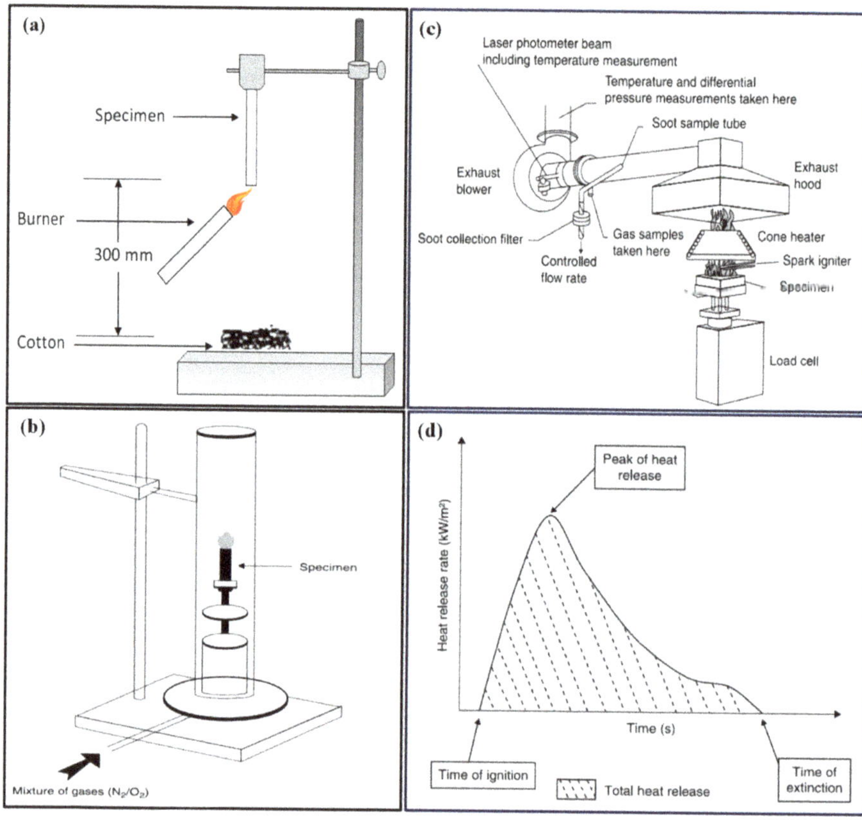

Figure 3. Schematic diagram of (**a**) the UL-94 vertical burning test [98]. Copyright 2020, reproduced with permission from the authors, MDPI, Basel. (**b**) The cone calorimeter, (**c**) the limiting oxygen index test, and (**d**) a typical cone calorimeter curve [100]. Copyright 2011, reproduced with permission from Woodhead Publishing Limited.

Cone calorimetry is a highly effective test for evaluating the fire behavior of medium-sized polymer samples. This method measures the decrease in oxygen concentration in the combustion gases of a sample exposed to a specific heat flux, generally between 10 and 100 kW/m^2. In the United States, it is standardized under ASTM E 1354 [102] and covered by the international standard ISO 5660 [103]. In this test, a sample measuring $100 \times 100 \times 4$ mm^3 is placed on a load cell to monitor mass loss throughout the experiment. The sample is irradiated uniformly from above by a conical radiant electric heater, and combustion is initiated using an electric spark (Figure 3c). The resulting combustion gases pass through the heating cone and are captured by an exhaust system equipped with a centrifugal fan and hood. This system measures gas flow, oxygen, CO, CO_2 concentrations, and smoke density.

Table 3. Classification of UL-94 V testing.

UL-94 Vo	Each specimen must have the first flame (t_1) and the second flame (t_2) less than 10 s. The total time for the first and second flames (t_1 + t_2) across all five specimens must be less than 50 s. Additionally, the second and third flames (t_2 + t_3) must be less than 30 s for each specimen. There should be no after-flame or afterglow up to the holding clamp, and no burning drops are allowed.
UL-94 V1	Each specimen must have t_1 and t_2 less than 30 s. The total time across all five specimens must be less than 250 s. Additionally, the t_2 and t_3 flames (t_2 + t_3) must be less than 60 s for each specimen. There should be no after-flame or afterglow up to the holding clamp, and no burning drops are allowed.
UL-94 V2	Each specimen must have t_1 and t_2 less than 30 s. The total time (t_1 + t_2) across all five specimens must be less than 250 s. Additionally, the second and third flames (t_2 + t_3) must be less than 60 s for each specimen. There should be no after-flame or afterglow up to the holding clamp, but burning drops are allowed.

The data from gas flow and oxygen concentration are used to calculate the heat release rate (HRR), expressed in kW/m^2, indicating the amount of heat released per unit time and surface area. The progression of HRR over time, especially its peak value (pHRR or HRRmax), is critical for assessing fire properties. The total heat release (THR), expressed in kJ/m^2, is obtained by integrating the HRR over time. Additionally, this test provides information on the time to ignition (TTI), duration of combustion or extinction (TOF), mass loss during combustion, quantities of CO and CO_2 produced, and total smoke released (TSR), as depicted in Figure 3d [100].

4. Mechanism of Flame Retardancy

The flame retardancy mechanisms of fire-retardant polymer nanocomposites involve gas phase inhibition, char formation or heat sink effect, and cooling through the generation of insulating layers. Gas phase inhibition occurs when flame retardants added to gases produced during polymer heating trap free radicals, halting the combustion process, typically seen with halogenated flame retardants (Figure 4a) [104]. Char formation involves flame-retardant chemicals reacting with the material's surface to create a carbonaceous layer, insulating the polymer, and reducing pyrolysis and gas release during burning, often used with non-halogen systems utilizing phosphorous and nitrogen chemistries [105]. The formation of carbonaceous char reduces the release of volatile by-products. The mechanism depicting the reactivity of phosphorus-based flame retardants in the gas phase is shown in this paper [26]. The cooling mechanism involves endothermic reactions releasing water molecules, which cool the polymer and dilute combustion, with hydrated metal salts, such as aluminum trihydroxide, utilized for this purpose. Additionally, synergistic approaches combining different flame retardants, such as antimony oxide, with halogen-containing ones, enhance flame retardancy by inhibiting vapor phase combustibility and generating heat sink behavior in polymer composites (Figure 4b) [105].

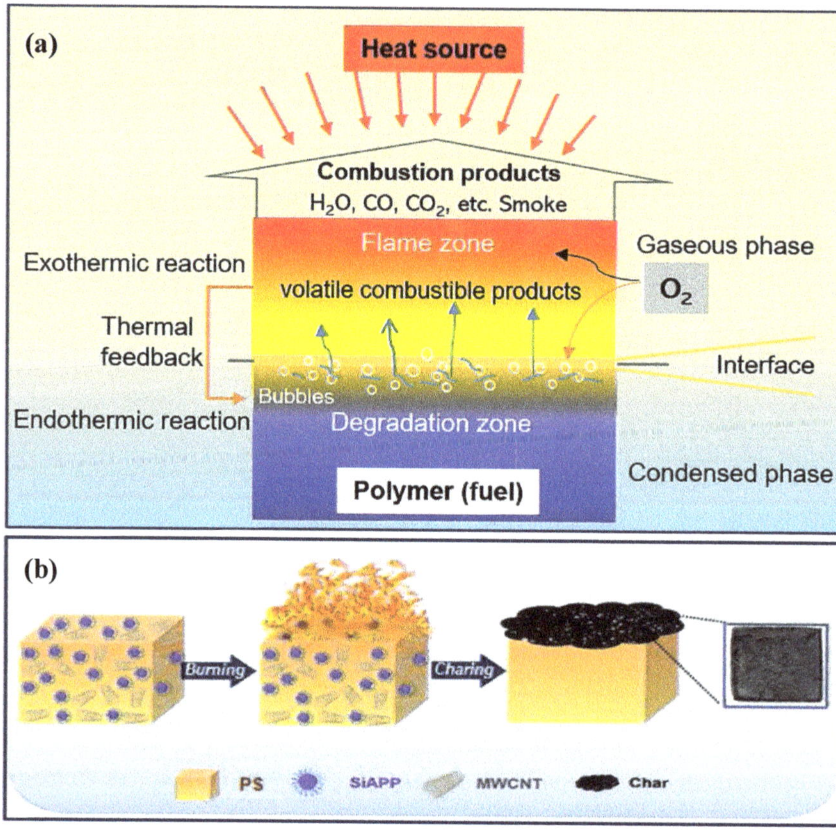

Figure 4. (**a**) The typical combustion process of polymers [104]. Copyright 2020, reproduced with permission from Published by Elsevier Ltd. (**b**) A schematic illustration of the flame-retardant nature [106]. Copyright 2023, reproduced with permission from Wiley-VCH GmbH.

5. Research on Flame-Retardant Chemicals

Research on flame-retardant treatment and combustion characteristics of wood has been evolving over the years, with studies from different years providing valuable insights into the effectiveness of various flame-retardant formulations and treatment methods. For instance, Park et al. [107] conducted a thermal analysis to examine the combustion characteristics of fire-retardant-treated wood. Their findings indicated that flame-retardant treatment significantly influences wood's thermal decomposition and combustion properties. Zhang Zhi-jun et al. [108] conducted a fire retardation performance test on a wood flour/polystyrene composite (WF-PS) treated with ammonium polyphosphate (APP) and tested using a cone calorimeter. The study found that the heat release rate was 35 kW/m^2, significantly reducing the total heat release. Additionally, the treatment with APP extended the ignition time of the composite, indicating improved fire-retardant properties.

Jinxue Jiang et al. [109] found that flame-retardant-treated wood exhibited the highest limiting oxygen index (LOI) values, indicating synergistic interactions between phosphorus and nitrogen (P–N). As the degree of degradation increased, the activation energies of the treated wood decreased by 19.6–50.4% compared to untreated wood. This led to higher char formation and reduced production of combustible products during degradation. These findings highlight the effectiveness of the P–N flame-retardant treatment in enhancing the fire resistance of wood by promoting char formation and inhibiting combustion. Lin Zhou et al. [110] investigated the effects of ammonium polyphosphate

(APP) and 3-(methylacryloxyl) propyltrimethoxy silane-modified APP (M-APP) on wood flour/polypropylene composites (WF/PP). M-APP significantly improved the mechanical properties of WF/PP composites and acted as an effective flame retardant, surpassing the performance of APP according to cone calorimetry results. Moreover, M-APP enhanced char formation, as evidenced by SEM analysis, indicating its potential in enhancing the fire resistance of WF/PP composites through improved char formation capabilities. Using cone calorimetry and thermogravimetric analysis (TGA), the study found significant reductions in the peak heat release rate (HRR) by 21% and total heat release (THR) by 44.2%.

Dong Won Son et al. [111] treated Japanese red pine, hemlock, and radiate pine with inorganic chemicals, such as 50% sodium silicate, 3% boric acid, 3% ammonium phosphate, and 3% ammonium borate, using a vacuum/pressure (vacuum 78 kPa, 30 min, pressure 18 kg/cm^2, time 2 h) impregnated method. The study reported that the ignition time of the treated wood was effectively delayed by these treatments, particularly with sodium silicate, ammonium phosphate, and ammonium borate, demonstrating an improvement in fire resistance. Seo et al. [112] analyzed the combustion and thermal properties of wood used indoors, such as the heat release rate, total heat release, and gas generation, using TGA and a cone calorimeter (KS F ISO 5660-1 [103]). They found that wood's material properties significantly impacted its combustion behavior, and the formation of a carbonization layer notably varied by tree species, showing a high correlation between total heat release and weight loss. The study emphasized that ignition time and total heat release are crucial data for imparting fire resistance performance to wood.

Chai, et al. [113] evaluated the fire retardation effect on Cryptomeria fortunei wood treated with a boric acid-urea-formaldehyde (MUF) resin mixed with borax. The treatment increased the oxygen index and time to ignition (TTI). Park et al. [114] compared the flame-retardant performance and combustion characteristics of cypress wood and particle board. Cypress wood injected with flame-retardant resin via vacuum pressurization outperformed specimens treated with surface flame-retardant paint. In their study, this group demonstrated that water-soluble phosphate flame retardants, when mixed with poly ammonium phosphate, guanylurea phosphate, phosphoric acid, and resin, effectively impregnated perforated Hinkoi plywood used as a sound absorber. They found that the frequency of perforations influenced the impregnation process, with narrower hole spacing leading to increased impregnation. This resulted in a 15% improvement in flame-retardant performance compared to untreated samples [115].

Sathasivam Pratheep Kumar et al. [116] applied a composite coating of sodium silicate and clay minerals to wood as an innovative inorganic flame retardant. Cone calorimeter tests revealed that the composite-coated wood significantly reduced the heat release rate, delayed total heat release and ignition, and exhibited superior flame retardation compared to other tested coatings due to the dense surface layer. The addition of vermiculite to sodium silicate enhanced ignition resistance and resulted in lower HRR values than the wood coated solely with sodium silicate. Ribeiro et al. [117] developed an unsaturated polyester-based composite with enhanced fire retardancy using nano/micro-oxide particles and common flame retardants. Results showed that hybrid-flame-retardant systems improved fire properties but sometimes decreased mechanical properties due to poor matrix-filler adhesion.

Rocha et al. [118] engineered a high-density polyethylene composite reinforced with lignocellulosic fibers as a potential substitute for natural pine wood. The HDPE/sponge gourd fiber composite demonstrated the best impact resistance, and with the addition of magnesium hydroxide, it showed improved flammability and thermal stability. Sheng Li et al. [119] developed a biomass-based flame-retardant additive derived from renewable chitosan, melamine formaldehyde resin-coated ammonium polyphosphate, and organic montmorillonite, incorporated into waterborne epoxy resin (WBEP) for wood coatings. The resulting coated wood composites achieved a UL-94 V-0 rating, a limiting oxygen index of 31.8%, and maintained excellent flame-retardant performance even after water resistance tests. The biomass-based additive enhanced the carbonization capability, with residual char reaching 23.9 wt.% at 800 °C. Cone calorimeter tests showed reduced heat and smoke release, forming an effective char layer that protected the wood substrate. The WBEP coating demonstrated superior water resistance and flame-retardant efficiency, making it suitable for flame-retardant wood composites.

Lu et al. [61] impregnated melamine (MEL) with organic phosphoric acid (AP) into the porous structure of wood. The limiting oxygen index (LOI) and cone calorimetry tests showed that MEL/AP presence significantly improved fire resistance. The LOI value increased from 21.0% to 68.5%, and the peak heat release rate and total heat release amount decreased by 41.7% and 80.2%, respectively, compared to the control sample (as shown in Figure 5). This indicates that MEL/AP in a porous wood structure improves flame retardancy. Price et al. [120] developed tannic acid (TA)-based composites for fire safety, comparing them to pentaerythritol (PER) composites. TA composites significantly extended the time to failure from seconds or minutes to over 15 min, with a maximum of 27 min. They exhibited better fire performance, with lower peak heat release values (211 vs. 108 kW/m^2), lower total heat release values (37.2 vs. 24.4 MJ/m^2; as shown in Figure 6), and slower fire growth rates (2.43 vs. 1.27 kW/m^{2}s^{-1}).

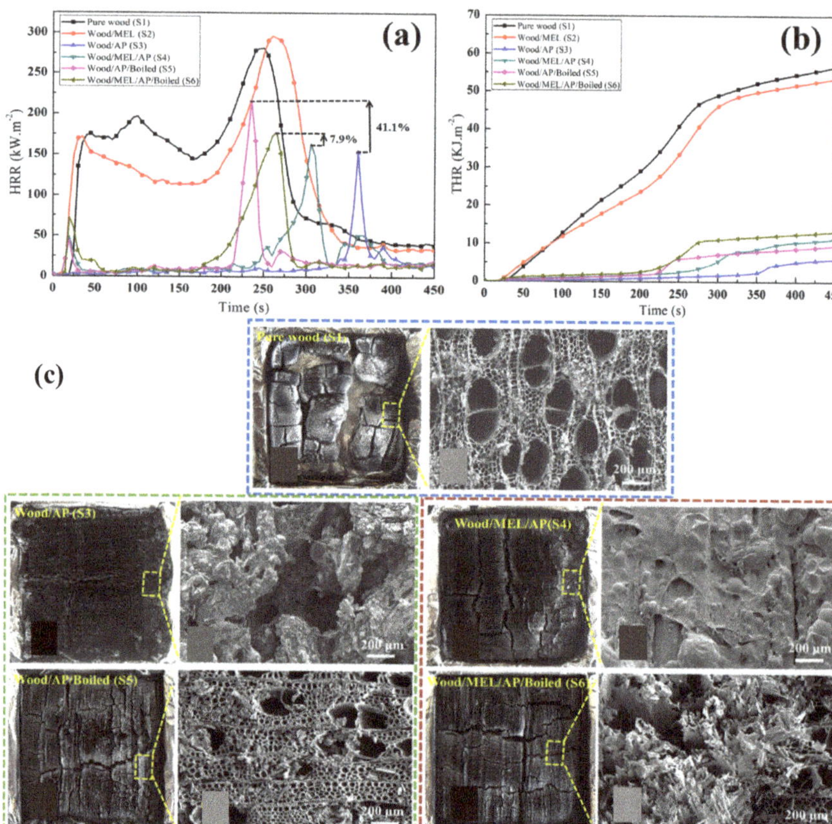

Figure 5. (**a**) Heat release rate (HRR) and (**b**) total heat release (THR) curves for both pure and modified wood. (**c**) Digital photographs and SEM images of pure wood and modified wood after cone calorimeter testing (CCT) [61]. Copyright 2020, reproduced with permission from Elsevier Ltd.

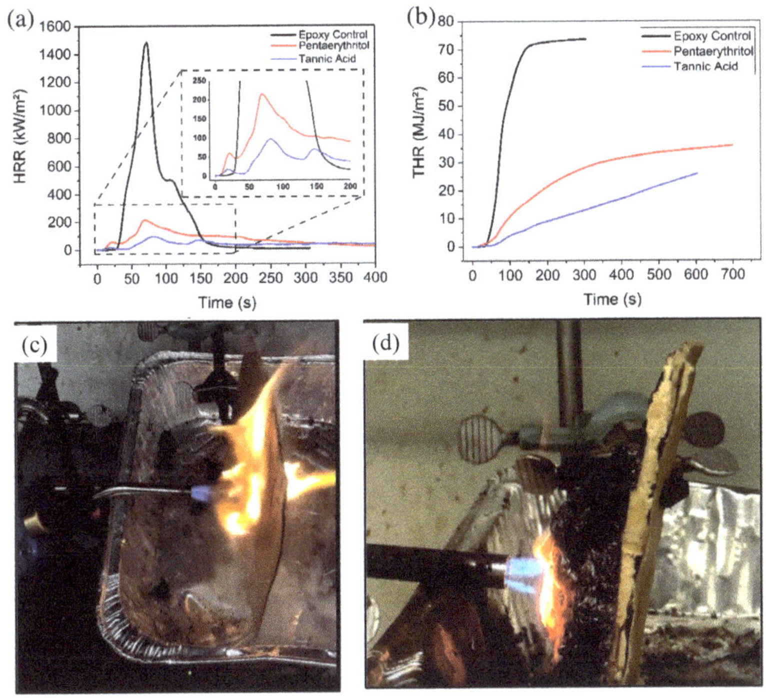

Figure 6. Cone calorimetry data illustrating (**a**) heat release rate (HRR) versus time and (**b**) total heat release (THR) versus time for cardboard coatings exposed to a propane blow torch. The uncoated cardboard was compromised after 7 s (**c**) and the 1 mm coated cardboard burned for 27 min (**d**). The coated sample exhibited a final expansion 25 times its original thickness [120]. Copyright 2020, reproduced with permission from the authors. SPE Polymers published by Wiley Periodicals LLC on behalf of the Society of Plastics Engineers.

X-ray photoelectron spectroscopy showed that TA char was more carbonaceous (54.71 at.% C vs. 39.63 at.% C in PER char). These findings demonstrate that TA composites provide superior fire protection, offering significant advancements for fire safety applications [120].

Özkan et al. [121] treated with fire retardants, including di-ammonium phosphate (DAP), borax, boric acid, and glucose, in aqueous solutions of 10%, 20%, and 30%. Post-heat treatment at 120 °C, 150 °C, and 180 °C improved water resistance, dimensional stability, mechanical strength, and thermal properties. The DAP/glucose complex enhanced phosphorus fixation, reducing leaching and providing long-term fire protection. These findings suggest potential for using treated wood in structural applications, offering enhanced durability and fire safety.

Yutao Yan et al. [122] developed a durable flame-retardant coating on wood using a layer-by-layer self-assembly method with chitosan (CS), graphene oxide (GO), and ammonium polyphosphate (APP; as shown in Figure 7). The coating enhanced thermal stability by lowering decomposition temperatures and increasing char residue due to the effective heat barrier properties of GO. Fire resistance significantly improved, with the LOI increasing from 22 to 42 and the heat release rate decreasing from 105.50 kW/m^2 to 57.51 kW/m^2 after 15 layers of CS-GO-APP were applied. The coating showed excellent durability in immersion and abrasion tests, forming a protective char layer that inhibited flame spread on wood surfaces.

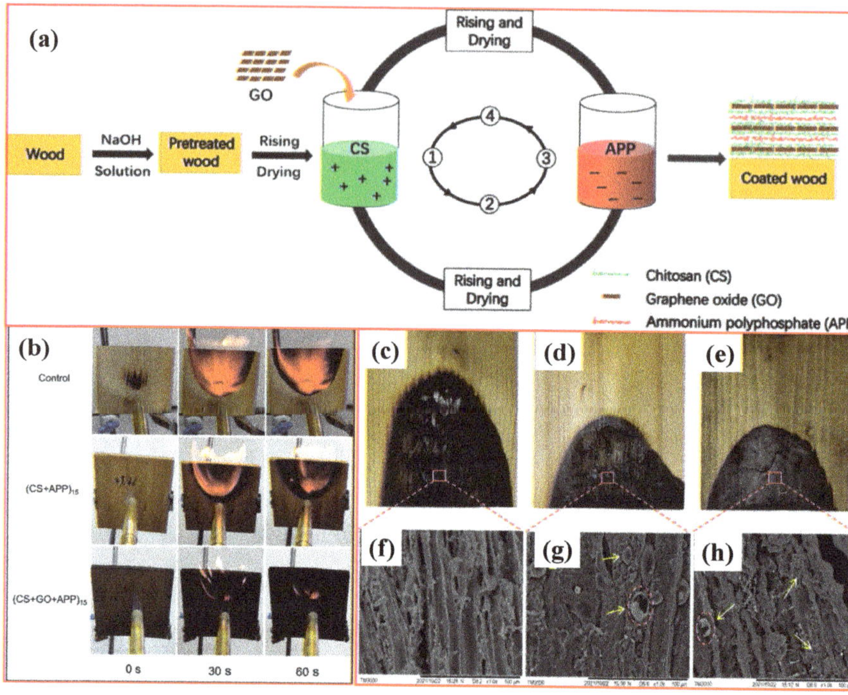

Figure 7. (**a**) Schematic illustration of the layer-by-layer (LBL) self-assembly CS-GO-APP coating on wood. (**b**) Fire behavior comparison of coated and uncoated wood samples at various burning times. Digital photographs and SEM images after combustion test: (**c**,**f**) uncoated wood, (**d**,**g**) (CS-APP)15-coated wood, and (**e**,**h**) (CS-GO-APP)15-coated wood [122]. Copyright 2022, reproduced with permission from the authors. Published by the American Chemical Society.

Recently, Rantuch et al. [24] treated spruce wood with a furfuryl alcohol solution enriched with biochar via vacuum infiltration. The research aimed to assess the suitability of this treatment and evaluate its impact on thermal degradation properties. Thermal gravimetric analysis revealed that the biochar-furfurylated wood bio-composite (BFW) exhibited enhanced thermal stability compared to untreated wood (W) and furfurylated wood (FW). BFW also demonstrated improved fire characteristics, including decreased effective heat of combustion and carbon monoxide yield, highlighting its potential for enhancing wood's fire resistance properties [24]. Besides, mineral fillers, such as aluminum and magnesium hydroxide, and natural mixtures, such as huntite and hydro-magnesite, are increasingly used as eco-friendly fire retardants. They act through endothermic decomposition, increasing the heat capacity of polymer residues, and enhancing the gas phase heat capacity with water or carbon dioxide. Despite the complexities in application across polymers, these fillers reduce flammability by up to 70%, as evidenced by LOI, UL-94, and cone calorimeter tests. Quantifying their effects reveals their nuanced impacts, crucial for sustainable fire safety solutions [123].

Liu et al. [124] explored the use of industrial lignin modified with phosphorus, nitrogen, and copper as a bio-based flame-retardant additive for wood-plastic composites (WPCs). The modified lignin (F-lignin) significantly enhanced the thermal stability and flame retardancy of WPCs compared to unmodified lignin (O-lignin). It reduced the heat release rate, total heat release, and smoke production during combustion, while promoting the formation of a dense, protective char layer. This innovative approach demonstrates the potential of utilizing industrial lignin in green flame-retardant strategies for sustainable WPC applications.

Moreover, Yang et al. [125] developed high-performance bio-composite materials from recyclable forestry waste, using lignin and cellulose as a natural adhesive matrix. Pretreatment with hydrogen peroxide, sodium hydroxide, and sodium silicate enhanced the material properties significantly, increasing tensile and bending strengths by over 90%. The bio-composite exhibited hydrophobicity with a water contact angle of 99.96° and maintained thermal stability up to 1300 °C without disintegration. These attributes make it highly suitable for eco-friendly construction applications, offering sustainable alternatives to petroleum-based materials. Yu et al. [126] introduced an eco-friendly method for creating flame-retardant wood composites using carboxymethylated alkali lignin, phytic acid, and melamine-urea-glyoxal resin. The resulting modified wood (MW/MPUC) showed significant improvements in flame retardancy, with a 56.8% reduction in total heat release, a 92.3% decrease in total smoke production, and an increase in the limiting oxygen index from 23.6% to 41.5% (as shown in Figure 8). Additionally, all modified wood samples passed the UL-94 V-1 flammability test and exhibited enhanced mechanical properties and dimensional stability.

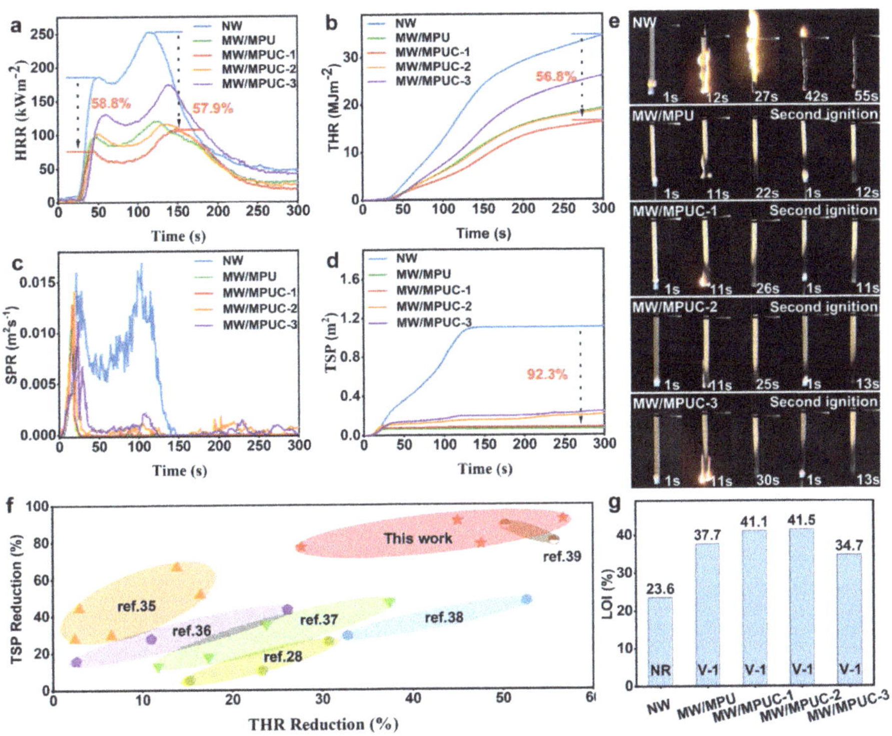

Figure 8. (**a**) The HRR curves, (**b**) the THR curves, (**c**) the SPR curves, and (**d**) the TSP curves of the specimens. (**e**) Digital photographs of specimens during UL-94 testing. (**f**) Summary of flame-retardant efficiency of PA and lignin, and (**g**) LOI and UL-94 testing results of the specimens [126].

From the above literature study, a summary table focusing on eco-friendly flame retardants is shown in Table 4.

Table 4. Summary of some eco-friendly flame retardants from the above literature review.

Study	Flame Retardant	Key Results	Thermal Properties	Flammability Test Results	Ref.
Study on TA-based composites	Tannic acid (TA)	Time to failure: 15–27 min	Peak HRR: 211 vs. 108 kW/m^2; total HRR: 37.2 vs. 24.4 MJ/m^2	Lower fire growth rates: 2.43 vs. 1.27 kW/m^2s^{-1}	[120]
Study on biochar-furfurylated wood	Furfuryl alcohol and biochar	Enhanced thermal stability and reduced flammability up to 70%	Decreased effective heat of combustion; higher char residue	Higher mass loss at low temperatures	[121]
Study on mineral fillers	Aluminum/magnesium hydroxide and magnesium carbonate	Endothermic decomposition, increased heat capacity, and reduced flammability by up to 70%	Quantified fire-retardant effects	Improved LOI, UL-94, and cone calorimeter results	[123]
Study on bio-composites with lignin	Lignin with P, N, and Cu elements	Reduced heat release rate, total heat release, and smoke production	Increased char residues	Enhanced flame retardancy	[124]
Study on MPUC flame retardant	Carboxymethylation alkali lignin, phytic acid, and melamine-urea-glyoxal resin	Total heat release reduction: 56.8%; total smoke production decrease: 92.3%	LOI: 23.6% to 41.5%	Passed UL-94 V-1 rating	[126]
Study on CS-GO-APP coating	Chitosan, graphene oxide, and ammonium polyphosphate	LOI: 22 to 42; HRR decrease: 105.50 to 57.51 kW/m^2; THR decrease: 62.43 to 34.31 MJ/m^2	Decreased initial and maximum thermal decomposition temperature	Excellent durability in water resistance and abrasion tests	[122]
This study introduces a PVA composite enhanced with graphene oxide and phytic acid	Graphene oxide and phytic acid	Achieved exceptional flame retardancy	pHRR reduction of 88.6%; THR reduction of 66.5%	Maintained structural integrity for over 2400 s	[72]
Study on polycarbonate (PC) hybridization in wood flour/high-density polyethylene (HDPE) composites	Boric acid and polycarbonate	Improved fire retardancy and mechanical properties	Char residue rate increased by 6.7% at 28% PC content	Heat release rate reduced upon combustion	[75]
Phenolic resins based on two natural products, namely, lignin and tannins, were implemented as bio-based fireproofing coatings for wood	Lignin and tannins with inorganic nanoparticles	Reduced heat release during combustion, improved wood integrity, and delayed flame propagation	Improved thermal resistance with TGA	Comparable performance to top commercial coatings	[79]

5.1. Discussions

The trend in fire-retardant research has been shifting significantly toward eco-friendly polymer nanocomposite coatings, reflecting an increasing emphasis on sustainability and environmental safety. Early studies focused primarily on inorganic chemicals and conventional flame retardants, such as ammonium polyphosphate (APP), boric acid, and sodium silicate [108–111]. However, the growing awareness of the environmental impact and the potential health risks associated with these traditional retardants have driven researchers to explore green alternatives [7]. Recent advancements have seen the development of bio-based and eco-friendly flame retardants, utilizing natural materials such as chitosan, lignin, and cellulose, often enhanced with nanotechnology [122,123]. For instance, the use of carboxymethylated alkali lignin, phytic acid, and melamine-urea-glyoxal resin to create flame-retardant wood composites exemplifies this shift, achieving significant improvements in fire resistance and reduced smoke production while maintaining mechanical properties and dimensional stability [125]. Nanocomposite coatings incorporating materials such as graphene oxide, clay minerals, and biochar are becoming prominent, leveraging their superior thermal stability and char-forming capabilities to enhance fire retardancy [126]. These innovations not only meet the stringent fire safety standards but also align with the principles of green chemistry, paving the way for sustainable applications in construction and material science. This eco-conscious direction marks a significant evolution in the field, aiming to mitigate the environmental impact while ensuring effective fire protection.

5.2. Limitation, Challengs, and Opportunities

The shift toward eco-friendly polymer nanocomposite coatings for flame retardancy presents challenges and opportunities in the field. Challenges include higher production costs and the need to scale manufacturing processes to meet industrial demands. There are also concerns about potential compromises in mechanical properties and the long-term durability of these coatings under various environmental conditions. Standardizing testing protocols and gaining acceptance from industries and consumers are additional hurdles. However, these challenges are accompanied by significant opportunities. Advances in nanotechnology offer the potential to develop highly effective flame retardants with minimal environmental impact. Integrating renewable resources, such as chitosan, lignin, and cellulose, supports sustainability goals and enhances the value of agricultural by-products. Innovation in multifunctional coatings that offer additional benefits beyond fire resistance is another promising avenue. Moreover, increasing regulatory pressures and consumer demand for greener products create a favorable market environment for developing and adopting eco-friendly flame-retardant technologies. A probable future roadmap for eco-friendly flame-retardant materials is illustrated in Figure 9.

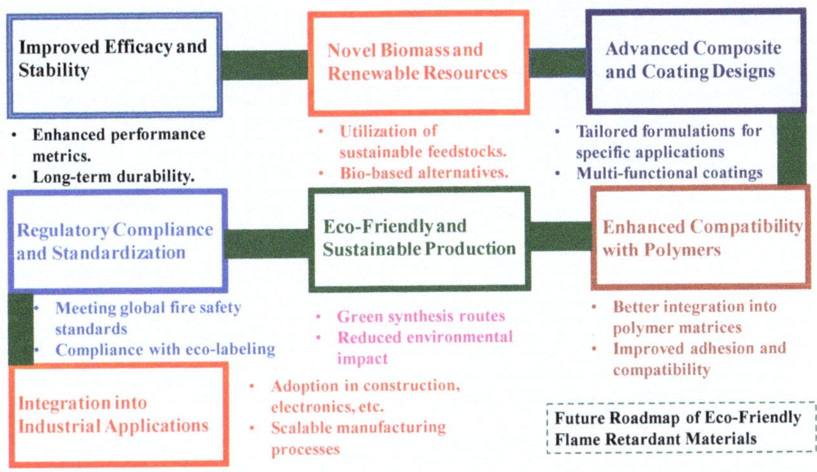

Figure 9. A probable future roadmap for the development and application of eco-friendly flame-retardant materials.

6. Conclusions

The evolution toward eco-friendly polymer nanocomposite coatings represents a pivotal advancement in fire-retardant technology for building materials. These coatings offer substantial opportunities, while facing challenges such as production costs, scalability, and ensuring mechanical integrity. Advances in nanotechnology enable the development of highly efficient flame retardants using renewable resources, aligning with sustainability objectives and regulatory requirements. Innovations in multifunctional coatings enhance fire resistance and provide added benefits, such as antimicrobial properties and thermal insulation. With growing market demand and regulatory support for greener solutions, the future holds promising prospects for integrating eco-friendly polymer nanocomposite coatings into mainstream building materials, ensuring safer environments while minimizing the environmental impact.

Author Contributions: H.K., conceptualization, writing—original draft preparation, writing—review and editing; C.-W.K., project administration, funding acquisition, writing—review and editing. All authors have read and agreed to the published version of the manuscript.

Funding: The Basic Science Research Program supported this research through the National Research Foundation of Korea (NRF), funded by the Ministry of Education (NRF-2019R1I1A3A02059471).

Institutional Review Board Statement: Not applicable.

Data Availability Statement: Not applicable.

Conflicts of Interest: The authors declare no conflicts of interest.

References

1. Bakhtiari, F. International cooperative initiatives and the United Nations Framework Convention on Climate Change. *Clim. Policy* **2018**, *18*, 655–663. [CrossRef]
2. Rogelj, J.; Geden, O.; Cowie, A.; Reisinger, A. Net-zero emissions targets are vague: Three ways to fix. *Nature* **2021**, *591*, 365–368. [CrossRef]

3. Costa, C.; Wollenberg, E.; Benitez, M.; Newman, R.; Gardner, N.; Bellone, F. Roadmap for achieving net-zero emissions in global food systems by 2050. *Sci. Rep.* **2022**, *12*, 15064. [CrossRef]
4. Nässén, J.; Hedenus, F.; Karlsson, S.; Holmberg, J. Concrete vs. wood in buildings—An energy system approach. *Build. Environ.* **2012**, *51*, 361–369. [CrossRef]
5. Webster, M.D.; Arehart, J.; Ruthwik, C.; Aloisio, J.; Karineh, G.; Gryniuk, M.; Hogroian, J.; Jezeritz, C.; Johnson, L.; Kestner, D. *Achieving Net Zero Embodied Carbon in Structural Materials by 2050*; American Society of Civil Engineers: Reston, VA, USA, 2020.
6. Zang, X.; Liu, W.; Wu, D.; Pan, X.; Zhang, W.; Bian, H.; Shen, R. Contemporary Fire Safety Engineering in Timber Structures: Challenges and Solutions. *Fire* **2024**, *7*, 2. [CrossRef]
7. Hull, T.R.; Law, R.J.; Bergman, Å. Chapter 4—Environmental drivers for replacement of halogenated flame retardants. In *Polymer Green Flame Retardants*; Papaspyrides, C.D., Kiliaris, P., Eds.; Elsevier: Amsterdam, The Netherlands, 2014; pp. 119–179. ISBN 978-0-444-53808-6.
8. Shaw, S. Halogenated Flame Retardants: Do the Fire Safety Benefits Justify the Risks? *Rev. Environ. Health* **2010**, *25*, 261–306. [CrossRef]
9. Sabet, M. Advancements in halogen-free polymers: Exploring flame retardancy, mechanical properties, sustainability, and applications. *Polym. Technol. Mater.* **2024**, *63*, 1794–1818. [CrossRef]
10. Popescu, C.-M.; Pfriem, A. Treatments and modification to improve the reaction to fire of wood and wood based products—An overview. *Fire Mater.* **2020**, *44*, 100–111. [CrossRef]
11. Sandberg, D. Additives in Wood products—today and future development. In *Environmental Impacts of Traditional and Innovative Forest-Based Bioproducts*; Kutnar, A., Muthu, S.S., Eds.; Springer: Singapore, 2016; pp. 105–172. ISBN 978-981-10-0655-5.
12. Hazarika, A.; Baishya, P.; Maji, T.K. Bio-based Wood Polymer Nanocomposites: A sustainable high-performance material for future. In *Eco-Friendly Polymer Nanocomposites: Processing and Properties*; Thakur, V.K., Thakur, M.K., Eds.; Springer: New Delhi, India, 2015; pp. 233–257. ISBN 978-81-322-2470-9.
13. Ling, M.; Yin, N.; Chen, Y.; Zhou, Z.; Chen, H.; Dai, C.; Huang, J.; Zhang, W. Construction of polylactic acid-based flame retardant composites by zinc oxide and bamboo carbon. *Carbon Lett.* **2024**, *34*, 665–675. [CrossRef]
14. Liu, M.; Wang, J.; Yan, Q.; Lyu, J.; Lei, Y.; Lyu, S.; Yan, L. Green bio-derived epoxidized linseed-oil plasticizer improves the toughness, strength, and dimensional stability of furfuryl alcohol-modified wood. *Ind. Crops Prod.* **2024**, *217*, 118886. [CrossRef]
15. Du, X.; Li, Z.; Zhang, J.; Li, X.; Du, G.; Deng, S. Development of environmentally friendly glyoxal-based adhesives with outstanding water repellency utilizing wheat gluten protein. *Int. J. Biol. Macromol.* **2024**, *273*, 133081. [CrossRef] [PubMed]
16. Chen, Y.; Rao, Y.; Liu, P.; Wu, L.; Zhang, G.; Zhang, J.; Xie, F. High-amylose starch-based gel as green adhesive for plywood: Adhesive property, water-resistance, and flame-retardancy. *Carbohydr. Polym.* **2024**, *339*, 122247. [CrossRef] [PubMed]
17. Nine, M.J.; Tran, D.N.H.; Tung, T.T.; Kabiri, S.; Losic, D. Graphene-Borate as an Efficient Fire Retardant for Cellulosic Materials with Multiple and Synergetic Modes of Action. *ACS Appl. Mater. Interfaces* **2017**, *9*, 10160–10168. [CrossRef]
18. Hu, X.; Sun, Z. Nano CaAlCO3-layered double hydroxide-doped intumescent fire-retardant coating for mitigating wood fire hazards. *J. Build. Eng.* **2021**, *44*, 102987. [CrossRef]
19. Guo, G.; Park, C.B.; Lee, Y.H.; Kim, Y.S.; Sain, M. Flame retarding effects of nanoclay on wood–fiber composites. *Polym. Eng. Sci.* **2007**, *47*, 330–336. [CrossRef]
20. Zhang, Y.; Huang, Y.; Li, M.-C.; Zhang, S.; Zhou, W.; Mei, C.; Pan, M. Bioinspired, stable adhesive Ti3C2Tx MXene-based coatings towards fire warning, smoke suppression and VOCs removal smart wood. *Chem. Eng. J.* **2023**, *452*, 139360. [CrossRef]
21. Kawalerczyk, J.; Walkiewicz, J.; Dziurka, D.; Mirski, R. Nanomaterials to improve fire properties in wood and wood-based composite panels. In *Emerging Nanomaterials: Opportunities and Challenges in Forestry Sectors*; Taghiyari, H.R., Morrell, J.J., Husen, A., Eds.; Springer: Cham, Germany, 2023; pp. 65–96. ISBN 978-3-031-17378-3.
22. Qiu, S.; Wang, X.; Yu, B.; Feng, X.; Mu, X.; Yuen, R.K.K.; Hu, Y. Flame-retardant-wrapped polyphosphazene nanotubes: A novel strategy for enhancing the flame retardancy and smoke toxicity suppression of epoxy resins. *J. Hazard. Mater.* **2017**, *325*, 327–339. [CrossRef] [PubMed]
23. Zhang, S.; Li, S.-N.; Wu, Q.; Li, Q.; Huang, J.; Li, W.; Zhang, W.; Wang, S. Phosphorus containing group and lignin toward intrinsically flame retardant cellulose nanofibril-based film with enhanced mechanical properties. *Compos. Part B Eng.* **2021**, *212*, 108699. [CrossRef]
24. Rantuch, P.; Kvorková, V.; Wachter, I.; Martinka, J.; Štefko, T. Is biochar a suitable fire retardant for furfurylated wood? *Compos. Part C Open Access* **2024**, *14*, 100454. [CrossRef]
25. Kolya, H.; Mondal, S.; Kang, C.-W.; Nah, C. 18—The use of polymer-graphene composites in catalysis. In *Woodhead Publishing Series in Composites Science and Engineering*; Rahaman, M., Nayak, L., Hussein, I.A., Das, N.C.B.T.-P.N.C.G., Eds.; Woodhead Publishing: Sawston, UK, 2022; pp. 537–556. ISBN 978-0-12-821639-2.
26. Taib, M.N.A.M.; Antov, P.; Savov, V.; Fatriasari, W.; Madyaratri, E.W.; Wirawan, R.; Osvaldová, L.M.; Hua, L.S.; Ghani, M.A.A.; Al Edrus, S.S.A.O.; et al. Current progress of biopolymer-based flame retardant. *Polym. Degrad. Stab.* **2022**, *205*, 110153. [CrossRef]
27. Padil, V.V.T.; Akshay Kumar, K.P.; Murugesan, S.; Torres-Mendieta, R.; Wacławek, S.; Cheong, J.Y.; Černík, M.; Varma, R.S. Sustainable and safer nanoclay composites for multifaceted applications. *Green Chem.* **2022**, *24*, 3081–3114. [CrossRef]
28. Green, J. An overview of the fire retardant chemicals industry, Past—Present—Future. *Fire Mater.* **1995**, *19*, 197–204. [CrossRef]
29. Zhang, M.; Buekens, A.; Li, X. Brominated flame retardants and the formation of dioxins and furans in fires and combustion. *J. Hazard. Mater.* **2016**, *304*, 26–39. [CrossRef] [PubMed]

30. Beyer, G. Chapter 1—Introduction to flame retardant systems. In *Flame Retardant Nanocomposites*; Thomas, S., Vahabi, H., Somasekharan, L., Eds.; Woodhead Publishing: Sawston, UK, 2024; pp. 1–22. ISBN 978-0-443-15421-8.
31. Kaur, A.; Kapoor, K.; Mandot, A.; Godara, S.K.; Sood, A.K.; Singh, M. Fire-Retardant coatings for modern lightweight materials. In *Functional Coatings: Innovations and Challenges*; Davim, J.P., Arya, R.K., Verros, G.D., Eds.; Wiley: Hoboken, New Jersey, USA, 2024; pp. 202–230. ISBN 9781394207305.
32. Purser, D. Toxicity of fire retardants in relation to life safety and environmental hazards. *Fire Retard. Mater.* **2001**, 69–127.
33. Zhu, K.; Yang, Y.; Lin, C.; Wang, Q.; Ye, D.; Jiang, H.; Wu, K. Effect of Compounded Aluminum Hydroxide Flame Retardants on the Flammability and Smoke Suppression Performance of Asphalt Binders. *ACS Omega* **2024**, *9*, 2803–2814. [CrossRef] [PubMed]
34. Dey, R.; Bhakare, M.A.; Some, S. One-pot synthesis of aluminum phosphate-supported, chitosan-linked expandable graphite as a novel flame retardant for textile. *J. Appl. Polym. Sci.* **2024**, *141*, e55581. [CrossRef]
35. Liu, K.; Li, Y.; Xu, L.; Zhu, F.; Zhang, Y.; Meng, Y.; Xia, X. Preparation of ethyl cellulose microencapsulated ammonium polyphosphate and its application in flame retardant cellulose paper. *Ind. Crops Prod.* **2024**, *210*, 118132. [CrossRef]
36. Goller, S.M.; Schartel, B.; Krüger, S. Phosphorus features halogen–calcium hypophosphite replaces antimony trioxide, reduces smoke, and improves flame retardancy. *Thermochim. Acta* **2024**, *737*, 179764. [CrossRef]
37. Lin, M.; Guo, X.; Xu, Y.; Zhang, X.; Hu, D. A Top-Down Approach to the Fabrication of Flame-Retardant Wood Aerogel with In Situ-Synthesized Borax and Zinc Borate. *Materials* **2024**, *17*, 2638. [CrossRef]
38. Zhu, J.; Wang, Y.; Zhao, X.; Li, N.; Guo, X.; Zhao, L.; Yin, Y. Anisotropic composite aerogel with thermal insulation and flame retardancy from cellulose nanofibers, calcium alginate and boric acid. *Int. J. Biol. Macromol.* **2024**, *267*, 131450. [CrossRef]
39. Levchik, S.; Bocchini, S.; Camino, G. Halogen-containing flame retardants. In *Fire Retardancy of Polymeric Materials*; Wilkie, C.A., Morgan, A.B., Eds.; CRC Press: Boca Raton, FL, USA, 2024; pp. 58–80.
40. Gao, L.; Zheng, G.; Zhou, Y.; Hu, L.; Feng, G.; Zhang, M. Synergistic effect of expandable graphite, diethyl ethylphosphonate and organically-modified layered double hydroxide on flame retardancy and fire behavior of polyisocyanurate-polyurethane foam nanocomposite. *Polym. Degrad. Stab.* **2014**, *101*, 92–101. [CrossRef]
41. Wang, Z.; Gao, S.J.; Che, X.X.; Shen, C.H. Synthesis and characterization of a flame retardant Dimethyl methyl Phosphonate (DMMP) and its application in FRP. *Adv. Mater. Res.* **2013**, *804*, 29–35. [CrossRef]
42. Kausar, A.; Anwar, Z.; Muhammad, B. Overview of nonflammability characteristics of graphene and graphene oxide-based polymeric composite and essential flame retardancy techniques. *Polym. Plast. Technol. Eng.* **2017**, *56*, 488–505. [CrossRef]
43. Wi, S.; Kim, Y.U.; Choi, J.Y.; Shin, B.; Kim, S. Active protection against fire: Enhancing the flame retardancy of sandwich panels using an expandable graphite layer formation. *Int. J. Therm. Sci.* **2024**, *195*, 108658. [CrossRef]
44. Wu, K.; Zhu, K.; Kang, C.; Wu, B.; Huang, Z. An experimental investigation of flame retardant mechanism of hydrated lime in asphalt mastics. *Mater. Des.* **2016**, *103*, 223–229. [CrossRef]
45. Wang, L.; Yan, W.-J.; Zhong, C.-Z.; Chen, C.-R.; Luo, Q.; Pan, Y.-T.; Tang, Z.-H.; Xu, S. Construction of TiO2-based decorated with containing nitrogen-phosphorus bimetallic layered double hydroxides for simultaneously improved flame retardancy and smoke suppression properties of EVA. *Mater. Today Chem.* **2024**, *36*, 101952. [CrossRef]
46. Dun, L.; Ouyang, Z.; Sun, Q.; Yue, X.; Wu, G.; Li, B.; Kang, W.; Wang, Y. A Simple and Efficient Magnesium Hydroxide Modification Strategy for Flame-Retardancy Epoxy Resin. *Polymers* **2024**, *16*, 1471. [CrossRef] [PubMed]
47. Wang, Z.; Wang, J.; Zhang, J.; Guo, W. Synergistic flame retardant modification of bio-based nylon 56 by graphitic carbon nitride and melamine cyanurate. *J. Vinyl Addit. Technol.* **2024**, *30*, 456–469. [CrossRef]
48. Wang, Y.; Qu, C.; Yu, K.; Si, Z.; Zhang, J. PTFE-based flame retardant coatings optimized by melamine polyphosphate/aluminum diethyl hypophosphite/anhydrous transparent powder through orthogonal experiment. *Prog. Org. Coat.* **2024**, *191*, 108423. [CrossRef]
49. Huang, G.; Liang, H.; Wang, Y.; Wang, X.; Gao, J.; Fei, Z. Combination effect of melamine polyphosphate and graphene on flame retardant properties of poly (vinyl alcohol). *Mater. Chem. Phys.* **2012**, *132*, 520–528. [CrossRef]
50. Bi, X.; Song, K.; Zhang, H.; Pan, Y.-T.; He, J.; Wang, D.-Y.; Yang, R. Dimensional change of red phosphorus into nanosheets by metal–organic frameworks with enhanced dispersion in flame retardant polyurea composites. *Chem. Eng. J.* **2024**, *482*, 148997. [CrossRef]
51. Chen, Q.; Huo, S.; Lu, Y.; Ding, M.; Feng, J.; Huang, G.; Xu, H.; Sun, Z.; Wang, Z.; Song, P. Heterostructured Graphene@ Silica@ Iron Phenylphosphinate for Fire-Retardant, Strong, Thermally Conductive Yet Electrically Insulated Epoxy Nanocomposites. *Small* **2024**, *20*, 2310724. [CrossRef] [PubMed]
52. Kanayama, K.; Nakamura, H.; Maruta, K.; Bodi, A.; Hemberger, P. Unimolecular Decomposition Mechanism of Trimethyl Phosphate. *Chem. Eur. J.* **2024**, *30*, e202401750. [CrossRef] [PubMed]
53. Gumus, N.; Doganci, E.; Aytac, A. Evaluations of the effects of different flame retardants combinations on particleboards produced using urea–formaldehyde resin. *Eur. J. Wood Wood Prod.* **2024**, *82*, 747–759. [CrossRef]
54. Van der Veen, I.; de Boer, J. Phosphorus flame retardants: Properties, production, environmental occurrence, toxicity and analysis. *Chemosphere* **2012**, *88*, 1119–1153. [CrossRef]
55. Wang, L.; Yang, Y.; Deng, H.; Duan, W.; Zhu, J.; Wei, Y.; Li, W. Flame retardant properties of a guanidine phosphate–zinc borate composite flame retardant on wood. *ACS Omega* **2021**, *6*, 11015–11024. [CrossRef] [PubMed]
56. Öhrn, O.; Sykam, K.; Gawusu, S.; Mensah, R.A.; Försth, M.; Shanmugam, V.; Babu, N.B.K.; Sas, G.; Jiang, L.; Xu, Q. Surface coated ZnO powder as flame retardant for wood: A short communication. *Sci. Total Environ.* **2023**, *897*, 165290. [CrossRef] [PubMed]

57. Rezaei Qazviniha, M.; Piri, F. Preparation, Identification, and Evaluation of the Thermal Properties of Novolac Resins Modified with TiO_2, MgO, and V_2O_5 Oxides. *Mech. Adv. Compos. Struct.* **2024**, *11*, 1–9.
58. Jin, E.; Chung, Y.-J. Evaluation of combustion characteristics for wood specimens coated with metal oxides of different oxidation states in the secondary stage of combustion (II). *Wood Sci. Technol.* **2024**, *58*, 253–271. [CrossRef]
59. Wen, M.-Y.; Kang, C.-W.; Park, H.-J. Impregnation and mechanical properties of three softwoods treated with a new fire retardant chemical. *J. Wood Sci.* **2014**, *60*, 367–375. [CrossRef]
60. Che, W.; Li, Z.; Huo, S.; Dinh, T.; Hong, M.; Maluk, C.; Yu, Y.; Xie, Y. Fire-retardant anti-microbial robust wood nanocomposite capable of fire-warning by graded-penetration impregnation. *Compos. Part B Eng.* **2024**, *280*, 111482. [CrossRef]
61. Lu, J.; Jiang, P.; Chen, Z.; Li, L.; Huang, Y. Flame retardancy, thermal stability, and hygroscopicity of wood materials modified with melamine and amino trimethylene phosphonic acid. *Constr. Build. Mater.* **2021**, *267*, 121042. [CrossRef]
62. Rabajczyk, A.; Zielecka, M.; Małozięć, D. Hazards Resulting from the Burning Wood Impregnated with Selected Chemical Compounds. *Appl. Sci.* **2020**, *10*, 6093. [CrossRef]
63. Mariappan, T. Fire retardant coatings. In *New Technologies in Protective Coatings*; Giudice, C., Canosa, G., Eds.; IntechOpen: London, UK, 2017; p. 28.
64. Lu, S.; Feng, Y.; Zhang, P.; Hong, W.; Chen, Y.; Fan, H.; Yu, D.; Chen, X. Preparation of flame-retardant polyurethane and its applications in the leather industry. *Polymers* **2021**, *13*, 1730. [CrossRef] [PubMed]
65. Ng, Y.H.; Dasari, A.; Tan, K.H.; Qian, L. Intumescent fire-retardant acrylic coatings: Effects of additive loading ratio and scale of testing. *Prog. Org. Coat.* **2021**, *150*, 105985. [CrossRef]
66. Mathews, L.D.; Capricho, J.C.; Peerzada, M.; Salim, N.V.; Parameswaranpillai, J.; Hameed, N. Recent progress and multifunctional applications of fire-retardant epoxy resins. *Mater. Today Commun.* **2022**, *33*, 104702. [CrossRef]
67. Wang, F.; Pan, S.; Zhang, P.; Fan, H.; Chen, Y.; Yan, J. Synthesis and application of phosphorus-containing flame retardant plasticizer for polyvinyl chloride. *Fibers Polym.* **2018**, *19*, 1057–1063. [CrossRef]
68. Temane, L.T.; Ray, S.S.; Orasugh, J.T. Review on Processing, Flame-Retardant Properties, and Applications of Polyethylene Composites with Graphene-Based Nanomaterials. *Macromol. Mater. Eng.* **2024**, 2400104. [CrossRef]
69. Kang, F.; Han, H.; Wang, H.; He, D.; Zhou, M. Construction of a flame retardant three-dimensional network structure in sisal/polypropylene composites. *Ind. Crops Prod.* **2024**, *209*, 117973. [CrossRef]
70. Kassaun, B.B.; Fatehi, P. Solvent-Free Lignin-Silsesquioxane wood coating formulation with superhydrophobic and Flame-Retardant functionalities. *Chem. Eng. J.* **2024**, *493*, 152582. [CrossRef]
71. Trifeldaite-Baranauskiene, G.; Stankute, E.; Aniskevich, A.; Zeleniakiene, D.; Zukiene, K. Preparation and Characterisation of Composites from Industrial Waste: Wood Flour and Expanded Ethylene Vinyl Acetate. *Mech. Compos. Mater.* **2024**, *60*, 1–16. [CrossRef]
72. Zhang, Z.; Zhou, Z.; Huang, J.; Wang, Y. A flame retardant poly vinyl alcohol/graphene oxide/phytic acid composite for a quick response and ultra-long fire alarm. *J. Mater. Chem. A* **2024**, *12*, 6050–6066. [CrossRef]
73. Chen, R.; Lu, S.; Li, C.; Li, M.; Lo, S. Characterization of thermal decomposition behavior of commercial flame-retardant ethylene–propylene–diene monomer (EPDM) rubber. *J. Therm. Anal. Calorim.* **2015**, *122*, 449–461. [CrossRef]
74. Gadhave, R.V.I.; Dhawale, P. V State of research and trends in the development of polyvinyl acetate-based wood adhesive. *Open J. Polym. Chem.* **2022**, *12*, 13–42. [CrossRef]
75. Zhang, J.; Koubaa, A.; Xing, D.; Wang, H.; Wang, F.; Wang, X.-M.; Wang, Q. Flammability, thermal stability, and mechanical properties of wood flour/polycarbonate/polyethylene bio-based composites. *Ind. Crops Prod.* **2021**, *169*, 113638. [CrossRef]
76. Çelen, U.; Balçik Tamer, Y.; Berber, H. The potential use of natural expanded perlite as a flame retardant additive for acrylonitrile-butadiene-styrene based composites. *J. Vinyl Addit. Technol.* **2024**, *30*, 277–293. [CrossRef]
77. Ren, G.; Fang, Y.; Yang, R.; Zhu, J.; Fu, Y.; Wang, W.; Ou, R.; Song, Y.; Wang, Q. Creation of a high strength, hydrophobic and fireproof surface on wood by polyamide acid under mild and simple conditions. *Prog. Org. Coat.* **2024**, *189*, 108313. [CrossRef]
78. Jimenez, M.; Gallou, H.; Duquesne, S.; Jama, C.; Bourbigot, S.; Couillens, X.; Speroni, F. New routes to flame retard polyamide 6, 6 for electrical applications. *J. Fire Sci.* **2012**, *30*, 535–551. [CrossRef]
79. De Hoyos-Martínez, P.L.; Issaoui, H.; Herrera, R.; Labidi, J.; Charrier-El Bouhtoury, F. Wood fireproofing coatings based on biobased phenolic resins. *ACS Sustain. Chem. Eng.* **2021**, *9*, 1729–1740. [CrossRef]
80. Wei, A.; Ou, M.; Wang, S.; Zou, Y.; Xiang, C.; Xu, F.; Sun, L. Preparation of a Highly Flame-Retardant Urea–Formaldehyde Resin and Flame Retardance Mechanism. *Polymers* **2024**, *16*, 1761. [CrossRef]
81. Wu, M.; Emmerich, L.; Kurkowiak, K.; Militz, H. Combined treatment of wood with thermosetting resins and phosphorous flame retardants. *Eur. J. Wood Wood Prod.* **2024**, *82*, 167–174. [CrossRef]
82. Ma, X.; Wang, J.; Li, L.; Wang, X.; Gong, J. Co-pyrolysis model for polylactic acid (PLA)/wood composite and its application in predicting combustion behaviors. *Renew. Energy* **2024**, *225*, 120267. [CrossRef]
83. Al-Mosawi, A.I.; Abdulsada, S.A.; Rijab, M.A.; Hashim, A. Flame retardancy of biopolymer polyhydroxyalkanoate composite. *Int. J.* **2015**, *3*, 883–886.
84. Passauer, L.P. P–N-modified starch: A polymeric flame retardant for wood-based materials. In *Bio-Based Flame-Retardant Technology for Polymeric Materials*; Hu, Y., Nabipour, H., Wang, X., Eds.; Elsevier: Amsterdam, The Netherlands, 2022; pp. 339–368.

85. Breuer, R.; Zhang, Y.; Erdmann, R.; Vernaez Hernandez, O.E.; Kabasci, S.; Kostka, M.; Reinhardt, N.; Facklam, M.; Hopmann, C. Development and processing of flame retardant cellulose acetate compounds for foaming applications. *J. Appl. Polym. Sci.* **2020**, *137*, 48863. [CrossRef]
86. Yang, H.; Yu, B.; Xu, X.; Bourbigot, S.; Wang, H.; Song, P. Lignin-derived bio-based flame retardants toward high-performance sustainable polymeric materials. *Green Chem.* **2020**, *22*, 2129–2161. [CrossRef]
87. Wei, A.; Wang, S.; Zou, Y.; Xiang, C.; Xu, F.; Sun, L. Preparation of a Flame-Retardant Curing Agent Based on Phytic Acid–Melamine Ion Crosslinking and Its Application in Wood Coatings. *Polymers* **2024**, *16*, 1557. [CrossRef] [PubMed]
88. Shukla, V. Review of electromagnetic interference shielding materials fabricated by iron ingredients. *Nanoscale Adv.* **2019**, *1*, 1640–1671. [CrossRef] [PubMed]
89. Guo, Q.; Ghadiri, R.; Weigel, T.; Aumann, A.; Gurevich, E.L.; Esen, C.; Medenbach, O.; Cheng, W.; Chichkov, B.; Ostendorf, A. Comparison of in Situ and ex Situ Methods for Synthesis of Two-Photon Polymerization Polymer Nanocomposites. *Polymers* **2014**, *6*, 2037–2050. [CrossRef]
90. Dobkowski, Z. Thermal analysis techniques for characterization of polymer materials. *Polym. Degrad. Stab.* **2006**, *91*, 488–493. [CrossRef]
91. Venkateshaiah, A.; Padil, V.V.T.; Nagalakshmaiah, M.; Waclawek, S.; Černík, M.; Varma, R.S. Microscopic Techniques for the Analysis of Micro and Nanostructures of Biopolymers and Their Derivatives. *Polymers* **2020**, *12*, 512. [CrossRef]
92. Praharaj, S.; Rout, D. Structural investigation of carbon nanotube-polymer composites by FTIR, UV, NMR, and Raman Spectroscopy. In *Handbook of Carbon Nanotubes*; Abraham, J., Thomas, S., Kalarikkal, N., Eds.; Springer: Cham, Germany, 2020; pp. 1–24. ISBN 978-3-319-70614-6.
93. Danilova, S.N.; Okhlopkova, A.A.; Yarusova, S.B.; Dyakonov, A.A.; Gordienko, P.S.; Papynov, E.K.; Shichalin, O.O.; Buravlev, I.Y.; Vasilev, A.P.; Zhevtun, I.G.; et al. Study on the Impact of a Combination of Synthetic Wollastonite and 2-Mercaptobenzothiazole-Based Fillers on UHMWPE Polymeric Matrix. *J. Compos. Sci.* **2023**, *7*, 431. [CrossRef]
94. Abhilash, V.; Rajender, N.; Suresh, K. Chapter 14—X-ray diffraction spectroscopy of polymer nanocomposites. In *Spectroscopy of Polymer Nanocomposites*; Thomas, S., Rouxel, D., Ponnamma, D., Eds.; William Andrew Publishing: Norwich, NY, USA, 2016; pp. 410–451. ISBN 978-0-323-40183-8.
95. Shojaeiarani, J.; Bajwa, D.S.; Stark, N.M. Green esterification: A new approach to improve thermal and mechanical properties of poly(lactic acid) composites reinforced by cellulose nanocrystals. *J. Appl. Polym. Sci.* **2018**, *135*, 46468. [CrossRef]
96. Gupta, P.; Ruzicka, E.; Benicewicz, B.C.; Sundararaman, R.; Schadler, L.S. Dielectric Properties of Polymer Nanocomposite Interphases Using Electrostatic Force Microscopy and Machine Learning. *ACS Appl. Electron. Mater.* **2023**, *5*, 794–802. [CrossRef] [PubMed]
97. Chakraborty, I.; Rongpipi, S.; Govindaraju, I.; B, R.; Mal, S.S.; Gomez, E.W.; Gomez, E.D.; Kalita, R.D.; Nath, Y.; Mazumder, N. An insight into microscopy and analytical techniques for morphological, structural, chemical, and thermal characterization of cellulose. *Microsc. Res. Tech.* **2022**, *85*, 1990–2015. [CrossRef] [PubMed]
98. Maqsood, M.; Seide, G. Biodegradable Flame Retardants for Biodegradable Polymer. *Biomolecules* **2020**, *10*, 1038. [CrossRef] [PubMed]
99. Crossley, R.; Schubel, P.; Stevenson, A. Furan matrix and flax fibre as a sustainable renewable composite: Mechanical and fire-resistant properties in comparison to phenol, epoxy and polyester. *J. Reinf. Plast. Compos.* **2013**, *33*, 58–68. [CrossRef]
100. Dewaghe, C.; Lew, C.Y.; Claes, M.; Belgium, S.A.; Dubois, P. 23—Fire-retardant applications of polymer–carbon nanotubes composites: Improved barrier effect andsynergism. In *Polymer–Carbon Nanotube Composites*; McNally, T., Pötschke, P., Eds.; Woodhead Publishing: Sawston, UK, 2011; pp. 718–745. ISBN 978-1-84569-761-7.
101. *STN EN ISO 4589-2: 2017*; Plastics. Determination of Burning Behaviour by Oxygen Index. Part 2: Ambient Temperature Test. ISO: Geneva, Switzerland, 2017.
102. *ASTM E 1354*; Standard Test Method for Heat and Visible Smoke Release Rates for Materials Using an OxygenConsumption Calorimeter. ASTM International: West Conshohocken, PA, USA,, 2004.
103. *ISO 5660-1*; Reaction to Fire Tests—Heat Release, Smoke Production and Mass Loss Rate - Part 1: Heat Release(Cone calorimeter method). ISO: Geneva, Switzerland, 2002.
104. He, W.; Song, P.; Yu, B.; Fang, Z.; Wang, H. Flame retardant polymeric nanocomposites through the combination of nanomaterials and conventional flame retardants. *Prog. Mater. Sci.* **2020**, *114*, 100687. [CrossRef]
105. Mouritz, A.P.; Gibson, A.G. *Fire Properties of Polymer Composite Materials*; Springer: Berlin/Heidelberg, Germany, 2007; Volume 143, ISBN 1402053568.
106. Kumar, S.; Dhawan, R.; Shukla, S.K. Flame Retardant Polymer Nanocomposites: An Overview. *Macromol. Symp.* **2023**, *407*, 2200089. [CrossRef]
107. Park, H.-J.; Kang, Y.-G.; Kim, H. A study on combustion characteristics of fire retardant treated wood. *J. Korean Wood Sci. Technol.* **2005**, *33*, 38–44.
108. Fang, Y.Q.; Wang, Q.W.; Song, Y.M.; Zhang, Z.J. The fire retardancy of wood flour/PS composites treated with APP-starch. *Polym. Mater. Sci. Eng* **2008**, *24*, 83–86.
109. Jiang, J.; Li, J.; Hu, J.; Fan, D. Effect of nitrogen phosphorus flame retardants on thermal degradation of wood. *Constr. Build. Mater.* **2010**, *24*, 2633–2637. [CrossRef]

110. Zhou, L.; Guo, C.; Li, L. Influence of ammonium polyphosphate modified with 3-(methylacryloxyl) propyltrimethoxy silane on mechanical and thermal properties of wood flour–polypropylene composites. *J. Appl. Polym. Sci.* **2011**, *122*, 849–855. [CrossRef]
111. Son, D.-W.; Kang, M.-R.; Kim, J.-I.; Park, S.-B. Fire performance of the wood treated with inorganic fire retardants. *J. Korean Wood Sci. Technol.* **2012**, *40*, 335–342. [CrossRef]
112. Seo, H.J.; Kang, M.R.; Son, D.W. Combustion properties of woods for indoor use (II). *J. Korean Wood Sci. Technol.* **2015**, *43*, 478–485. [CrossRef]
113. Chai, Y.B.; Liu, J.L.; Zhen, X. Dimensional stability, mechanical properties and fire resistance of MUF-boron treated wood. *Adv. Mater. Res.* **2012**, *341*, 80–84. [CrossRef]
114. Park, S.-H.; Baek, E.-S. A Study on the combustion characteristics of wood according to flame resistant treatment. *Fire Sci. Eng.* **2015**, *29*, 12–18. [CrossRef]
115. Jang, E.-S.; Yong, W.-J.; Jo, S.-U.; Kang, C.-W.; Park, H.-J. Evaluation of flame retardant impregnation in perforated Hinoki (*Chamaecyparis obtusa*) plywood: Flame retardant impregnation in perforated Hinoki plywood. *Wood Fiber Sci.* **2024**, *56*, 43–50.
116. Kumar, S.P.; Takamori, S.; Araki, H.; Kuroda, S. Flame retardancy of clay–sodium silicate composite coatings on wood for construction purposes. *RSC Adv.* **2015**, *5*, 34109–34116. [CrossRef]
117. Ribeiro, M.C.S.; Sousa, S.P.B.; Nóvoa, P.R.O. An Investigation on Fire and Flexural Mechanical Behaviors of Nano and Micro Polyester Composites Filled with SiO_2 and Al_2O_3 Particles. *Mater. Today Proc.* **2015**, *2*, 8–19. [CrossRef]
118. Rocha, J.d.S.; Escócio, V.A.; Visconte, L.L.Y.; Pacheco, É.B.A. V Thermal and flammability properties of polyethylene composites with fibers to replace natural wood. *J. Reinf. Plast. Compos.* **2021**, *40*, 726–740. [CrossRef]
119. Li, S.; Wang, X.; Xu, M.; Liu, L.; Wang, W.; Gao, S.; Li, B. Effect of a biomass based waterborne fire retardant coating on the flame retardancy for wood. *Polym. Adv. Technol.* **2021**, *32*, 4805–4814. [CrossRef]
120. Price, E.J.; Covello, J.; Paul, R.; Wnek, G.E. Tannic acid based super-intumescent coatings for prolonged fire protection of cardboard and wood. *SPE Polym.* **2021**, *2*, 153–168. [CrossRef]
121. Özkan, O.E.; Temiz, A.; Tor, Ö.; Vurdu, H. Effect of post-heat treatment on fire retardant treated wood properties. *Holzforschung* **2022**, *76*, 645–657. [CrossRef]
122. Yan, Y.; Dong, S.; Jiang, H.; Hou, B.; Wang, Z.; Jin, C. Efficient and Durable Flame-Retardant Coatings on Wood Fabricated by Chitosan, Graphene Oxide, and Ammonium Polyphosphate Ternary Complexes via a Layer-by-Layer Self-Assembly Approach. *ACS Omega* **2022**, *7*, 29369–29379. [CrossRef] [PubMed]
123. Hull, T.R.; Witkowski, A.; Hollingbery, L. Fire retardant action of mineral fillers. *Polym. Degrad. Stab.* **2011**, *96*, 1462–1469. [CrossRef]
124. Liu, L.; Qian, M.; Song, P.; Huang, G.; Yu, Y.; Fu, S. Fabrication of Green Lignin-based Flame Retardants for Enhancing the Thermal and Fire Retardancy Properties of Polypropylene/Wood Composites. *ACS Sustain. Chem. Eng.* **2016**, *4*, 2422–2431. [CrossRef]
125. Yang, Y.; Zhang, L.; Zhang, J.; Ren, Y.; Huo, H.; Zhang, X.; Huang, K.; Rezakazemi, M.; Zhang, Z. Fabrication of environmentally, high-strength, fire-retardant biocomposites from small-diameter wood lignin in situ reinforced cellulose matrix. *Adv. Compos. Hybrid Mater.* **2023**, *6*, 140. [CrossRef]
126. Yu, F.; Ba, Z.; Gao, Z.; Wang, Y.; Xie, Y.; Wang, H.; Qiu, Z.; Xiao, Z. Modification with lignin-based N-P flame retardant to improve the flame retardancy and smoke suppression of wood. *Chem. Eng. J.* **2024**, *493*, 152827. [CrossRef]

MDPI AG
Grosspeteranlage 5
4052 Basel
Switzerland
Tel.: +41 61 683 77 34

Polymers Editorial Office
E-mail: polymers@mdpi.com
www.mdpi.com/journal/polymers

www.ingramcontent.com/pod-product-compliance
Lightning Source LLC
Chambersburg PA
CBHW061513210926
51850CB00008B/1904
9783725853090